Aperiodic Crystals

Siegbert Schmid • Ray L. Withers • Ron Lifshitz
Editors

Aperiodic Crystals

Springer

Editors
Siegbert Schmid
School of Chemistry
The University of Sydney
Sydney, New South Wales, Australia

Ron Lifshitz
School of Physics & Astronomy
Tel Aviv University
Tel Aviv, Israel

Ray L. Withers
Research School of Chemistry
The Australian National University
Canberra, Australia

ISBN 978-94-007-6430-9
ISBN 978-94-007-6431-6 (eBook)
DOI 10.1007/978-94-007-6431-6
Springer Dordrecht Heidelberg New York London

Library of Congress Control Number: 2013936945

Printed on acid-free paper

Springer is part of Springer Science+Business Media (www.springer.com)

Preface

Aperiodic Crystals collects 37 selected papers from the scientific contributions presented at the Seventh International Conference on Aperiodic Crystals, *Aperiodic 2012*, held in Cairns, Australia from the 2nd to the 7th of September, 2012 and organized under the auspices of the Commission on Aperiodic Crystals of the International Union of Crystallography (IUCr). It followed Aperiodic'94 (Les Diablerets, Switzerland), Aperiodic'97 (Alpe d'Huez, France), Aperiodic 2000 (Nijmegen, The Netherlands), Aperiodic 2003 (Belo Horizonte, Brazil), Aperiodic 2006 (Zao, Japan) and Aperiodic 2009 (Liverpool, U.K.). The *Aperiodic* series of conferences in turn followed on four earlier conferences held under the title of Modulated Structures, Polytypes and Quasicrystals (MOSPOQ). The eighth conference in the *Aperiodic* series will be held in Prague in 2015.

The program was wonderfully diverse, covering a wide range of topics including: the mathematics of aperiodic long-range order and the fascinating types of tilings resulting from it; the synthesis, growth and stability of metallic aperiodic crystals and related complex metallic alloys; new methods and associated structural characterisation studies of aperiodic crystals; theoretical and experimental studies of the electronic, magnetic and other physical properties of aperiodic crystals; partial order, correlated disorder, and structured diffuse scattering; modulated structures, quasicrystals and approximants; soft-matter quasicrystals, and aperiodic ordering in bio-molecules and proteins; the dynamics of aperiodic crystals; as well as aperiodic surfaces, thin films and adsorbates. This impressive diversity in subject matter is well reflected in the contributions to this volume.

The conference was attended by more than 110 delegates from 23 different countries, including Dan Shechtman from Israel, Laureate of the 2011 Nobel Prize in Chemistry. Prof. Shechtman delivered a special celebratory Nobel lecture on the 30th anniversary year of his pioneering electron-diffraction characterization of the first quasicrystal on April 8, 1982. The introductory tutorial talk by Ted Janssen highlighted the fact that we also celebrated a half a century of work on aperiodic crystals, which could be considered as dating back to the pioneering work of Pim de Wolff on γ-Na_2CO_3 around 1962 and published in 1964. Presentations at the conference included 3 invited introductory tutorials, 11 invited talks, 46 contributed talks, and 43 poster presentations, all discussing state-of-the-art research in this fascinating field of scientific endeavour. What we know and what we still don't know about aperiodic order was carefully examined and hotly debated throughout this conference.

We would like to thank all the participants for coming the very long way to Australia as well as for their enthusiastic and considered contributions to, and participation in, the conference. Special thanks are due to the International Program Committee for their work in the organisation of the conference program and to the members of the Local Organizing Committee for making Aperiodic 2012 the very successful and highly stimulating meeting it was. We would also like to thank the wonderful staff from Springer for their help in the production of this volume. Finally, we gratefully acknowledge financial and other support from our sponsors.

Contents

Contributors

N. Abe Department of Advanced Materials Science, The University of Tokyo, Tokyo, Japan

R. Al Ajlouni College of Architecture, Texas Tech University, Lubbock, USA

T. Arima Department of Advanced Materials Science, The University of Tokyo, Tokyo, Japan

T. Asaka Nagoya Institute of Technology, Nagoya, Japan

M. Baake Fakultät für Mathematik, Universität Bielefeld, Bielefeld, Germany

K. Barkan Raymond and Beverly Sackler School of Physics and Astronomy, Tel Aviv University, Tel Aviv, Israel

B. Bauer Department of Earth and Environmental Sciences, Crystallography Section, Ludwig-Maximilians-Universität München, Munich, Germany

N. Bédaride LATP, Univ. Aix-Marseille, Marseille, France

P.J. Bereciartua Institute of Physics of the Academy of Sciences of the Czech Republic, Praha, Czech Republic

W.R. Brant School of Chemistry, The University of Sydney, Sydney, NSW, Australia

P.D. Carr Research School of Chemistry, Australian National University, Canberra, Australia

A. Castro Instituto de Ciencia de Materiales de Madrid (ICMM, CSIC), Madrid, Spain

I. Černičková Faculty of Materials Science and Technology in Trnava, Slovak University of Technology in Bratislava, Trnava, Slovak Republic

R. Čička Faculty of Materials Science and Technology in Trnava, Slovak University of Technology in Bratislava, Trnava, Slovak Republic

M. de Boissieu Sciences et Ingénierie des Matériaux et Procédés, Grenoble-INP, CNRS, UJF, St Martin d'Hères Cedex, France

S. Doyle Institute for Synchrotron Radiation (ISS), ANKA, Eggenstein-Leopoldshafen, Germany

M. Drienovský Faculty of Materials Science and Technology in Trnava, Slovak University of Technology in Bratislava, Trnava, Slovak Republic

C. Ecolivet Institut de Physique de Rennes, UMR UR1-CNRS 6251, Université de Rennes 1, Rennes, France

H. Euchner Institut für Theoretische und Angewandte Physik, Universität Stuttgart, Stuttgart, Germany; Sciences et Ingénierie des Matériaux et Procédés, Grenoble-INP, CNRS, UJF, St Martin d'Hères Cedex, France

T. Fernique LIPN, CNRS & Univ. Paris 13, Paris, France

F. Frey Department of Earth and Environmental Sciences, Crystallography Section, Ludwig-Maximilians-Universität München, Munich, Germany

K. Fukuda Nagoya Institute of Technology, Nagoya, Japan

V. Fung School of Chemistry, The University of Sydney, Sydney, NSW, Australia

F. Gähler Faculty of Mathematics, Bielefeld University, Bielefeld, Germany

U. Grimm Department of Mathematics and Statistics, The Open University, Milton Keynes, UK

L. Guérin Institut de Physique de Rennes, UMR UR1-CNRS 6251, Université de Rennes 1, Rennes, France

J. Hafner Faculty for Physics and Center for Computational Materials Science, Vienna University, Vienna, Austria

A.P. Heerdegen Research School of Chemistry, Australian National University, Canberra, Australia

K. Hiraga Institute for Materials Research, Tohoku University, Sendai, Japan

M.D. Hollingsworth Department of Chemistry, Kansas State University, Manhattan, KS, USA

C. Huck Fakultät für Mathematik, Universität Bielefeld, Bielefeld, Germany

E. Illeková Institute of Physics, Slovak Academy of Science, Bratislava, Slovak Republic

M. Inukai Toyota Technological Institute, Hisakata, Nagoya, Japan

Y. Ishii Department of Physics, Chuo University, Tokyo, Japan

T. Ishimasa Division of Applied Physics, Graduate School of Engineering, Hokkaido University, Sapporo, Japan

N. Ishizawa Nagoya Institute of Technology, Nagoya, Japan

D. Janičkovič Institute of Physics, Slovak Academy of Sciences, Bratislava, Slovak Republic; Institute of Physics, Slovak Academy of Science, Bratislava, Slovak Republic

J. Janovec Faculty of Materials Science and Technology in Trnava, Slovak University of Technology in Bratislava, Trnava, Slovak Republic

T. Janssen University of Nijmegen, Nijmegen, The Netherlands

I. Kanazawa Department of Physics, Tokyo Gakugei University, Tokyo, Japan

I. Kanazawa Department of Physics, Tokyo Gakugei University, Tokyo, Japan

L. Korkidi Raymond and Beverly Sackler School of Physics and Astronomy, Tel Aviv University, Tel Aviv, Israel

B. Kozakowski Faculty of Physics and Applied Computer Science, AGH University of Science and Technology, Krakow, Poland

M. Krajčí Institute of Physics, Slovak Academy of Sciences, Bratislava, Slovakia

P. Kuczera Faculty of Physics and Applied Computer Science, AGH University of Science and Technology, Krakow, Poland

T. Kudo Nagoya Institute of Technology, Nagoya, Japan

R. Lifshitz Raymond and Beverly Sackler School of Physics and Astronomy, Tel Aviv University, Tel Aviv, Israel

T.Y. Liu Faculty of Materials Science and Technology in Trnava, Slovak University of Technology in Bratislava, Trnava, Slovak Republic

M. Lowe Surface Science Research Centre and Department of Physics, The University of Liverpool, Liverpool, UK

C. Mariette Institut de Physique de Rennes, UMR UR1-CNRS 6251, Université de Rennes 1, Rennes, France

R. McGrath Surface Science Research Centre and Department of Physics, The University of Liverpool, Liverpool, UK

M. Mihalkovič Institute of Physics, Slovak Academy of Sciences, Bratislava, Slovakia

U. Mizutani Nagoya Industrial Science Research Institute, Nagoya, Japan

M. Moll Fakultät für Mathematik, Universität Bielefeld, Bielefeld, Germany

D.T. Murphy School of Chemistry, The University of Sydney, Sydney, NSW, Australia

K. Nozawa Department of Physics, Chuo University, Tokyo, Japan

B. Pedersen Forschungs-Neutronenquelle Heinz Maier-Leibnitz (FRM II), TU München, Garching, Germany

J.M. Perez-Mato Department of Condensed Matter Physics, University of the Basque Country, Bilbao, Spain

V. Petříček Institute of Physics of the Academy of Sciences of the Czech Republic, Praha, Czech Republic

P. Priputen Faculty of Materials Science and Technology in Trnava, Slovak University of Technology in Bratislava, Trnava, Slovak Republic

L. Pytlik Faculty of Physics and Applied Computer Science, AGH University of Science and Technology, Krakow, Poland

P. Rabiller Institut de Physique de Rennes, UMR UR1-CNRS 6251, Université de Rennes 1, Rennes, France

J. Rodríguez-Carvajal Institut Laue-Langevin, Grenoble, France

S. Rols Institut Laue Langevin, Grenoble, France

K. Saito Department of Materials Science and Engineering, Akita University, Akita, Japan

M. Saito Department of Physics, Tokyo Gakugei University, Tokyo, Japan

M. Sandbrink Institut für Theoretische Physik 2: Weiche Materie, Heinrich-Heine-Universität Düsseldorf, Düsseldorf, Germany

T. Sasaki Department of Physics, Tokyo Gakugei University, Tokyo, Japan

H. Sato Aichi University of Education, Aichi, Japan

S. Schmid School of Chemistry, The University of Sydney, Sydney, NSW, Australia

M. Schmiedeberg Institut für Theoretische Physik 2: Weiche Materie, Heinrich-Heine-Universität Düsseldorf, Düsseldorf, Germany

H. Schober Institut Laue Langevin, Grenoble, France; UFR de Physique, Université Joseph Fourier, Grenoble, France

M. Schreiber Institut für Physik, Technische Universität Chemnitz, Chemnitz, Germany

H.R. Sharma Surface Science Research Centre and Department of Physics, The University of Liverpool, Liverpool, UK

R. Simura Institute for Materials Research, Tohoku University, Sendai, Japan

R. Strzalka Faculty of Physics and Applied Computer Science, AGH University of Science and Technology, Krakow, Poland

K. Sugiyama Institute for Materials Research, Tohoku University, Sendai, Japan

S. Suzuki Institute for Materials Research, Tohoku University, Sendai, Japan

P. Švec Faculty of Materials Science and Technology in Trnava, Slovak University of Technology in Bratislava, Trnava, Slovak Republic; Institute of Physics, Slovak Academy of Sciences, Bratislava, Slovak Republic; Faculty of Materials Science and Technology in Trnava, Slovak University of Technology in Bratislava, Trnava, Slovak Republic; Institute of Physics, Slovak Academy of Science, Bratislava, Slovak Republic

R. Tamura Department of Materials Science and Technology, Tokyo University of Science, Noda, Japan

V. Terauds Fakultät für Mathematik, Universität Bielefeld, Bielefeld, Germany

S. Thiem Institut für Physik, Technische Universität Chemnitz, Chemnitz, Germany

B. Toudic Institut de Physique de Rennes, UMR UR1-CNRS 6251, Université de Rennes 1, Rennes, France

E. Vila Instituto de Ciencia de Materiales de Madrid (ICMM, CSIC), Madrid, Spain

T.R. Welberry Research School of Chemistry, Australian National University, Canberra, Australia

T.A. Whittle School of Chemistry, The University of Sydney, Sydney, NSW, Australia

J. Wolny Faculty of Physics and Applied Computer Science, AGH University of Science and Technology, Krakow, Poland

T.P. Yadav Surface Science Research Centre and Department of Physics, The University of Liverpool, Liverpool, UK

K. Yamada Department of Physics, Tokyo Gakugei University, Tokyo, Japan

T. Yamada Sciences et Ingénierie des Matériaux et Procédés, Grenoble-INP, CNRS, UJF, St Martin d'Hères Cedex, France; Department of Materials Science and Technology, Tokyo University of Science, Tokyo, Japan

A. Yasuhara EM Application Group, EM Business Unit, JEOL Ltd., Akishima, Japan; EM Application Group, EM Business Unit, JEOL Ltd., Tokyo, Japan; EM Application Group, JEOL Ltd., Tokyo, Japan

K. Yubuta Institute for Materials Research, Tohoku University, Sendai, Japan

P. Zeiner Faculty of Mathematics, Bielefeld University, Bielefeld, Germany

E.S. Zijlstra Theoretical Physics, University of Kassel, Kassel, Germany

F.J. Zuñiga Department of Condensed Matter Physics, University of the Basque Country, Bilbao, Spain

Chapter 1
A Brief History of Aperiodic Crystals: 1962–2012

T. Janssen

Abstract About 50 years ago, it was shown that there are solid state systems with perfect order but without lattice periodicity. These systems were called crystalline phases because of the order and incommensurate because of the lack of periodicity. They formed a challenge for crystallographers and physicists to understand the structure, the physical properties and the reason for their appearance. Later other classes of this type were found (occupation modulated crystals, incommensurate magnetic systems, incommensurate composites), the most important one being that of quasicrystals. The discovery of the latter class in 1982 caused a huge increase in interest. The first conferences on this new type of materials were called Modulated Crystals, later polytypes and quasicrystals were included in the title MOSPOQ. Nowadays these conferences continue under the name Aperiodic (Crystals). The field has become very active worldwide, and our insight into structure and properties has increased impressively. A brief sketch of the development of the field is given in this chapter.

1.1 History

On April 21st, 1912, Friedrich, Knipping and Laue [1] followed a proposal by Max Laue (after 1913 von Laue) and performed an experiment throwing X-rays onto a crystal of copper sulfate [2]. They found the sharp diffraction spots foreseen by Laue showing that the crystal has lattice periodicity. Shortly afterwards the Braggs developed the fundamental techniques of crystallography. For half a century the paradigm was that ideal crystals are built of identical unit cells. Later the International Union of Crystallography incorporated this property into the definition of a 'crystal'. This idea remained intact for half a century. Then systems with also sharp diffraction spots but without lattice periodicity were found. The idea to consider these materials also as crystals reached the larger crystallographic community much later, after Dan Shechtman had discovered quasicrystals on April 8th, 1982. The first aperiodic crystals, however, were discovered earlier, but this did not attract so much attention.

T. Janssen (✉)
University of Nijmegen, Nijmegen, The Netherlands
e-mail: T.Janssen@science.ru.nl

S. Schmid et al. (eds.), *Aperiodic Crystals*, DOI 10.1007/978-94-007-6431-6_1,

It is more difficult to state when exactly the first aperiodic crystal was found. In 1960, a spin wave was found with a period that did not fit the periodicity of the underlying lattice [3], but one did not observe an effect on the positions of the atoms. In 1963, satellites were seen in $NaNO_2$ [4], next to the main reflections. These were interpreted as micro-domains, in a very small (1.5 degree centigrade) temperature interval, around the ferroelectric phase transition and there was no discussion of the incommensurate character. But in 1964 the first incommensurately modulated structure was found. So there is some reason to consider 1962 (between 1960 and 1964) as the beginning of the field.

1.1.1 Incommensurate Modulated Phases and Composites

In 1964, Pim de Wolff and collaborators [5] found satellite peaks in the γ-phase of anhydrous Na_2CO_3: next to the main reflections of the monoclinic basic structure there were peaks at positions $h\mathbf{a}^* + k\mathbf{b}^* + \ell c^* + m(\alpha\mathbf{a}^* + \beta\mathbf{b}^*)$. First they were found in powder, later also in a single crystal. The interpretation was a periodic displacement of the atoms, with wave vector $\mathbf{q}$ which has irrational indices α and β with respect to the reciprocal lattice vectors $\mathbf{a}^*$ and $\mathbf{b}^*$ of the basic structure. The conclusion is that the structure is not lattice periodic. Mathematically, it is a quasiperiodic structure, but to stress the aperiodicity these phases were called *incommensurately modulated phases*. Soon other examples of such structures followed, e.g. thiourea and K_2SeO_4, to mention two examples which were studied extensively. For such structures the positions of the atoms can be given as

$$\mathbf{r}_{\mathbf{n}j} = \mathbf{n} + \mathbf{r}_j + \mathbf{u}_j\big(\mathbf{q}.(\mathbf{n} + \mathbf{r}_j)\big), \tag{1.1}$$

where $\mathbf{n}$ are the lattice points of the lattice periodic basic structure, $\mathbf{r}_j$ the position of the jth atom in the unit cell, $\mathbf{u}$ a periodic function with period 1. Later, also structures with 2 or more modulation vectors $\mathbf{q}$ were found. The diffraction spots are at positions

$$\mathbf{H} = h\mathbf{a}^* + k\mathbf{b}^* + \ell\mathbf{c}^* + \sum_{s=1}^{d} m_s\mathbf{q}_s. \tag{1.2}$$

d is the number of independent modulation wave vectors. Main reflections are the peaks with $m_s = 0$. Because main reflections are mapped onto main reflections, the symmetry of the diffraction must be one of the three-dimensional crystallographic point groups.

A new type of aperiodic crystal was found in 1975. Tetrathiafulvalene (TTF) pentaiodide has a subsystem consisting of TTF molecules with as second subsystem iodide in channels, the compound having composition $TTF_7I_{5-\delta}$ [6]. The basic structure of the TTF system is C-centered monoclinic, that of the iodine system A-centered monoclinic, but the lattice constants are incommensurate. Therefore, one needs more than 3 basis vectors to index the diffraction spots with integers: one additional vector is needed. A second example is $Hg_{3-\delta}AsF_6$ [7], consisting of 3 subsystems: the AsF_6 host lattice, and two systems of mercury chains, one in the a-direction, and one in the b-direction. Each subsystem has a basic structure with 3 basis vectors for their reciprocal lattices. Nevertheless, instead of nine, one only needs four basis vectors for indexing the diffraction peaks with integers. Such structures with mutually incommensurate subsystems are called incommensurate composites. Since these first examples many other have been found. A particular class is that of misfit structures, layered structures for which the layers do not have the same translation symmetry, not even for the basic structure one has to consider, because generally the layers are modulated by the interaction with the other layers.

The diffraction pattern shows reflections common to several subsystems, reflections corresponding to main peaks for one of the subsystems and summation reflections corresponding to modulations of one subsystem caused by the interaction with other subsystems. The symmetry of the diffraction pattern consists of all orthogonal transformations mapping each spot to another of the same intensity. Because subsystems may be mapped on each other, the crystallographic condition no longer holds: in principle, the symmetry may contain non-crystallographic elements, i.e. elements which are impossible for a three-dimensional lattice periodic structure, such as a five-fold rotation. However, such symmetries have not been observed for composites.

1.1.2 Aperiodic Tilings and Quasicrystals

Non-crystallographic symmetries play a role in mathematical constructions, aperiodic tilings of the plane. The best known example is the Penrose tiling, a tiling of the plane with copies of two different tiles, without gaps or overlaps. One realization of a Penrose tiling is by means of rhombs, one with an angle of 36° and with an angle of 72°. An early overview was given by Martin Gardiner in the January 1977 issue of *Scientific American*. Crystallographers played with the idea that there could be crystals with a comparable structure. Alan Mackay [8] showed experimentally that the diffraction pattern has ten-fold symmetry. Later this could be proven mathematically. However, such structures, aperiodic and quasiperiodic, were not known by then.

Therefore, it was a big surprise when Dan Shechtman found real structures with sharp diffraction peaks and ten-fold symmetry in the diffraction pattern. He studied rapidly cooled AlMn particles and observed this phenomenon. Unfortunately, the referees were not aware of the existence of tilings with these properties, or they did not believe that this could happen in nature and did not accept the explanations for the findings. However, other explanations could all be proven to be false. It took two and a half years before the results could be published [9]. The new material did not only show ten-fold symmetry in the diffraction, but even the symmetry of an icosahedron. It became known as a quasicrystal. This discovery can be considered as the most important event in the history of aperiodic crystals.

Mackay already had used the term ‘quasi-lattice’ for the diffraction pattern of the Penrose tiling. For the new material Levine and Steinhardt proposed the term ‘quasicrystal’. It is supposed to mean ‘quasiperiodic crystal’, but the latter term is actually too broad. First, according to the original mathematical definition, any periodic function is also quasiperiodic; and second, there are aperiodic and quasiperiodic structures that are usually not considered to be quasicrystals, like the incommensurate phases. Actually, there is not yet a consensus about the term ‘quasicrystal’. ‘Aperiodic crystal’ is the general term for a structure with sharp diffraction peaks that is quasiperiodic and not lattice periodic.

After the discovery of the quasicrystalline structure in AlMn, new examples were rapidly found. Next to the icosahedral quasicrystals, there are the decagonal quasicrystals with quasiperiodicity in planes perpendicular to an axis along which the crystal is periodic. Also new classes of icosahedral quasicrystals were discovered which turned out to be of better quality and stable (the very small AlMn quasicrystals are unstable). Several families of ternary alloys (AlCuFe, AlMnPd, etc.) and binary alloys (e.g. YbCd) were developed. The larger size and higher quality were essential for the study of crystallographic and physical properties.

1.1.3 Incommensurate Magnetic Structures

As mentioned before, aperiodicity in crystals was found as an incommensurate spin wave in $MnAu_2$. There no influence of the spin wave on the crystal structure was reported. Such an interaction was found in chromium [10]. Below the Curie temperature, satellites are observed with a modulation wave vector equal to the spin wave vector. The coupling between spin and lattice causes a modulation, a mechanism comparable to the coupling between charge density waves and the lattice, leading to a modulation of the latter. Analogously to the displacive modulation (Eq. (1.1)), the spin wave at discrete positions may be written as

$$\mathbf{S}(\mathbf{n}, j) = \sum_{\mathbf{H}} \hat{\mathbf{S}}(\mathbf{H}) \exp\bigl(i\mathbf{H}.(\mathbf{n} + \mathbf{r}_j)\bigr). \tag{1.3}$$

Whereas displacive modulations are usually transversal or longitudinal, spin structures may show complicated spiral structures, especially in systems with rare-earth elements.

1.2 Superspace Treatment

A quasiperiodic function $f(x)$ is a function that is the restriction of a function g, periodic in each of its n variables, to a line in the n-dimensional space: $f(x) = g(\alpha_1 x, \dots, \alpha_n x)$, where the numbers α_i are irrational. Notice that for $n = 1$ the function is periodic: a periodic function is also quasiperiodic. It may be shown that the projection of the Fourier transform of $f(x)$ on the line consists of sharp peaks at positions $q = \sum_{i=1}^{n} m_i q_i$. This corresponds to the observation by de Wolff that the peaks for γ-Na_2CO_3 can be seen as the projection points of a regular lattice in four dimensions.

The diffraction spots of a quasiperiodic crystal can be labeled with n indices. If $n = 3$ (or the dimension of the crystal which could also be 1 or 2), the crystal has lattice periodicity and the spots belong to the reciprocal lattice. In general, the spots are given as points on a *Fourier module*: linear combinations of n basis vectors $\mathbf{a}_i^*$. The number n is the *rank* of the Fourier module. If the orthogonal 3-dimensional transformation R leaves the pattern (inclusive intensities) invariant, one has

$$R\mathbf{a}_i^* = \sum_{j=1}^{n} \Gamma_{ij}(R)\mathbf{a}_j^*. \tag{1.4}$$

Because there are only a finite number of peaks above a certain intensity around the origin, the matrices $\Gamma(R)$ form a finite group, and group theory then tells us that this group on another basis consists of orthogonal matrices. Because the pattern in 3 dimensions is left invariant, the elements R correspond to pairs (R_E, R_I) of orthogonal transformations, in 3 and $(n-3)$ dimensions, respectively. The n-dimensional group then leaves an n-dimensional (reciprocal) lattice invariant, and its direct lattice as well. This is the general idea of constructing a periodic n-dimensional structure for which the restriction to 3 dimensions gives the physical, aperiodic crystal. An alternative way is the following. Suppose the aperiodic crystal has a density $\rho(\mathbf{r})$. Its Fourier component is non-zero only for points of the Fourier module, and these correspond to points of the n-dimensional reciprocal lattice. Then one can construct a lattice periodic function in n-dimensions

$$\rho(\mathbf{r}) = \sum_{\mathbf{H}} \hat{\rho}(\mathbf{H}) \exp(i\mathbf{H}.\mathbf{r}) \rightarrow \rho(\mathbf{r}, \mathbf{r}_I) = \sum_{\mathbf{H}} \hat{\rho}(\mathbf{H}) \exp\bigl(i(\mathbf{H}.\mathbf{r} + \mathbf{H}_I.\mathbf{r}_I)\bigr). \tag{1.5}$$

In the case of point atoms, the function $\rho(\mathbf{r}_E, \mathbf{r}_I)$ is restricted to $(n-3)$-dimensional hypersurfaces, called *atomic surfaces*. These may extend to infinity or be of finite volume, when they are disjunct. It is the goal of structure determination to find position and shape of these atomic surfaces. On them an occupation function may be defined which determines the probability of finding a certain chemical species there.

Because the projection of the lattice on internal space is a dense set, the energy of every 3-plane through a point of this set is the same, and this corresponds to a global translation along internal space. In many cases, local shifts do not cost much energy. Statically, this corresponds to phason disorder, dynamically to phason excitations (either jumps or collective motions). At finite temperature, there will be phason disorder which shows itself in points outside the atomic surfaces which may occur with a certain probability. The disorder could contribute to the entropy of the system, and in turn this might influence the balance between periodic and aperiodic structures.

Since the function $\rho(\mathbf{r}_E, \mathbf{r}_I)$ is lattice periodic, its symmetry group is an n-dimensional space group, a *superspace group*. An element g is a combination of a separable orthogonal transformation (R_E, R_I), where the two components are orthogonal transformations in physical and internal space, respectively, and a pair of transformations $(\mathbf{v}_E, \mathbf{v}_I)$ of translations in these 2 subspaces. Its action on $\rho(\mathbf{r}_E, \mathbf{r}_I)$ is given by

$$T_g\rho(\mathbf{r}_E, \mathbf{r}_I) = \rho\big(R_E^{-1}(\mathbf{r}_E - \mathbf{v}_E), R_I^{-1}(\mathbf{r}_I - \mathbf{v}_I)\big) \tag{1.6}$$

and on its Fourier transform by

$$T_g\hat{\rho}(\mathbf{k}) = \exp\big(i(\mathbf{k}_E.\mathbf{v}_E + \mathbf{k}_I.\mathbf{v}_I)\big)\hat{\rho}\big(R_E^{-1}\mathbf{k}\big). \tag{1.7}$$

This formula gives the systematic extinctions associated with the superspace group: if $R\mathbf{k} = \mathbf{k}$, then $\hat{\rho}(\mathbf{k}) = 0$, unless the argument of exp is a multiple of 2π.

Let us consider this procedure for the various families.

1.2.1 Incommensurate Modulated Phases and Composites

De Wolff [11] noticed that the positions of the diffraction pattern (Eq. (1.2)) for $d = 1$ can be seen as the projection of a reciprocal lattice in 4 dimensions. As 4th dimension one can consider the phase of the modulation wave (Eq. (1.1)). Then one gets

$$\mathbf{r}_{\mathbf{n}j}(\phi) = \mathbf{n} + \mathbf{r}_j + \mathbf{u}_j\big(\mathbf{q}.(\mathbf{n} + \mathbf{r}_j) + \phi\big), \tag{1.8}$$

which is equivalent to the expression to obtain a quasiperiodic function from a periodic function in a higher-dimensional space. The symmetry of the pattern (1.8) is a four-dimensional space group. Such space groups had been used by Janner and Janssen for the study of space-time symmetries of electrodynamic systems. Then the phase ϕ is taken over by the time t. Together the approach could be generalized to more general modulated phases, and later to general quasiperiodic crystals. This has been dealt with in the beginning of this section (Eq. (1.5)).

For incommensurate composites each subsystem (ν) has a lattice periodic basic structure with a reciprocal lattice on a basis $\mathbf{a}_i^{\nu*}$. A basis for the Fourier module, the set generated by all diffraction spot positions, is given by $\mathbf{a}_i^*$ $(n = 1, \ldots, n)$). The former can be expressed in the latter by $\mathbf{a}_i^{\nu*} = \sum_j Z_{ij}^{\nu}\mathbf{a}_j^*$. The basis $\mathbf{a}_i^*$ can be embedded into an n-dimensional space, and consequently the reciprocal and the direct lattice of each subsystem is embedded in an n-dimensional space. Each subsystem is modulated by the interaction with the other subsystems. This means that the main peaks of one may coincide with satellite peaks of other subsystems. In addition, there may be other modulations as well, which would increase the dimension of the superspace, but we shall disregard that possibility here.

1.2.2 Aperiodic Tilings and Quasicrystals

With some effort it is possible to index the diffraction pattern made by Mackay [8]. It has rank four, the symmetry group is the group $\bar{5}$m corresponding to a finite group of integer matrices, for which one can determine the invariant lattice in 4 dimensions and the unit cell. The problem is the determination of the atomic surfaces in this unit cell. It turns out that one has 4 different and disjunct atomic surfaces, all pentagons, of different size and orientation.

This illustrates the problems for the description of quasicrystals. The trivial part is the determination of the unit cell (or the asymmetric unit cell). This follows easily from the positions of the diffraction spots and the symmetry of the pattern. For decagonal (and octagonal or dodecagonal) quasicrystals the dimension of the superspace is five, for icosahedral quasicrystals six. Like for the Penrose tiling, the atomic surfaces here are disjunct. So one has to determine the position of these atomic surfaces and their shape. Moreover, there may be chemical order or disorder on the atomic surfaces. In the determination, the knowledge of the superspace group may help. Moreover, there is a closeness condition. This means the following. Two atomic surfaces with nearby positions in physical space do not overlap in projection on internal space. But if one changes the position of the n-dimensional crystal in internal space and an intersection point of an atomic surface with physical space vanishes, a new intersection point on another atomic surface should appear. This means that the projection of a border of an atomic surface should coincide with that of another atomic surface. This is the closeness condition which poses limitations on the shape of the atomic surfaces. Another helpful fact can be the existence of an approximant, a lattice periodic structure with similar chemical composition and similar local ordering. There are a number of structure models for ternary systems. Using the knowledge of the structure of an approximant, one has been able to determine the structure of the binary icosahedral YbCd [12].

1.2.3 Incommensurate Magnetic Structures

The description of incommensurate magnetic structures (Eq. (1.3)) and their effect on the nuclear structure in superspace follows the same lines. Both the spin waves and modulation waves may be embedded in superspace. The superspace group of the nuclear structure consists of all elements leaving it invariant. For the spin system one may introduce the time reversal operator θ. Then the action of the combination of a superspace group element g and θ on the spin wave in superspace is

$$T_g\theta\mathbf{S}(\mathbf{r}_E,\mathbf{r}_I) = -\mathrm{Det}(R_E)\mathbf{S}\big(g^{-1}(\mathbf{r}_E,\mathbf{r}_I)\big), \tag{1.9}$$

and for g itself the same expression holds without the minus-sign. The group of all elements g and $g\theta$ which leave $\mathbf{S}$ invariant is the magnetic superspace group. This group and the corresponding structure have been determined for several aperiodic magnetic structures such as chromium, rare-earth compounds and multiferroics [13–15].

1.3 Phase Transitions

As in conventional 3-dimensional crystals, aperiodic crystals may show phase transitions in the composition–pressure–temperature space. However, the variety is bigger for aperiodic systems. In the first place, one has to distinguish transitions where the dimension of superspace does not change, and those where there is a difference in dimension at the phase transition. An example of the former, is a transition to a superstructure, for example, a change in centering. If the transition is 2nd order, the order parameter is related to irreducible representations of the high-symmetry group, and one can apply Landau's theory of phase transitions.

Phase transitions where the dimension changes, are more typical for aperiodic crystals. Examples are the transition from an unmodulated to a modulated structure. These are often of 2nd order and related to a soft mode, a vibration mode becoming unstable at the transition. The appearing modulated structure may be described using irreducible representations of the symmetry of the unmodulated phase, at least near the phase transition. There is a relatively simple connection between such irreducible representations and the superspace group of the modulated phase [16]. Also modulations of quasicrystals may be described in this way, e.g. the icosahedral modulation of an icosahedral quasicrystal is a transition from a six- to a 12-dimensional structure.

For incommensurate composites the phase transition may correspond to a change in the relationship between the subsystems. Examples of such phase transitions have been observed in nonadecane-urea [17]. In the phase diagram, one finds structures with rank 3, 4 and 5.

1.4 Conclusion

In 50 years the field of aperiodic crystals has grown to a rich and important topic. There is a very large variety of systems, and these are interesting from various points of view. The number of systems is large, and so is the total amount of such crystals, because many minerals belong to this class of materials. Although many mathematical, physical and chemical questions have been answered, there still remain many fundamental open questions concerning the origin of aperiodic order, the growth of the order, the reason for the stability, and the character of elementary excitations. Finally, the question remains in how far this class of materials may lead to new applications.

Additional information "http://www.janssenweb.net/ted/janssen.htm" and the book "Aperiodic Crystals" by Ted Janssen, Gervais Chapuis and Marc de Boissieu, Oxford University Press 2007.

References

1. Forman P (1969) The discovery of the diffraction of X-rays by crystals. Arch Hist Exact Sci 6:38–71
2. Friedrich W, Knipping P, Laue M (1912) Interferenz-Erscheinungen bei Röntgenstrahlen. Bayer Akad Wiss München, Sitzungsber Math-Phys Kl August:303–322
3. Herpin A, Mériel P, Villain J (1960) Antiferromagnétisme hélicoïdal. J Phys Radium 21:67

4. Tanisaki S (1963) X-ray study on the ferroelectric phase transition of $NaNO_2$. J Phys Soc Jpn 18:1181–1191
5. Brouns E, Visser JW, de Wolff PM (1964) An anomaly in the crystal structure of γ-Na_2CO_3. Acta Crystallogr 17:614
6. Johnson CK, Watson CR (1976) Superstructure and modulation wave analysis for the unidimensional conductor TTF pentaiodide. J Chem Phys 64:2271–2286
7. Pouget JP, Shirane G, Hastings JM, Heeger AJ, Miro ND, MacDiarmid AG (1978) Elastic neutron scattering of the 'phase ordering' phase transition in $Hg_{3-\delta}AsF_6$. Phys Rev B 18:3645–3656
8. Mackay A (1982) Crystallography and the Penrose tiling. Physica 114A:609–613
9. Shechtman D, Blech I, Gratias D, Cahn JW (1984) Metallic phase with long-range orientational order and no translational symmetry. Phys Rev Lett 53:1951–1953
10. Tsunoda Y, Mori M, Kunitomi N, Teraoka Y, Kanamori J (1974) Strain wave in pure chromium. Solid State Commun 14:287–289
11. de Wolff PM (1974) The pseudo-symmetry of modulated crystal structures. Acta Crystallogr, Ser A 30:777–785
12. Takakura H, Gomez CP, Yamamoto A, de Boissieu M, Tsai AP (2007) Atomic structure of the binary icosahedral Yb–Cd quasicrystal. Nat Mater 6:58–63
13. Janner A, Janssen T (1980) Symmetry of incommensurate crystal phases, I: commensurate basic structure. Acta Crystallogr, Ser A 36:399–408
14. Schobinger-Papamantellos P, Rodriguez-Carvajal J, Ritter C, Buschow KHJ (2010) Competing magnetic structures in the DySi FeB-type phase diagram. J Magn Magn Mater 322:119–132
15. Perez-Mato JM, Ribeiro JL, Petricek V, Aroyo MI (2012) Magnetic superspace groups and symmetry constraints in incommensurate magnetic phases. J Phys Condens Matter 24:163201
16. Janssen T, Janner A (1984) Superspace groups and representations of ordinary space groups: alternative approaches to the symmetry of incommensurate crystal phases. Physica A 126:163–176
17. Toudic B, Rabiller P, Bourgeois L, Huard M, Ecolivet C, McIntyre GJ, Bourges P, Breczewski T, Janssen T (2011) Europhys Lett 93:16003

Chapter 2
Squiral Diffraction

Uwe Grimm and Michael Baake

Abstract The Thue–Morse system is a paradigm of singular continuous diffraction in one dimension. Here, we consider a planar generalisation, constructed by a bijective block substitution rule, which is locally equivalent to the squiral inflation rule. For balanced weights, its diffraction is purely singular continuous. The diffraction measure is a two-dimensional Riesz product that can be calculated explicitly.

2.1 Introduction

The diffraction of (fully) periodic systems and of aperiodic structures based on cut and project sets (or model sets) is well understood; see [4, 5] and references therein. These systems (in the case of model sets under suitable assumptions on the window) are pure point diffractive, and the diffraction can be calculated explicitly.

The picture changes for structures with continuous diffraction. Not much is known in general, in particular for the case of singular continuous diffraction, even though both absolutely and singular continuous diffraction show up in real systems [13, 14]. The paradigm of singular continuous diffraction is the Thue–Morse chain, which in its balanced form (constructed via the primitive inflation rule $1 \mapsto 1\bar{1}$, $\bar{1} \mapsto \bar{1}1$ with weights 1 and $\bar{1} = -1$, say) shows purely singular continuous diffraction. This was shown by Kakutani [10], see also [1], and the result can be extended to an entire family of generalised Thue–Morse sequences [3].

Here, we describe a two-dimensional system which, in its balanced form, has purely singular continuous diffraction. For mathematical details, we refer to [6]. Again, it is possible to obtain an explicit formula for the diffraction measure in terms of a Riesz product [12, Sect. 1.3], with convergence in the vague topology.

U. Grimm (✉)
Department of Mathematics and Statistics, The Open University, Walton Hall, Milton Keynes MK7 6AA, UK
e-mail: u.g.grimm@open.ac.uk

M. Baake
Fakultät für Mathematik, Universität Bielefeld, Postfach 100131, 33501 Bielefeld, Germany
e-mail: mbaake@math.uni-bielefeld.de

S. Schmid et al. (eds.), *Aperiodic Crystals*, DOI 10.1007/978-94-007-6431-6_2,

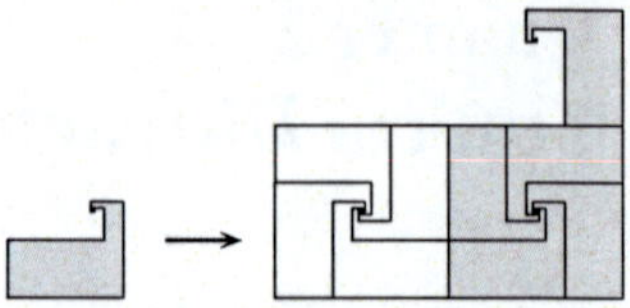

Fig. 2.1 The primitive inflation rule for the squiral tiling of the Euclidean plane

Fig. 2.2 Patch of the squiral tiling obtained by two inflation steps from the central seed

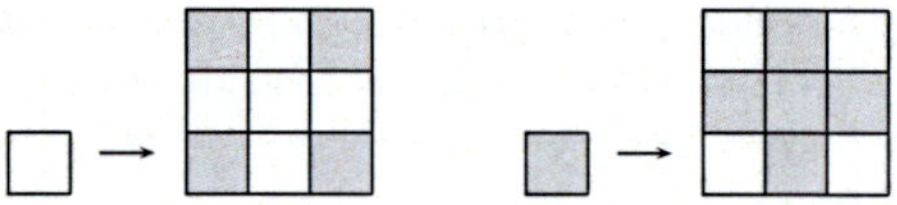

Fig. 2.3 Equivalent block inflation rule for the squiral tiling of Fig. 2.2

2.2 The Squiral Block Inflation

The squiral tiling (a name that comprises 'square' and 'spiral') was introduced in [8, Fig. 10.1.4] as an example of an inflation tiling with prototiles comprising infinitely many edges. The inflation rule is shown in Fig. 2.1; it is compatible with reflection symmetry, so that the reflected prototile is inflated accordingly.

A patch of the tiling is shown in Fig. 2.2. Clearly, the tiling consists of a two-colouring of the square lattice, with each square comprising four squiral tiles of the same chirality. The two-colouring can be obtained by the simple block inflation rule shown in Fig. 2.3, which is bijective in the sense of [11]. Again, the rule is compatible with colour exchange. The corresponding hull has D_4 symmetry, and also contains an element with exact individual D_4 symmetry; see [6] for details and an illustration.

Due to the dihedral symmetry of the inflation tiling, it suffices to consider a tiling of the positive quadrant. Using the lower left point of the square as the reference point, the induced block inflation ρ produces a two-cycle of configurations v and

ρv. They satisfy, for all $m, n \geq 0$ and $0 \leq r, s \leq 2$, the fixed point equations

$$(\rho v)_{3m+r,3n+s} = \begin{cases} \overline{v}_{m,n}, & \text{if } r \equiv s \equiv 0 \bmod 2, \\ v_{m,n}, & \text{otherwise.} \end{cases} \tag{2.1}$$

2.3 Autocorrelation and Diffraction Measure

For a fixed point tiling under ρ^2, we mark each (coloured) square by a point at its lower left corner $z \in \mathbb{Z}^2$. For the balanced version, each point carries a weight $w_z = 1$ (for white) or $w_z = \bar{1} = -1$ (for grey). Consider the corresponding Dirac comb

$$\omega = w\delta_{\mathbb{Z}^2} = \sum_{z \in \mathbb{Z}^2} w_z \delta_z, \tag{2.2}$$

which also is a special decoration of the original squiral tiling. Following the approach pioneered by Hof [9], the natural autocorrelation measure γ of ω is defined as

$$\gamma = \omega \circledast \widetilde{\omega} := \lim_{N \to \infty} \frac{(\omega|_{C_N}) * \widetilde{(\omega|_{C_N})}}{(2N+1)^2}, \tag{2.3}$$

where C_N stands for the closed centred square of side length $2N$. Here, $\widetilde{\mu}$ denotes the measure defined by $\widetilde{\mu}(g) = \overline{\mu(\widetilde{g})}$ for $g \in C_{\mathsf{c}}(\mathbb{R}^2)$, with $\widetilde{g}(x) := \overline{g(-x)}$ (and where the bar denotes complex conjugation). The autocorrelation measure γ is of the form $\gamma = \eta\delta_{\mathbb{Z}^2}$ with autocorrelation coefficients

$$\eta(m,n) = \lim_{N \to \infty} \frac{1}{(2N+1)^2} \sum_{k,\ell=-N}^{N} w_{k,\ell} w_{k-m,\ell-n}. \tag{2.4}$$

All limits exists due to the unique ergodicity of the underlying dynamical system [6], under the action of the group $\mathbb{Z}^2$.

Clearly, one has $\eta(0,0) = 1$, while Eq. (2.1) implies the nine recursion relations

$$\begin{aligned} \eta(3m, 3n) &= \eta(m,n), \\ \eta(3m, 3n+1) &= -\frac{2}{9}\eta(m,n) + \frac{1}{3}\eta(m,n+1), \\ \eta(3m, 3n+2) &= \frac{1}{3}\eta(m,n) - \frac{2}{9}\eta(m,n+1), \\ \eta(3m+1, 3n) &= -\frac{2}{9}\eta(m,n) + \frac{1}{3}\eta(m+1,n), \\ \eta(3m+1, 3n+1) &= -\frac{2}{9}\big(\eta(m+1,n) + \eta(m,n+1)\big) + \frac{1}{9}\eta(m+1,n+1), \end{aligned} \tag{2.5}$$

$$\eta(3m+1,3n+2) = -\frac{2}{9}\big(\eta(m,n)+\eta(m+1,n+1)\big)+\frac{1}{9}\eta(m+1,n),$$

$$\eta(3m+2,3n) = \frac{1}{3}\eta(m,n)-\frac{2}{9}\eta(m+1,n),$$

$$\eta(3m+2,3n+1) = -\frac{2}{9}\big(\eta(m,n)+\eta(m+1,n+1)\big)+\frac{1}{9}\eta(m,n+1),$$

$$\eta(3m+2,3n+2) = \frac{1}{9}\eta(m,n)-\frac{2}{9}\big(\eta(m+1,n)+\eta(m,n+1)\big),$$

which hold for all $m,n \in \mathbb{Z}$ and determine all coefficients uniquely [6]. The autocorrelation coefficients show a number of remarkable properties, which are interesting in their own right, and useful for explicit calculations.

Since the support of ω is the lattice $\mathbb{Z}^2$, the diffraction measure $\hat{\gamma}$ is $\mathbb{Z}^2$-periodic [2], and can thus be written as

$$\hat{\gamma} = \mu * \delta_{\mathbb{Z}^2},$$

where μ is a positive measure on the fundamental domain $\mathbb{T}^2 = [0,1)^2$ of $\mathbb{Z}^2$. One can now analyse $\hat{\gamma}$ via the measure μ, which, via the Herglotz–Bochner theorem, is related to the autocorrelation coefficients by Fourier transform

$$\eta(k) = \int_{\mathbb{T}^2} e^{2\pi i k z}\,\mathrm{d}\mu(z),$$

where $k=(m,n)\in\mathbb{Z}^2$ and kz denotes the scalar product. We now sketch how to determine the spectral type of μ, and how to calculate it.

Defining $\Sigma(N) := \sum_{m,n=0}^{N-1}\eta(m,n)^2$, the recursions (2.5) lead to the estimate

$$\Sigma(3N) \le \frac{319}{81}\Sigma(N),$$

so that $\Sigma(N)/N^2 \longrightarrow 0$ as $N\to\infty$. An application of Wiener's criterion in its multidimensional version [6, 7] implies that μ, and hence also the diffraction measure $\hat{\gamma}$, is continuous, which means that it comprises no Bragg peaks at all.

Since $\eta(0,1)=\eta(1,0)=-1/3$, which follows from Eq. (2.5) by a short calculation, the first recurrence relation implies that $\eta(0,3^j)=\eta(3^j,0)=-1/3$ for all integer $j\ge 0$. Consequently, the coefficients cannot vanish at infinity. Due to the linearity of the recursion relations, the Riemann–Lebesgue lemma implies [6] that μ cannot have an absolutely continuous component (relative to Lebesgue measure). The measure μ, and hence $\hat{\gamma}$ as well, must thus be purely singular continuous.

2.4 Riesz Product Representation

Although the determination of the spectral type of $\hat{\gamma}$ is based on an abstract argument, the recursion relations (2.5) hold the key to an explicit, iterative calculation of μ (and hence $\hat{\gamma}$). One defines the distribution function $F(x,y) :=$

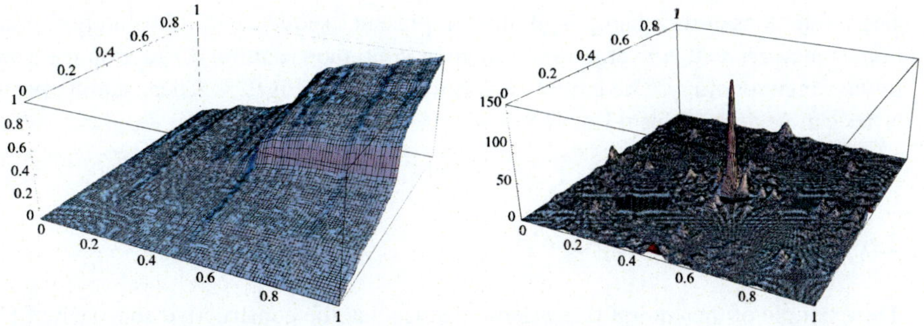

Fig. 2.4 The distribution function $F^{(3)}$ of Eq. (2.7) (*left*) and the corresponding density $f^{(3)}$ of Eq. (2.8) (*right*), approximating the diffraction measure $\hat{\gamma}$ of the squiral tiling on $[0, 1]^2$

$\mu([0, x] \times [0, y])$ for rectangles with $0 \le x, y < 1$, which is then extended to the positive quadrant as

$$F(x, y) = \hat{\gamma}\big([0, x] \times [0, y]\big).$$

This can finally be extended to $\mathbb{R}^2$ via $F(-x, y) = F(x, -y) = -F(x, y)$ and hence $F(-x, -y) = F(x, y)$. In particular, one has $F(0, 0) = 0$ as well as $F(0, y) = F(x, 0) = 0$, and F is continuous on $\mathbb{R}^2$. The latter property is non-trivial, and follows from the continuity of certain marginals; see [6] and references therein for details.

One can show that, as a result of Eq. (2.5), F satisfies the functional relation

$$F(x, y) = \frac{1}{9} \int_0^{3x} \int_0^{3y} \vartheta\left(\frac{x}{3}, \frac{y}{3}\right) \mathrm{d}F(x, y), \tag{2.6}$$

written in Lebesgue–Stieltjes notation, with the trigonometric kernel function

$$\vartheta(x, y) = \frac{1}{9}\big(1 + 2\cos(2\pi x) + 2\cos(2\pi y) - 4\cos(2\pi x)\cos(2\pi y)\big)^2.$$

The functional relation (2.6) induces an iterative approximation of F as follows. Starting from $F^{(0)}(x, y) = xy$ (which corresponds to Lebesgue measure, $\mathrm{d}F^{(0)} = \lambda$) and continuing with the iteration

$$F^{(N+1)}(x, y) = \frac{1}{9} \int_0^{3x} \int_0^{3y} \vartheta\left(\frac{x}{3}, \frac{y}{3}\right) \mathrm{d}F^{(N)}(x, y), \tag{2.7}$$

one obtains a uniformly (but not absolutely) converging sequence of distribution functions, each of which represents an absolutely continuous measure. With $\mathrm{d}F^{(N)}(x, y) = f^{(N)}(x, y)\,\mathrm{d}x\,\mathrm{d}y$, where $f^{(N)}(x, y) = \frac{\partial^2}{\partial x\,\partial y} F^{(N)}(x, y)$, one finds the Riesz product

$$f^{(N)}(x, y) = \prod_{\ell=0}^{N-1} \vartheta\big(3^\ell x, 3^\ell y\big). \tag{2.8}$$

These functions of increasing ‘spikiness’ represent a sequence of (absolutely continuous) measures that converge to the singular continuous squiral diffraction measure in the vague topology. The case $N = 3$ is illustrated in Fig. 2.4. Local scaling properties can be derived from Eq. (2.8).

2.5 Summary and Outlook

The example of the squiral tiling demonstrates that the constructive approach of [1, 3, 10] can be extended to more than one dimension. The result is as expected, and analogous arguments apply to a large class of binary block substitutions that are bijective in the sense of [11]. This leads to a better understanding of binary systems with purely singular continuous diffraction.

It is desirable to extend this type of analysis to substitution systems with larger alphabets. Although the basic theory is developed in [12], there is a lack of concretely worked-out examples. Moreover, there are various open questions in this direction, including the (non-) existence of bijective constant-length substitutions with absolutely continuous spectrum (the celebrated example from [12, Ex. 9.3] was recently recognised to be inconclusive by Alan Bartlett and Boris Solomyak).

Acknowledgements We thank Tilmann Gneiting and Daniel Lenz for discussions. This work was supported by the German Research Council (DFG), within the CRC 701.

References

1. Baake M, Grimm U (2008) The singular continuous diffraction measure of the Thue–Morse chain. J Phys A, Math Theor 41:422001. arXiv:0809.0580
2. Baake M (2002) Diffraction of weighted lattice subsets. Can Math Bull 45:483–498. arXiv:math.MG/0106111
3. Baake M, Gähler F, Grimm U (2012) Spectral and topological properties of a family of generalised Thue–Morse sequences. J Math Phys 53:032701. arXiv:1201.1423
4. Baake M, Grimm U (2011) Kinematic diffraction from a mathematical viewpoint. Z Kristallogr 226:711–725. arXiv:1105.0095
5. Baake M, Grimm U (2012) Mathematical diffraction of aperiodic structures. Chem Soc Rev 41:6821–6843. arXiv:1205.3633
6. Baake M, Grimm U (2012) Squirals and beyond: substitution tilings with singular continuous spectrum. Ergod Theory Dyn Syst, to appear. arXiv:1205.1384
7. Baake M, Grimm U (2013) Theory of aperiodic order: a mathematical invitation. Cambridge University Press, Cambridge, to appear
8. Grünbaum B, Shephard GC (1987) Tilings and patterns. Freeman, New York
9. Hof A (1995) On diffraction by aperiodic structures. Commun Math Phys 169:25–43
10. Kakutani S (1972) Strictly ergodic symbolic dynamical systems. In: LeCam LM, Neyman J, Scott EL (eds) Proceedings of the 6th Berkeley symposium on mathematical statistics and probability. University of California Press, Berkeley, pp 319–326
11. Frank NP (2005) Multi-dimensional constant-length substitution sequences. Topol Appl 152:44–69

12. Queffélec M (2010) Substitution dynamical systems—spectral analysis, 2nd edn. LNM, vol 1294 Springer, Berlin
13. Welberry TR, Withers RL (1991) The rôle of phase in diffuse diffraction patterns and its effect on real-space structure. J Appl Crystallogr 24:18–29
14. Withers RL (2005) Disorder, structured diffuse scattering and the transmission electron microscope. Z Kristallogr 220:1027–1034

Chapter 3
Random Noble Means Substitutions

Michael Baake and Markus Moll

Abstract The random local mixture of a family of primitive substitution rules with noble mean inflation multiplier is investigated. This extends the random Fibonacci example that was introduced by Godrèche and Luck in (J. Stat. Phys. 55:1–28, 1989). We discuss the structure of the corresponding dynamical systems, and determine the entropy, an ergodic invariant measure and diffraction spectra.

3.1 Introduction

Despite many open problems (including the famous Pisot substitution conjecture), the structure of systems with pure point diffraction is rather well understood [5, 13]. Due to recent progress [1, 2], also the situation for various systems with diffraction spectra of mixed type has improved, especially from a computational point of view. In particular, one can explicitly calculate the diffraction measure in closed form for certain classes of examples. Still, the understanding of spectra in the presence of entropy is only at its beginning; compare [1, 4] and references therein. The purpose of this contribution is a further step into 'disordered territory', here via the analysis of mixed substitutions that are randomly applied at a *local* level. This is in contrast to global mixtures (which leads to S-adic systems), which have no entropy. Local mixtures were investigated in [8, 9], where the essential properties of the Fibonacci case were derived, along with first results on planar systems based on triangle inflation rules. Here, we extend the random Fibonacci system to the noble means family, and present the results from the point of view of dynamical systems. The entire family is still relatively simple because each individual member of a fixed noble mean family defines the same (deterministic) hull. Various generalisations are possible, but not discussed here.

M. Baake (✉) · M. Moll
Fakultät für Mathematik, Universität Bielefeld, Postfach 100131, 33501 Bielefeld, Germany
e-mail: mbaake@math.uni-bielefeld.de

M. Moll
e-mail: mmoll@math.uni-bielefeld.de

S. Schmid et al. (eds.), *Aperiodic Crystals*, DOI 10.1007/978-94-007-6431-6_3,

3.2 Construction

Let $\mathcal{A} = \{a, b\}$ be our two letter alphabet. For any fixed integer $m \geq 1$, we define a family $\mathcal{H}_m$ of substitution rules by

$$\zeta_{m,i}: \begin{cases} a \mapsto a^i b a^{m-i}, \\ b \mapsto a, \end{cases}$$

with $0 \leq i \leq m$, and refer to each $\zeta_{m,i}$ as a *noble means substitution (NMS)*. Each member of $\mathcal{H}_m$ is a primitive substitution with Pisot inflation multiplier $\lambda_m = \frac{1}{2}(m + \sqrt{m^2+4})$ and algebraic conjugate $\lambda_m' = \frac{1}{2}(m - \sqrt{m^2+4})$. Each substitution possesses a reflection symmetric and aperiodic two-sided discrete (or symbolic) hull $\mathbb{X}_{m,i}$, where the hull, as usual, is defined as the orbit closure of a fixed point in the local topology. Moreover, all elements of $\mathcal{H}_m$ are pairwise conjugate to each other which implies that, for fixed m, the hulls $\mathbb{X}_{m,i}$ are equal for $0 \leq i \leq m$.

We now fix a probability vector $(p_0, \dots, p_m)$, that is $p_i \geq 0$ and $\sum_{j=0}^{m} p_j = 1$. We define the random substitution rule

$$\zeta_m: \begin{cases} a \mapsto \begin{cases} \zeta_{m,0}(a) = ba^m, & \text{with probability } p_0, \\ \quad\vdots & \quad\vdots \\ \zeta_{m,m}(a) = a^m b, & \text{with probability } p_m, \end{cases} \\ b \mapsto a, \end{cases} \qquad \text{where } M := \begin{pmatrix} m & 1 \\ 1 & 0 \end{pmatrix}$$

is its substitution matrix. We refer to ζ_m as the *random noble means substitution (RNMS)*. The application of ζ_m occurs *locally*, which means that we decide separately on each letter a which of the $m+1$ possible realisations we choose. In particular, for each $k \in \mathbb{N}$, $\zeta_m^k(a)$ is a random variable. As there is no direct analogue of a fixed point in the stochastic situation, we have to slightly adjust the notion of the two-sided discrete hull in this context. Note that aa is a legal word (see below for more) for all m, and consider

$$X_m := \{ w \in \mathcal{A}^{\mathbb{Z}} \mid w \text{ is an accumulation point of } (S^{j_n} \zeta_m^n(a|a))_{n \in \mathbb{N}_0} \},$$

where S denotes the shift. The two-sided discrete hull $\mathbb{X}_m$ is defined as the smallest closed and shift-invariant subset of $\mathcal{A}^{\mathbb{Z}}$ with $X_m \subset \mathbb{X}_m$. It is immediate that $\mathbb{X}_m$ is a superset of $\mathbb{X}_{m,i}$. Note that typical elements of $\mathbb{X}_m$ contain the subword bb, which is absent in $\mathbb{X}_{m,i}$. The hull $\mathbb{X}_m$ is characterised by the property that it contains all elements of $\{a, b\}^{\mathbb{Z}}$ that contain ζ_m-legal subwords only (see below for more).

3.3 Topological Entropy

In this section, we assume that all probabilities p_i are strictly positive. We call a finite word w *legal* with respect to ζ_m if there is a power $k \in \mathbb{N}$ such that w is a subword of some realisation of $\zeta_m^k(a)$. Furthermore, let $\mathcal{D}_{m,\ell}$ be the set of all legal words of length ℓ with respect to ζ_m. We refer to the function $C\colon \mathbb{N} \to \mathbb{N}$, $\ell \mapsto |\mathcal{D}_{m,\ell}|$ as the *complexity function* of ζ_m. It is known that the discrete hull of each member of $\mathcal{H}_m$ has linear complexity, which implies that the topological entropy vanishes here. In the stochastic setting, the picture changes; see [4] for background.

Let $m \in \mathbb{N}$ be arbitrary but fixed. The sets $\mathcal{G}_0 := \varnothing$, $\mathcal{G}_1 := \{b\}$, $\mathcal{G}_2 := \{a\}$ and

$$\mathcal{G}_n := \bigcup_{i=0}^{m} \prod_{j=0}^{m} \mathcal{G}_{n-1-\delta_{ij}}, \tag{3.1}$$

with δ_{ij} denoting the Kronecker symbol, are called the *generation sets* of ζ_m. The product in (3.1) is meant to be the set-theoretic product with respect to concatenation of words. Moreover, we define $\mathcal{G} := \lim_{n\to\infty} \mathcal{G}_n$ and refer to $\mathcal{G}_n$ as the *nth generation set*. The length ℓ_n of words in $\mathcal{G}_n$ is given by the sequence $\ell_0 = 0$, $\ell_1 = 1$, $\ell_2 = 1$ and $\ell_{n+1} = m\ell_n + \ell_{n-1}$, for $n \geq 2$. The set $\mathcal{G}_n$ consists of all possible exact realisations $\zeta_m^{n-1}(b)$. Since not all legal words result from an exact substitution, which can again be seen from the example bb, it is clear that $|\mathcal{G}_n| < C(\ell_n)$ for $n \geq 2$.

In [8], Godrèche and Luck computed the topological entropy of ζ_1 under the implicit assumption that

$$\lim_{n\to\infty} \frac{1}{\ell_n} \log\bigl(C(\ell_n)\bigr) = \lim_{n\to\infty} \frac{1}{\ell_n} \log\bigl(|\mathcal{G}_n|\bigr),$$

which was recently proved by J. Nilsson [12]. This asymptotic identity is crucial because the exact computation of the complexity function of ζ_m is still an open problem. It is easy to compute $|\mathcal{D}_{m,\ell}|$ for $\ell \leq m+2$ and it is known [10] that

$$|\mathcal{D}_{m,\ell}| = \sum_{i=0}^{3} \binom{\ell}{i} - \frac{1}{6} m(m+1)(3\ell - 2m - 4)$$

if $m+3 \leq \ell \leq 2m+2$, while an extension to arbitrary word lengths seems difficult. In [8], the entropy per letter for $m = 1$ is computed to be

$$h_1 = \lim_{n\to\infty} \frac{\log(|\mathcal{G}_n|)}{\ell_n} = \sum_{i=2}^{\infty} \frac{\log(i)}{\lambda_1^{i+2}} \approx 0.444399 > 0,$$

whereas a convenient representation for arbitrary m reads

$$h_m = \lim_{n\to\infty} \frac{\log(|\mathcal{G}_n|)}{\ell_n} = \frac{\lambda_m - 1}{1 - \lambda_m'} \cdot \sum_{i=2}^{\infty} \frac{\log(m(i-1)+1)}{\lambda_m^i}. \tag{3.2}$$

The result computed by Godrèche and Luck for $m = 1$ can be recovered by the observation that $(\lambda_1 - 1)/(1 - \lambda_1') = 1/\lambda_1^2$ in this case. Some numerical values are given in Table 3.1. It is not difficult to prove [10] that $\lim_{m\to\infty} h_m = 0$, which can be verified by estimating the logarithm in (3.2) via the square root and using the fact that λ_m/m tends to 1 as $m \to \infty$.

Table 3.1 Numerical values of the topological entropy for RNMS with $1 \le m \le 7$

m	1	2	3	4	5	6	7
h_m	0.444399	0.408549	0.371399	0.338619	0.310804	0.287298	0.267301

3.4 Frequencies of Subwords

We adopt the method of computing the frequencies of subwords via induced substitutions on words of length ℓ (with $\ell \in \mathbb{N}$), which was introduced in [13, Sect. 5.4.1], and modify it to fit the stochastic setting. To this end, we again assume that all probabilities p_i in the definition of ζ_m are strictly positive.

If $w = w_0 w_1 \cdots w_{\ell-1}$ is a word of length ℓ, we define $w_{[i,j]}$ to be the subword $w_i \cdots w_j$ of w of length $j - i + 1$. For $\ell \ge 2$, we denote $\zeta_{m,(\ell)} \colon \mathcal{D}_{m,\ell} \to \mathcal{D}_{m,\ell}$ as the *induced substitution* defined by

$$\zeta_{m,(\ell)} \colon w^{(i)} \mapsto \begin{cases} (v^{(i,1)}_{[k,k+\ell-1]})_{0 \le k \le |\zeta_m(w^{(i)}_0)|-1}, & \text{with probability } p_{i,1}, \\ \vdots & \vdots \\ (v^{(i,n_i)}_{[k,k+\ell-1]})_{0 \le k \le |\zeta_m(w^{(i)}_0)|-1}, & \text{with probability } p_{i,n_i}, \end{cases} \tag{3.3}$$

where $w^{(i)} \in \mathcal{D}_{m,\ell}$ and $v^{(i,j)}$ is a realisation of $w^{(i)}$ under ζ_m with probability $p_{i,j}$. This way, we ensure that we are neither under- nor overcounting subwords of a given length. Similar to the case of ζ_m, the result is a random variable.

The action of $\zeta_{m,(\ell)}$ on words in $\mathcal{D}_{m,\ell}$ is illustrated in Table 3.2 for $m = 1$ and $\ell = 2$. Applying the lexicographic order to the words in $\mathcal{D}_{m,\ell}$ leads to the corresponding substitution matrix $M_{m,\ell} := M(\zeta_{m,(\ell)})$. For any fixed $m \in \mathbb{N}$ and $\ell = 2$, we get

$$M_{m,2} = \begin{pmatrix} (m-1) + p_0 p_m & (m-1) + p_0 & 1 - p_0 & 1 \\ 1 - p_0 p_m & 1 - p_0 & p_0 & 0 \\ 1 - p_0 p_m & 1 & 0 & 0 \\ p_0 p_m & 0 & 0 & 0 \end{pmatrix}.$$

This matrix has the spectrum $\sigma(M_{m,2}) = \{\lambda_m, \lambda'_m, -p_0, p_0 p_m\}$. Furthermore, it is interesting to observe that the spectrum of the matrices $M_{m,\ell}$ for $\ell \ge 3$ is the same as that of $M_{m,2}$, except for the addition of zeros. Note that $\zeta_{m,(1)}$ agrees with ζ_m which implies that $M_{m,1} = M$.

The substitution matrix $M_{m,\ell}$ is primitive for all m and ℓ, which allows an application of Perron–Frobenius theory; see [14] for general background. This implies that there is a strictly positive right eigenvector $\phi^{(\ell)}$ to the eigenvalue λ_m. Note that λ_m does not depend on any of the p_i at all, whereas this is not the case for $\phi^{(\ell)}$.

We define a measure on the discrete hull $\mathbb{X}_m$ as follows. For any word $v \in \mathcal{D}_{m,\ell}$ and $k \in \mathbb{N}$, let $Z_k(v) := \{w \in \mathbb{X}_m \mid w_{[k,k+\ell-1]} = v\}$ be the cylinder set of v that starts at position k. Then, the family $\{Z_k(v)\}_{k \in \mathbb{N}}$ generates the product topology and

Table 3.2 The action of $\zeta_{1,(2)}$ on ζ_1-legal words of length 2

$w \in \mathcal{D}_{1,2}$	$\zeta_1(w)$	$v^{(i,j)}$	$\mathbb{P}$	$w \in \mathcal{D}_{1,2}$	$\zeta_1(w)$	$v^{(i,j)}$	$\mathbb{P}$
aa	$abab$	$(ab)(ba)$	p_1^2	ab	aba	$(ab)(ba)$	p_1
	$abba$	$(ab)(bb)$	$p_0 p_1$		baa	$(ba)(aa)$	p_0
	$baab$	$(ba)(aa)$	$p_0 p_1$	ba	aab	(aa)	p_1
	$baba$	$(ba)(ab)$	p_0^2		aba	(ab)	p_0
				bb	aa	(aa)	1

we define the measure $\mu\colon \mathbb{X}_m \to \mathbb{R}_{\geq 0}$ on the cylinder sets as $\mu(Z_k(v)) = \phi^{(\ell)}(v)$, where $\phi^{(\ell)}(v)$ is the entry of $\phi^{(\ell)}$ with respect to v. This is a proper (and consistent) definition of a measure on $\mathbb{X}_m$, which can also be found in [13, Sect. 5.4.2]. By construction, the measure is shift-invariant.

The following theorem [10] shows that, similar to the deterministic setting [13], it is possible to interpret the entries of $\phi^{(\ell)}$ as the frequencies of legal subwords with respect to ζ_m as follows:

Theorem 3.1 *Let $\mathbb{X}_m \subset \mathcal{A}^{\mathbb{Z}}$ be the hull of the random noble means substitution for $m \in \mathbb{N}$ and μ the shift-invariant probability measure on $\mathbb{X}_m$ as defined above. For any $f \in L^1(\mathbb{X}_m, \mu)$ and for an arbitrary but fixed $s \in \mathbb{Z}$, the identity*

$$\lim_{N\to\infty} \frac{1}{N} \sum_{i=s}^{N+s-1} f(S^i x) = \int_{\mathbb{X}_m} f \, \mathrm{d}\mu$$

holds for μ-almost every $x \in \mathbb{X}_m$.

The proof can be accomplished by inspecting the family of random variables $R = \{f(S^i w)\}_{i\in\mathbb{N}}$, where f is a *patch recognition function* that evaluates to 1 if $(S^i w)_{[s,s+\ell-1]} = v$ for an arbitrary but fixed word $v \in \mathcal{D}_{m,\ell}$ and to 0 otherwise. By observing that, given any $i \in \mathbb{Z}$, the sets

$$\mathcal{I}_i := \{(S^{i+k(\ell+m)} w)_{[s,s+\ell-1]} \mid k \in \mathbb{N}\}$$

consist of pairwise independent words, we can split up the summation over R appropriately and apply Etemadi's version of the strong law of large numbers [7, Theorem 1] to each sum over $\mathcal{I}_i$ separately. This, in conjunction with an application of the Stone–Weierstrass theorem, implies the assertion.

3.5 Diffraction Measure

The symbolic situation is turned into a geometric one as follows. In view of the left PF eigenvector of M, a and b are turned into intervals of lengths λ_m and 1, respectively. The left end points are the coordinates we use. The corresponding *continuous* hull $\mathbb{Y}_m$ is the orbit closure of all accumulation points of the geometric inflation rule under $\mathbb{R}$. Let $\varLambda \subset \mathbb{Z}[\lambda_m]$ be a coordinatisation of an element of $\mathbb{X}_m$ in $\mathbb{R}$. Then, $\varLambda$ is a discrete point set that fits into the same cut and project scheme as all elements of the family $\mathcal{H}_m$. With respect to $\varLambda$, the smallest interval that covers $\varLambda' = \{x' \mid x \in \varLambda\}$ in internal space is given by $[\lambda_m' - 1, 1 - \lambda_m']$. Then, $\varLambda$ is relatively dense with covering radius λ_m, and a subset of a model set, which implies that $\varLambda$ is a Meyer set by [11, Theorem 9.1]. Let $\varLambda_R = \varLambda \cap \overline{B_R}$ and consider the autocorrelation

$$\gamma := \lim_{R\to\infty} \frac{\delta_R * \widetilde{\delta_R}}{\operatorname{vol}(B_R)} \quad \text{with } \delta_R = \sum_{x\in\varLambda_R} \delta_x .$$

The limit almost surely exists due to the ergodicity of our system. By construction, γ is a positive definite measure which implies that its Fourier transform exists and is a positive measure. Regarding the Lebesgue decomposition $\widehat{\gamma} = (\widehat{\gamma})_{\mathrm{pp}} + (\widehat{\gamma})_{\mathrm{ac}} + (\widehat{\gamma})_{\mathrm{sc}}$, it is possible to compute the pure point part to be

$$(\widehat{\gamma})_{\mathrm{pp}} = \sum_{k\in L^{\circledast}} \big|\widehat{\eta_a}(-k') + \widehat{\eta_b}(-k')\big|^2 \delta_k ,$$

where $L^{\circledast} = \mathbb{Z}[\lambda_m]/\sqrt{m^2+4}$ is the Fourier module. In the case of $m = 1$, the invariant measures $\widehat{\eta_a}, \widehat{\eta_b}$ can be approximated via the recursion relation

$$\begin{pmatrix} \widehat{\eta_a}(y) \\ \widehat{\eta_b}(y) \end{pmatrix} = |\xi|^n \cdot \prod_{\ell=1}^{n} \left[p_0 \begin{pmatrix} \mathrm{e}^{-2\pi \mathrm{i} y \xi^{\ell-1}} & 1 \\ 1 & 0 \end{pmatrix} + p_1 \begin{pmatrix} 1 & 1 \\ \mathrm{e}^{-2\pi \mathrm{i} y \xi^{\ell}} & 0 \end{pmatrix} \right] \cdot \begin{pmatrix} \widehat{\eta_a}(y\xi^n) \\ \widehat{\eta_b}(y\xi^n) \end{pmatrix}, \tag{3.4}$$

with $n \in \mathbb{N}$ and $\xi := \lambda_1'$. As $\xi^n \to 0$ for $n \to \infty$, an appropriate choice of the eigenvector $(\widehat{\eta_a}(0), \widehat{\eta_b}(0))^T$ for the equation

$$\begin{pmatrix} 1 & 1 \\ 1 & 0 \end{pmatrix} \cdot \begin{pmatrix} \widehat{\eta_a}(0) \\ \widehat{\eta_b}(0) \end{pmatrix} = \lambda_1 \cdot \begin{pmatrix} \widehat{\eta_a}(0) \\ \widehat{\eta_b}(0) \end{pmatrix},$$

which results from (3.4) for $k = 0$ and $n = 1$, fixes the base of the recursion and provides the desired approximation. Since $\widehat{\eta_a}(0) + \widehat{\eta_b}(0)$ must be the density of $\varLambda$, which always is $\lambda_1/\sqrt{5}$, one finds $\widehat{\eta_a}(0) = 1/\sqrt{5}$ and $\widehat{\eta_b}(0) = (\lambda_1 - 1)/\sqrt{5}$. Let μ_s be the measure on $\mathbb{Y}_m$ induced by μ via suspension; see [6, Chap. 11] for general background. One consequence, due to a theorem of Strungaru [15] and an application of the methods of [3], is that our random dynamical system $D := (\mathbb{Y}_m, \mathbb{R}, \mu_s)$ is ergodic, but not weakly mixing. In particular, it has strong long-range order.

Due to the stochastic setting with positive entropy, one expects a non-trivial absolutely continuous part. For $m = 1$, the ergodicity of D almost surely yields a diffraction measure which can be represented as $\widehat{\gamma} = (\widehat{\gamma})_{\mathrm{pp}} + \alpha(k) \cdot \lambda$, where

$$\widehat{\gamma}(\{k\}) = \lim_{n\to\infty} \frac{1}{L_n^2} \cdot \left|\mathbb{E}\big(g_n(k)\big)\right|^2.$$

Here, $\mathbb{E}$ refers to averaging with respect to μ_s and

$$\alpha(k) := \lim_{n\to\infty} \frac{1}{L_n} \cdot \left(\mathbb{E}\big(|g_n(k)|^2\big) - \left|\mathbb{E}\big(g_n(k)\big)\right|^2\right),$$

with the random exponential sum $g_n(k) := \sum_{j=1}^{F_{n+1}} \mathrm{e}^{-2\pi \mathrm{i} k x_j}$ and $L_n := \lambda_1 F_n + F_{n-1}$, where F_n is the nth Fibonacci number. Now let

$$A_n(k) := \mathbb{E}\big(g_n(k)\big) \quad \text{and} \quad B_n(k) := \mathbb{E}\big(|g_n(k)|^2\big) - \left|\mathbb{E}\big(g_n(k)\big)\right|^2.$$

Godrèche and Luck [8] derived a recursion relation for the sequence $A_n(k)$,

$$A_n(k) = \big(p_1 + p_0 \mathrm{e}^{-2\pi \mathrm{i} k L_{n-2}}\big) \cdot A_{n-1}(k) + \big(p_0 + p_1 \mathrm{e}^{-2\pi \mathrm{i} k L_{n-1}}\big) \cdot A_{n-2}(k),$$

where $A_0(k) = \mathrm{e}^{-2\pi \mathrm{i} k}$ and $A_1(k) = \mathrm{e}^{-2\pi \mathrm{i} k \lambda_1}$. Analogously, one derives a recursion relation for the sequence $B_n(k)$,

$$B_n(k) = B_{n-1}(k) + B_{n-2}(k) + 2 p_0 p_1 \cdot \Delta_n(k),$$

with

$$\begin{aligned}\Delta_n(k) &= \big(1 - \cos(2\pi k L_{n-1})\big) \cdot |A_{n-2}(k)|^2 + \big(1 - \cos(2\pi k L_{n-2})\big) \cdot |A_{n-1}(k)|^2 \\ &\quad - \mathrm{Re}\Big[\big(1 - \mathrm{e}^{2\pi \mathrm{i} k L_{n-1}}\big) \cdot \big(1 - \mathrm{e}^{-2\pi \mathrm{i} k L_{n-2}}\big) \cdot A_{n-1}(k) \cdot \overline{A_{n-2}(k)}\Big]\end{aligned}$$

and $B_0(k) = B_1(k) = 0$. In [8], almost surely by way of a misprint, the authors applied complex conjugation on $A_{n-1}(k)$ instead of $A_{n-2}(k)$, which makes a huge difference, as the sequence $B_n(k)$ does not converge in that case. The recursion for $B_n(k)$ can be solved and the explicit representation reads

$$B_n(k) = 2 p_0 p_1 \cdot \sum_{i=2}^{n} F_{n+1-i} \Delta_i(k).$$

A detailed discussion of the continuous part of $\widehat{\gamma}$ can be found in [10].

The illustration of an approximation of the diffraction measure $\widehat{\gamma}$ in case of $m = 1$ and $p_0 = p_1 = \frac{1}{2}$, based on the sequences $A_n(k)$ and $B_n(k)$, is shown in Fig. 3.1, which agrees with the average over many realisations for the same length.

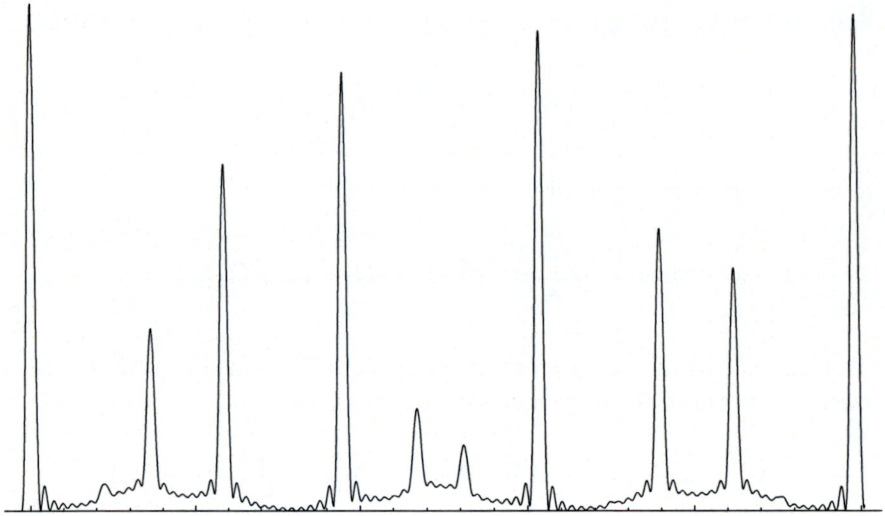

Fig. 3.1 Approximative calculation of the diffraction measure $\widehat{\gamma}$ for $m = 1$ and $p_0 = p_1 = \frac{1}{2}$, based on $A_n(k)$ and $B_n(k)$ with $n = 6$

Acknowledgements We thank Uwe Grimm and Johan Nilsson for discussions. This work was supported by the German Research Council (DFG), within the CRC 701.

References

1. Baake M, Birkner M, Moody RV (2010) Diffraction of stochastic point sets: explicitly computable examples. Commun Math Phys 293:611–660. arXiv:0803.1266
2. Baake M, Gähler F, Grimm U (2012) Spectral and topological properties of a family of generalised Thue–Morse sequences. J Math Phys 53:032701. arXiv:1201.1423
3. Baake M, Lenz D (2004) Dynamical systems on translation bounded measures: pure point dynamical and diffraction spectra. Ergod Theory Dyn Syst 24:1867–1893. arXiv:math.DS/0302231
4. Baake M, Lenz D, Richard C (2007) Pure point diffraction implies zero entropy for Delone sets with uniform cluster frequencies. Lett Math Phys 82:61–77. arXiv:0706.1677
5. Baake M, Moody RV (2004) Weighted Dirac combs with pure point diffraction. J Reine Angew Math 573:61–94. arXiv:math.MG/0203030
6. Cornfeld IP, Fomin SV, Sinai YG (1982) Ergodic theory. Springer, New York
7. Etemadi N (1981) An elementary proof of the strong law of large numbers. Z Wahrscheinlichkeitstheor Verw Geb 55:119–122
8. Godrèche C, Luck JM (1989) Quasiperiodicity and randomness in tilings of the plane. J Stat Phys 55:1–28
9. Lütkehölter C (2010) Diffraktion stochastischer Fibonacci-Mengen. Diplomarbeit, Univ Bielefeld
10. Moll M (2013) Random Noble means substitutions. PhD thesis, Univ Bielefeld, in preparation
11. Moody RV (1997) Meyer sets and their duals. In: The mathematics of long-range aperiodic order. NATO ASI series C, vol 489. Kluwer, Dordrecht, pp 403–441

12. Nilsson J (2012) On the entropy of a family of random substitution systems. Monatshefte Math 166:1–15. arXiv:1103.4777
13. Queffélec M (2010) Substitution dynamical systems—spectral analysis, 2nd edn. Springer, Berlin
14. Seneta E (2006) Non-negative matrices and Markov chains, 2nd revised edn. Springer, New York
15. Strungaru N (2005) Almost periodic measures and long-range order in Meyer sets. Discrete Comput Geom 33:483–505

Chapter 4
Magic Numbers in the Discrete Tomography of Cyclotomic Model Sets

Christian Huck

Abstract We report recent progress in the problem of distinguishing convex subsets of cyclotomic model sets Λ by (discrete parallel) X-rays in prescribed Λ-directions. It turns out that for any of these model sets Λ there exists a 'magic number' m_Λ such that any two convex subsets of Λ can be distinguished by their X-rays in any set of m_Λ prescribed Λ-directions. In particular, for pentagonal, octagonal, decagonal and dodecagonal model sets, the least possible numbers are in that very order 11, 9, 11 and 13.

4.1 Introduction

Discrete tomography is concerned with the inverse problem of retrieving information about some *finite* set in Euclidean space from (generally noisy) information about its slices. One important problem is the *unique reconstruction* of a finite point set in Euclidean 3-space from its (*discrete parallel*) *X-rays* in a small number of directions, where the *X-ray* of the finite set in a certain direction is the *line sum function* giving the number of points in the set on each line parallel to this direction.

The interest in the discrete tomography of planar Delone sets Λ with long-range order is motivated by the requirement in materials science for the unique reconstruction of solid state materials like *quasicrystals* slice by slice from their images under quantitative *high resolution transmission electron microscopy* (HRTEM). In fact, in [12, 14] a technique is described, which can, for certain crystals, effectively measure the number of atoms lying on densely occupied columns. Clearly, the aforementioned density condition forces us to consider only Λ-directions, i.e. directions parallel to lines through two different points of Λ.

In the *quasicrystallographic* setting, the positions to be determined form a finite subset of a nonperiodic Delone set with long-range order (more precisely, a *model set* [2, 13]). Model sets possess a dimensional hierarchy, i.e. they allow a slicing into planar model sets. In fact, many of the model sets that describe real quasicrystallographic structures allow a slicing such that each slice is an *n-cyclotomic model set*,

C. Huck (✉)
Fakultät für Mathematik, Universität Bielefeld, Postfach 100131, 33501 Bielefeld, Germany
e-mail: huck@math.uni-bielefeld.de

S. Schmid et al. (eds.), *Aperiodic Crystals*, DOI 10.1007/978-94-007-6431-6_4,

the latter being (planar) Delone sets contained in the additive subgroup of the Euclidean plane generated by the nth roots of unity; cf. [6, 7, 9] and [15] for details. It therefore suffices to study the discrete tomography of these cyclotomic model sets. In practice, the cases $n = 5, 8, 12$ are of particular interest. In the present text, we shall mainly focus on the larger class of *cyclotomic Delone sets*.

Since different finite subsets of an n-cyclotomic model set Λ may have the same X-rays in several directions, one is naturally interested in conditions to be imposed on the set of directions together with restrictions on the possible finite subsets of Λ such that the latter phenomenon cannot occur. Here, we consider the *convex subsets* of Λ and summarise recent results in the problem of distinguishing convex subsets of Λ by X-rays in prescribed Λ-directions. It turns that there are four prescribed Λ-directions such that any two convex subsets of Λ can be distinguished by the corresponding X-rays, whereas less than four Λ-directions never suffice for this purpose. Much more novel is the result obtained in collaboration with M. Spieß that there is a finite number m_Λ such that any two convex subsets of Λ can be distinguished by their X-rays in *any* set of m_Λ prescribed Λ-directions. Moreover, the least possible numbers m_Λ in the case of the practically most relevant examples of n-cyclotomic model sets Λ with $n = 5$, 8 and 12 only depend on n and are in that very order 11, 9 and 13. This extends a well-known result of R.J. Gardner and P. Gritzmann [5] on the corresponding problem for the crystallographic cases $n = 3, 4$ of the triangular (resp., square) lattice Λ (with least possible number $m_\Lambda = 7$ in both cases) to cases that are relevant in quasicrystallography.

The intention of this text is to provide an easy to read guide to the results of [10] with a view towards practical applications. Detailed proofs, related results and an extensive list of references can be found there. For the algorithmic reconstruction problem in the quasicrystallographic setting, the reader is referred to [3, 7].

4.2 Cyclotomic Delone Sets

Throughout the text, the Euclidean plane is identified with the complex numbers. For $z \in \mathbb{C}$, $\bar{z}$ denotes the complex conjugate of z. Further, we denote by K_Λ the smallest subfield of $\mathbb{C}$ that contains the rational numbers as well as the union of $\Lambda - \Lambda$ and its image $\overline{\Lambda - \Lambda}$ under complex conjugation. Recall that Λ is called a *Delone set* if it is both uniformly discrete and relatively dense. For $n \in \mathbb{N}$, we always let $\zeta_n = e^{2\pi i/n}$, a primitive nth root of unity in $\mathbb{C}$. Then, the smallest subfield of $\mathbb{C}$ that contains the rational numbers as well as ζ_n is the nth cyclotomic field denoted by $\mathbb{Q}(\zeta_n)$. The latter is just the $\mathbb{Q}$-span of the nth roots of unity and thus contains the $\mathbb{Z}$-span $\mathbb{Z}[\zeta_n]$ of the nth roots of unity. Recall that a *homothety* of the complex plane is given by $z \mapsto \lambda z + t$, where $\lambda \in \mathbb{R}$ is positive and $t \in \mathbb{C}$. For the purpose of this text, the following rather abstract definition provides a convenient framework.

Definition 4.1 Let $n \geq 3$. A Delone set $\Lambda \subset \mathbb{C}$ is called an *n-cyclotomic Delone set* if it satisfies the following properties:

(n-Cyc) K_Λ is contained in $\mathbb{Q}(\zeta_n)$.

(Hom) For any finite subset F of K_Λ, there is a homothety h of the complex plane that maps the elements of F to Λ.

Further, Λ is called a *cyclotomic Delone set* if it is an n-cyclotomic Delone set for a suitable $n \geq 3$.

Standard examples of n-cyclotomic Delone sets are the *n-cyclotomic model sets*, which were also called *cyclotomic model sets with underlying $\mathbb{Z}$-module* $\mathbb{Z}[\zeta_n]$ in [9, Sect. 4.5] and are defined via the canonical cut and project scheme that is given by the Minkowski representation of the $\mathbb{Z}$-module $\mathbb{Z}[\zeta_n]$; see [9, 10] for details. These sets are certain Delone subsets of the $\mathbb{Z}$-module $\mathbb{Z}[\zeta_n]$ and range from periodic examples like the fourfold square lattice ($n = 4$) or the sixfold triangular lattice ($n = 3$) to nonperiodic examples like the vertex set of the tenfold Tübingen triangle tiling ($n = 5$), the eightfold Ammann–Beenker tiling ($n = 8$) or the twelvefold shield tiling ($n = 12$); see [8, Fig. 1], [9, Fig. 2] and Fig. 1 below for illustrations. Note that the vertex sets of the famous Penrose tilings of the plane fail to be 5-cyclotomic model sets but can still be seen to be 5-cyclotomic Delone sets; see [1] and references therein.

4.3 Determination of Convex Subsets by X-Rays

Let (t_1, t_2, t_3, t_4) be an ordered tuple of four distinct elements of $\mathbb{R} \cup \{\infty\}$. Then, its *cross ratio* $\langle t_1, t_2, t_3, t_4 \rangle$ is the nonzero real number defined by

$$\langle t_1, t_2, t_3, t_4 \rangle = \frac{(t_3 - t_1)(t_4 - t_2)}{(t_3 - t_2)(t_4 - t_1)},$$

with the usual conventions if one of the t_i equals ∞.

The unit circle in $\mathbb{C}$ is denoted by $\mathbb{S}^1$ and its elements are also called *directions*. For a nonzero complex number z, we denote by $\mathrm{sl}(z)$ the slope of z, i.e. $\mathrm{sl}(z) = -i(z - \bar{z})/(z + \bar{z}) \in \mathbb{R} \cup \{\infty\}$. Let Λ be a subset of $\mathbb{C}$. A direction $u \in \mathbb{S}^1$ is called a *Λ-direction* if it is parallel to a nonzero element of the *difference set* $\Lambda - \Lambda = \{v - w \mid v, w \in \Lambda\}$ of Λ. By construction, the cross ratio of slopes of four pairwise nonparallel Λ-directions is an element of the field $K_\Lambda \cap \mathbb{R}$. In case of n-cyclotomic Delone sets Λ, these cross ratios are thus elements of the field $\mathbb{Q}(\zeta_n) \cap \mathbb{R}$.

Definition 4.2

- Let F be a finite subset of $\mathbb{C}$, let $u \in \mathbb{S}^1$ be a direction, and let $\mathcal{L}_u$ be the set of lines in the complex plane in direction u. Then the (*discrete parallel*) *X-ray* of F *in direction* u is the function $X_u F : \mathcal{L}_u \to \mathbb{N}_0 = \mathbb{N} \cup \{0\}$, defined by

$$X_u F(\ell) = \mathrm{card}(F \cap \ell).$$

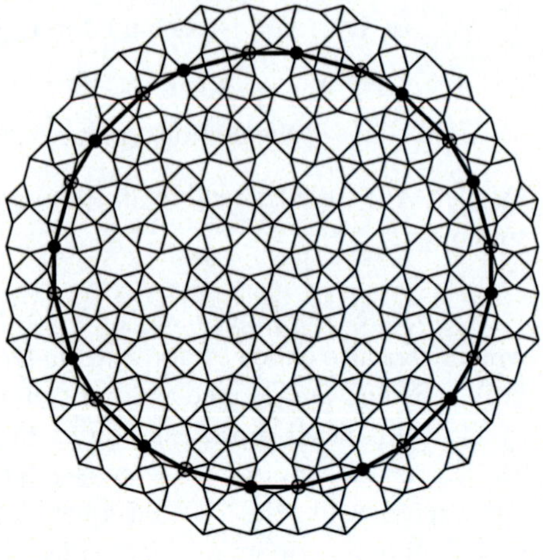

Fig. 1 The boundary of a U-polygon in the vertex set Λ of the twelvefold shield tiling, where U is the set of twelve pairwise nonparallel Λ-directions given by the edges and diagonals of the central regular dodecagon. The vertices of Λ in the interior of the U-polygon together with the vertices indicated by the *black* and *grey dots*, respectively, give two different convex subsets of Λ with the same X-rays in the directions of U

- Let $\mathcal{F}$ be a collection of finite subsets of $\mathbb{C}$ and let $U \subset \mathbb{S}^1$ be a finite set of directions. We say that the elements of $\mathcal{F}$ are *determined* by the X-rays in the directions of U if different elements of $\mathcal{F}$ cannot have the same X-rays in the directions of U.

One can easily see that no finite set of pairwise nonparallel Λ-directions suffices in order to determine the whole class of finite subsets of Λ by the corresponding X-rays [9]. It is therefore necessary to impose some restriction on the finite subsets of Λ to be determined. It has proven most fruitful to focus on the *convex subsets* of cyclotomic Delone sets. The latter are bounded (and thus finite) subsets C of Λ satisfying the equation $C = \mathrm{conv}(C) \cap \Lambda$, where $\mathrm{conv}(C)$ denotes the convex hull of C. One has the following fundamental result which shows that one has to choose the set U of Λ-directions in such a way that certain convex polygons cannot exist; cf. [9, Proposition 4.6 and Lemma 4.5]. Here, for a finite set $U \subset \mathbb{S}^1$ of directions, a nondegenerate convex polygon $P \subset \mathbb{C}$ is called a *U-polygon* if it has the property that whenever v is a vertex of P and $u \in U$, the line in the complex plane in direction u which passes through v also meets another vertex v' of P. P is called a *U-polygon in Λ*, if its vertices lie in Λ. Note that the proof of direction (ii)$\Rightarrow$(i) needs property (Hom) and see Fig. 1 for an illustration of the other direction (i)$\Rightarrow$(ii).

Theorem 4.1 *Let Λ be a cyclotomic Delone set and let $U \subset \mathbb{S}^1$ be a set of two or more pairwise nonparallel Λ-directions. The following statements are equivalent*:

(i) *The convex subsets of Λ are determined by the X-rays in the directions of U.*
(ii) *There is no U-polygon in Λ.*

In addition, if $\mathrm{card}(U) < 4$, *then there is a U-polygon in Λ.*

Employing Darboux's theorem on second midpoint polygons [4] together with a blend of sophisticated methods from the theory of cyclotomic fields and previous results obtained by Gardner and Gritzmann [5], one obtains the following deep result on U-polygons; cf. [10, Theorem 5.7]. Note that the proof heavily relies on property (n-Cyc).

Theorem 4.2 *Let $n \geq 3$ and let Λ be an n-cyclotomic Delone set. Further, let $U \subset \mathbb{S}^1$ be a set of four or more pairwise nonparallel Λ-directions and suppose the existence of a U-polygon. Then the cross ratio of slopes of any four directions of U, arranged in order of increasing angle with the positive real axis, is an element of the finite set $\mathcal{C}_{\mathrm{lcm}(2n,12)}(\mathbb{Q}(\zeta_n) \cap \mathbb{R})$ of numbers in the field $\mathbb{Q}(\zeta_n) \cap \mathbb{R}$ that can be written in the form*

$$\frac{(1-\zeta_{\mathrm{lcm}(2n,12)}^{k_1})(1-\zeta_{\mathrm{lcm}(2n,12)}^{k_2})}{(1-\zeta_{\mathrm{lcm}(2n,12)}^{k_3})(1-\zeta_{\mathrm{lcm}(2n,12)}^{k_4})},$$

where (k_1, k_2, k_3, k_4) is an element of the set

$$\{(k_1, k_2, k_3, k_4) \in \mathbb{N}^4 \mid k_3 < k_1 \leq k_2 < k_4 \leq \mathrm{lcm}(2n, 12) - 1 \text{ and } k_1 + k_2 = k_3 + k_4\}.$$

Moreover, $\mathrm{card}(U)$ is bounded above by a finite number $b_n \in \mathbb{N}$ that only depends on n. In particular, one can choose $b_3 = b_4 = 6$, $b_5 = 10$, $b_8 = 8$ and $b_{12} = 12$.

Theorems 4.1 and 4.2 now immediately imply our main result on the determination of convex subsets of cyclotomic Delone sets; cf. [10, Theorem 5.11].

Theorem 4.3 *Let $n \geq 3$ and let Λ be an n-cyclotomic Delone set.*

(a) *There are sets of four pairwise nonparallel Λ-directions such that the convex subsets of Λ are determined by the corresponding X-rays. In addition, less than four pairwise nonparallel Λ-directions never suffice for this purpose.*
(b) *There is a finite number $m_n \in \mathbb{N}$ that only depends on n such that the convex subsets of Λ are determined by the X-rays in any set of m_n pairwise nonparallel Λ-directions. In particular, one can choose $m_3 = m_4 = 7$, $m_5 = 11$, $m_8 = 9$ and $m_{12} = 13$.*

By Theorems 4.1 and 4.2 above, it suffices for Part (a) to take any set of four pairwise nonparallel Λ-directions such that the cross ratio of their slopes, arranged in order of increasing angle with the positive real axis, is *not* an element of the finite set $\mathcal{C}_{\mathrm{lcm}(2n,12)}(\mathbb{Q}(\zeta_n) \cap \mathbb{R})$; cf. [10, Corollary 4.10] for concrete results in the practically most important cases $n = 5, 8, 12$ of quasicrystallography.

4.4 Concluding Remarks

Our above analysis heavily relies on the assumption of ideal data and is therefore only a very first step towards a satisfactory tool for materials science. Further, it would certainly be interesting to abandon the slice by slice approach and work, for a Delone set Λ in Euclidean 3-space, with Λ-directions *in general position* instead; compare the approach to 3D reconstruction of atomic arrangements presented in [11]. In that case, it might well be that *seven* is a *universal* magic number for the determination of convex subsets by X-rays; cf. [4, Problem 2.1].

Acknowledgements This work was supported by the German Research Council (Deutsche Forschungsgemeinschaft), within the CRC 701.

References

1. Baake M, Huck C (2007) Discrete tomography of Penrose model sets. Philos Mag 87:2839–2846. arXiv:math-ph/0610056
2. Baake M, Moody RV (eds) (2000) Directions in mathematical quasicrystals. CRM monograph series, vol 13. AMS, Providence
3. Baake M, Gritzmann P, Huck C, Langfeld B, Lord K (2006) Discrete tomography of planar model sets. Acta Crystallogr 62:419–433. arXiv:math/0609393
4. Gardner RJ (2006) Geometric tomography, 2nd edn. Cambridge University Press, New York
5. Gardner RJ, Gritzmann P (1997) Discrete tomography: determination of finite sets by X-rays. Trans Am Math Soc 349:2271–2295
6. Huck C (2007) Discrete tomography of Delone sets with long-range order. PhD thesis, Universität Bielefeld. Logos, Berlin
7. Huck C (2009) Discrete tomography of icosahedral model sets. Acta Crystallogr 65:240–248. arXiv:0705.3005
8. Huck C (2009) On the existence of U-polygons of class $c \geq 4$ in planar point sets. Discrete Math 309:4977–4981 arXiv:0811.3546
9. Huck C (2009) Uniqueness in discrete tomography of Delone sets with long-range order. Discrete Comput Geom 42(4):740–758 arXiv:0711.4525
10. Huck C, Spieß M (2012) Solution of a uniqueness problem in the discrete tomography of algebraic Delone sets. J Reine Angew Math. doi:10.1515/crelle.2012.026. arXiv:1101.4149
11. Ishibashi Y, Sugiura H, Saitoh K, Tanaka N (2011) Three-dimensional reconstruction of the atomic arrangement of icosahedral quasicrystals by binary discrete tomography. Philos Mag 91:2519–2527
12. Kisielowski C, Schwander P, Baumann FH, Seibt M, Kim Y, Ourmazd A (1995) An approach to quantitative high-resolution transmission electron microscopy of crystalline materials. Ultramicroscopy 58:131–155
13. Moody RV (2000) Model sets: a survey. In: Axel F, Dénoyer F, Gazeau J-P (eds) From quasicrystals to more complex systems. EDP Sciences/Springer, Les Ulis/Berlin, pp 145–166. arXiv:math/0002020
14. Schwander P, Kisielowski C, Seibt M, Baumann FH, Kim Y, Ourmazd A (1993) Mapping projected potential, interfacial roughness, and composition in general crystalline solids by quantitative transmission electron microscopy. Phys Rev Lett 71:4150–4153
15. Steurer W (2004) Twenty years of structure research on quasicrystals, part I: pentagonal, octagonal, decagonal and dodecagonal quasicrystals. Z Kristallogr 219:391–446

Chapter 5
Some Comments on the Inverse Problem of Pure Point Diffraction

Venta Terauds and Michael Baake

Abstract In a recent paper arXiv:1111.3617, Lenz and Moody presented a method for constructing families of real solutions to the inverse problem for a given pure point diffraction measure. Applying their technique and discussing some possible extensions, we present, in a non-technical manner, some examples of homometric structures.

5.1 Introduction

Kinematic diffraction is concerned with the Fourier transform $\widehat{\gamma}$ of the autocorrelation γ of a given structure, the latter described by a measure ω, which is usually assumed to be translation bounded; see [2] for a recent summary of the state of affairs. Of particular relevance (for crystals and quasicrystals, say) are systems with pure point (or pure Bragg) diffraction, i.e. those where $\widehat{\gamma}$ is a pure point measure.

In considering the inverse problem, namely the problem of determining which structure or structures could have produced a given diffraction, one is naturally led to the concept of *homometry*, where structures that give rise to the same diffraction measure are said to be homometric. Via the autocorrelation measure, homometry is well-defined for certain classes of measures, and, accordingly, also for objects such as point sets and tilings (an overview of these concepts is given in [3, Chap. 9.6]). Various methods have been used (see [1, 3, 7, 8] and references therein) to construct different objects (mainly Dirac combs) that are homometric. However, a way of finding *all* objects with a given diffraction, which we shall refer to as the *diffraction solution class* of a given diffraction measure, has long remained elusive.

In [11], Lenz and Moody present a method for abstractly parametrising the real solution class of a given pure point diffraction measure. Their approach is based on the Halmos–von Neumann theorem in conjunction with concepts from the theory of

V. Terauds (✉) · M. Baake
Fakultät für Mathematik, Universität Bielefeld, Postfach 100131, 33501 Bielefeld, Germany
e-mail: terauds@math.uni-bielefeld.de

M. Baake
e-mail: mbaake@math.uni-bielefeld.de

S. Schmid et al. (eds.), *Aperiodic Crystals*, DOI 10.1007/978-94-007-6431-6_5,

(stochastic) point processes. The objects constructed via their method are in many cases measures, but as we shall see, even for a very simple periodic diffraction, one may construct objects with that diffraction that generically fail to be measures. Thus many open questions remain, in particular whether (or in what sense) non-measure solutions have a reasonable physical interpretation.

The mathematical formalism behind the construction method in [11] is based on a theory of (generalised) point processes and is quite formidable. In essence, it justifies the use of the lower path of the Wiener diagram

$$\begin{array}{ccc} \omega & \xrightarrow{\;\circledast\;} & \gamma \\ {\scriptstyle \mathcal{F}}\big\downarrow & & \big\downarrow{\scriptstyle \mathcal{F}} \\ \widehat{\omega} & \xrightarrow{\;|\cdot|^2\;} & \widehat{\gamma} \end{array}$$

in the reverse direction, where $\mathcal{F}$ denotes Fourier transform and $\circledast$ the volume-averaged (or Eberlein) convolution (so that $\gamma = \omega \circledast \tilde{\omega}$; compare [2, 10]). In particular, for a diffraction measure of the form $\widehat{\gamma} = \sum_{k \in L} |A(k)|^2 \delta_k$, with L a countable set, one may (formally, but consistently) invert the left as well as the bottom arrow of the diagram (the latter interpreted as $|\cdot|^2$ being applied to the coefficients or 'intensities' $I(k) = |A(k)|^2$ individually) to construct a measure (or at least a tempered distribution) ω with diffraction $\widehat{\gamma}$.

It is well known that for certain measures (for example, the Dirac comb of a lattice or, more generally, of a crystallographic structure [6, 9, 10]), one may proceed via the lower route in the forward, and thus also the reverse, direction in this way. However, the results of Lenz and Moody apply in a much more general situation.

The purpose of this brief contribution is to present, without the point process formalism of [11], some examples of object classes that display the same diffraction. Unless stated otherwise, all measures presented below are measures on the real line $\mathbb{R}$, which means that we illustrate everything with one-dimensional examples.

5.2 A Simple Diffraction Measure with Simple Origins

A *diffraction measure*, that is, a measure that represents the diffraction of some (physical) structure, must be real, positive and inversion symmetric. By *pure point*, we mean that the measure can be written as $\widehat{\gamma} = \sum_{k \in L} I(k) \delta_k$, with a countable set L (which may be finite) and locally summable intensities $I(k) > 0$. Let us begin then, with the 'simplest possible' pure point diffraction measure.

Example 5.1 Let $\widehat{\gamma} = \delta_0$. Proceeding backwards through the Wiener diagram via the 'bottom' route, we gain $\widehat{\omega} = A(0)\delta_0$, where, applying the method of [11], we must have $A(0) = 1$. Since $\widehat{\lambda} = \delta_0$ and $\widehat{\delta_0} = \lambda$, where λ is Lebesgue measure, we have $\omega = \lambda$ and the real diffraction solution class of δ_0 is $\{\lambda, -\lambda\}$ (as it is clear that $-\omega$ is homometric to ω). Of course, observing that $\lambda \circledast \tilde{\lambda} = \lambda$ gives us the same thing via the top route in the Wiener diagram.

It is not hard to deduce a bit more here: For $u \in \mathbb{S}^1$, the unit circle, we have $u\lambda \circledast \widetilde{u\lambda} = u\bar{u}\lambda = \lambda$, and so, via the top route, we see that $\{u\lambda \mid u \in \mathbb{S}^1\}$ is contained in the (complex) diffraction solution class of δ_0. Are there further measures with diffraction δ_0? Well, any measure of the form $\omega + \mu$, with μ a finite measure on $\mathbb{R}$, has the same diffraction as ω, as adding a finite measure to ω does not change the autocorrelation. This is well-known [3, 9] and a 'trivial' degree of freedom; in the framework of [11], the point process for $\omega + \mu$ is the same as that for ω.

In this example, we have a good idea of what the complex diffraction solution class is, and are certain that the real diffraction solution class contains only measures. In the next section, we shall see that the real diffraction solution class of a nice, periodic measure like $\delta_{\mathbb{Z}}$ contains both measures and non-measures. In fact, as will be shown in [12], the only pure point diffraction measures whose real solution class consists solely of measures are those supported on a finite set of points.

5.3 A Lattice Diffraction Measure with All Kinds of Origins

Let us now consider measures with diffraction $\widehat{\gamma} = \delta_{\mathbb{Z}} = \sum_{k\in\mathbb{Z}} \delta_k$. According to the Wiener diagram, objects ω with the diffraction $\delta_{\mathbb{Z}}$ must have the (possibly formal) Fourier transform $\widehat{\omega} = \sum_{k\in\mathbb{Z}} A(k)\delta_k$, with $|A(k)| = 1$ for all k. In the setting of [11], we have the further conditions that $A(0) = 1$ and $A(-k) = \overline{A(k)}$ for all k, and one constructs different objects with diffraction $\delta_{\mathbb{Z}}$ simply by choosing different sets of compliant coefficients $\{A(k) \mid k \in \mathbb{Z}\}$. An interpretation of ω and $\widehat{\omega}$ might need the theory of tempered distributions and their relations with measures.

Example 5.2 Choosing $A(k) = 1$ for all k, one gains $\omega = \delta_{\mathbb{Z}}$, as follows from the Poisson summation formula $\widehat{\delta_{\mathbb{Z}}} = \delta_{\mathbb{Z}}$; compare [3, 5].

Example 5.3 By splitting the set $\mathbb{Z}$ into subsets $n\mathbb{Z}, n\mathbb{Z}+1, \ldots, n\mathbb{Z}+(n-1)$, and choosing coefficients appropriately, one may construct an n-periodic measure with diffraction $\delta_{\mathbb{Z}}$. For example, to construct a 4-periodic measure, let

$$A(k) = \begin{cases} 1, & k \in 4\mathbb{Z}, \\ \alpha, & k \in 4\mathbb{Z}+1, \\ e, & k \in 4\mathbb{Z}+2, \\ \bar{\alpha}, & k \in 4\mathbb{Z}+3, \end{cases}$$

where $e = \pm 1$ and $\alpha \in \mathbb{S}^1$. Note that this exhausts all possibilities for compliant coefficients for this four-way splitting. We have

$$\widehat{\omega_{\alpha}^{\pm}} = (\delta_0 + \alpha\delta_1 + \bar{\alpha}\delta_{-1} \pm \delta_2) * \delta_{4\mathbb{Z}}.$$

Then, writing $\alpha = e^{2\pi i t_\alpha}$ and defining $\chi_s(t) := e^{2\pi i s t}$ for $t, s \in \mathbb{R}$, we obtain

$$\begin{aligned}\omega_\alpha^\pm &= \frac{1}{4}(\chi_0 + \alpha\chi_1 + \bar{\alpha}\chi_{-1} \pm \chi_2)\delta_{\frac{1}{4}\mathbb{Z}} \\ &= \frac{1}{4}\sum_{k\in\mathbb{Z}}\left(1 + 2\cos\left(2\pi\left(t_\alpha + \frac{k}{4}\right)\right) \pm \cos(\pi k)\right)\delta_{\frac{k}{4}},\end{aligned}$$

where we have used the Poisson summation formula and the identity $\widehat{\mu * \nu} = \widehat{\mu}\widehat{\nu}$. That is, we have a measure of the form $\omega_\alpha^\pm = (a\delta_0 + b\delta_{\frac{1}{4}} + c\delta_{\frac{1}{2}} + d\delta_{\frac{3}{4}}) * \delta_\mathbb{Z}$. For example, choosing $\alpha \in \{0, \mathrm{i}, 1, -\mathrm{i}\}$, we get $\omega_\alpha^\pm$ according to the following table.

$\omega_\alpha^\pm$	$e = 1$	$e = -1$
$t_\alpha = 0$	$\delta_\mathbb{Z}$	$\frac{1}{2}(\delta_0 + \delta_{\frac{1}{4}} - \delta_{\frac{1}{2}} + \delta_{\frac{3}{4}}) * \delta_\mathbb{Z}$
$t_\alpha = \frac{1}{4}$	$\frac{1}{2}(\delta_0 - \delta_{\frac{1}{4}} + \delta_{\frac{1}{2}} + \delta_{\frac{3}{4}}) * \delta_\mathbb{Z}$	$\delta_{\frac{3}{4}} * \delta_\mathbb{Z}$
$t_\alpha = \frac{1}{2}$	$\delta_{\frac{1}{2}} * \delta_\mathbb{Z}$	$\frac{1}{2}(-\delta_0 + \delta_{\frac{1}{4}} + \delta_{\frac{1}{2}} + \delta_{\frac{3}{4}}) * \delta_\mathbb{Z}$
$t_\alpha = \frac{3}{4}$	$\frac{1}{2}(\delta_0 + \delta_{\frac{1}{4}} + \delta_{\frac{1}{2}} - \delta_{\frac{3}{4}}) * \delta_\mathbb{Z}$	$\delta_{\frac{1}{4}} * \delta_\mathbb{Z}$

It is easy to verify (along the top route of the Wiener diagram) that these measures do indeed all have autocorrelation $\gamma = \delta_\mathbb{Z}$ and thus diffraction $\widehat{\gamma} = \delta_\mathbb{Z}$. All elements constructed are pure point measures; compare [8] for similar examples.

Example 5.4 We now construct a measure that is not itself pure point, but has diffraction $\delta_\mathbb{Z}$. Take $A(0) = 1$ and $A(k) = -1$ for all $k \in \mathbb{Z} \setminus \{0\}$. Then we have

$$\widehat{\omega} = \delta_0 - \sum_{k\in\mathbb{Z}\setminus\{0\}} \delta_k = 2\delta_0 - \delta_\mathbb{Z},$$

and see that $\omega = 2\lambda - \delta_\mathbb{Z}$ is a member of the (real) diffraction solution class of $\delta_\mathbb{Z}$.

Recalling that $\lambda \circledast \delta_\mathbb{Z} = \lambda$, one can easily verify the above example and construct similar ones via a little trial and error and the top route of the Wiener diagram. For example, via a quick calculation with the Eberlein convolution, one can see that the diffraction of $\delta_\mathbb{Z} - (1 + \mathrm{i})\lambda$ is also $\delta_\mathbb{Z}$, while $\delta_\mathbb{Z} - \lambda$ has diffraction $\delta_{\mathbb{Z}\setminus\{0\}}$. Such a trial and error method, however, would probably not lead to the following more elaborate system.

Example 5.5 To construct something a little different, we use an aperiodic set suggested by the period doubling sequence, which is limit-periodic; compare [3, 6]. From the set

$$\varLambda = \bigcup_{n\geq 0}(2.4^n\mathbb{Z} + (4^n - 1))$$

of left endpoints of 'a's in this sequence [4], we form the symmetric set

$$\varDelta := \varLambda \cup (-\varLambda) = 2\mathbb{Z} \,\dot{\cup} \dot{\bigcup_{n\geq 1}} \varDelta_n,$$

where $\varDelta_n := (2.4^n\mathbb{Z} + (4^n - 1)) \cup (2.4^n\mathbb{Z} + (1 - 4^n))$ for $n \geq 1$ and $\dot{\cup}$ denotes the disjoint union of sets. Then, we define $A(k) = 1$ for $k \in \varDelta$ and $A(k) = -1$ for $k \notin \varDelta$. Due to the symmetry of $\varDelta$, these coefficients satisfy the conjugacy condition given above. We get

$$\widehat{\omega} = \sum_{k\in\varDelta} \delta_k - \sum_{k\in\mathbb{Z}\setminus\varDelta} \delta_k = 2\delta_\varDelta - \delta_\mathbb{Z},$$

and thus $\omega = 2\widehat{\delta_\varDelta} - \delta_\mathbb{Z}$. Now, we have

$$\delta_L = \delta_{2\mathbb{Z}} + \sum_{n\geq 1} \delta_{\varDelta_n} = \delta_{2\mathbb{Z}} + \sum_{n\geq 1} \delta_{2.4^n\mathbb{Z}} * (\delta_{4^n-1} + \delta_{1-4^n}),$$

and hence may use the Poisson summation formula to derive the formal expression

$$\begin{aligned}\widehat{\delta_\varDelta} &= \frac{1}{2}\delta_{\frac{\mathbb{Z}}{2}} + \sum_{n\geq 1} \frac{\bar{\chi}_{4^n-1} + \bar{\chi}_{1-4^n}}{2.4^n} \delta_{\frac{\mathbb{Z}}{2.4^n}} \\ &= \frac{1}{2}\delta_{\frac{\mathbb{Z}}{2}} + \sum_{n\geq 1} \frac{\cos(2\pi(4^n - 1)k)}{4^n} \delta_{\frac{\mathbb{Z}}{2.4^n}}.\end{aligned}$$

As $\delta_\varDelta$ is a translation bounded measure, $\widehat{\delta_\varDelta}$ is a tempered distribution. It is not, however, a measure, as it is easy to find compact sets $K \subseteq \mathbb{R}$ (for example, take $K = [0, \frac{1}{4}]$) such that $|\widehat{\delta_\varDelta}|(K)$ is infinite. So, we have a tempered distribution, ω, that is not a measure, but nevertheless has diffraction $\delta_\mathbb{Z}$.

Of course, one can construct many such 'non-measures' (in [12], via a theorem of Cordoba [5], it is shown that the homometry class of $\delta_\mathbb{Z}$ contains uncountably many such objects), but this does not shed much light on the physical relevance of such constructions. A little more insight may be gained by noting that (in this case at least) our constructed distribution is the limit (in the weak-$*$ topology on the space $\mathcal{S}'(\mathbb{R})$ of tempered distributions) of a sequence of measures over $\mathbb{R}$.

For $\varepsilon > 0$, define

$$\rho_\varepsilon := \frac{1}{2}\delta_{\frac{\mathbb{Z}}{2}} + \sum_{n\geq 1} \frac{\cos(2\pi(4^n - 1)k)}{(4+\varepsilon)^n} \delta_{\frac{\mathbb{Z}}{2.4^n}}.$$

A short calculation reveals that, for $\varepsilon > 0$, $|\rho_\varepsilon|(K)$ is finite for all compact sets $K \subseteq \mathbb{R}$, so that ρ_ε is indeed a measure (it is even translation-bounded). Thus $\omega_\varepsilon := 2\rho_\varepsilon - \delta_\mathbb{Z}$ is also a measure. Moreover, for all Schwartz functions $g \in \mathcal{S}(\mathbb{R})$, one has $\omega_\varepsilon(g) \to \omega(g)$ as $\varepsilon \to 0^+$. This is a standard approach in Fourier analysis to enforce convergence of the series, which is sometimes referred to as 'regularisation' in physics. Such objects can still be given a reasonable physical meaning.

5.4 Further Remarks

The method of [11] may only be applied to diffraction measures $\widehat{\gamma}$ that are 'backward transformable', meaning that the (inverse) Fourier transform, γ (that is, the autocorrelation), is again a measure. Further contemplation of the previous example, however, makes this condition seem a little too restrictive. Using the scheme, we constructed the object $\omega = 2\widehat{\delta_\Delta} - \delta_{\mathbb{Z}}$, which is not a measure, but is the weak-$*$ limit (in $\mathcal{S}'(\mathbb{R})$) of measures. The object $\widehat{\delta_\Delta}$ is also a non-measure weak-$*$ limit of measures, but does not have the good fortune, as ω does, to have a measure-valued autocorrelation, so it is not covered by the scheme.

Presuming that one may extend the method of [11] to admit this case (or, in other words, that one may proceed via the lower path in the Wiener diagram to calculate the diffraction), one has $\omega = \widehat{\delta_\Delta}$, with $\widehat{\omega} = \delta_\Delta$, and thus diffraction $\widehat{\gamma} = \delta_\Delta$. The measure δ_Δ is positive, inversion symmetric and translation bounded, so has almost all of the properties that one expects from a diffraction measure.

The natural next step is to understand the classes of objects for which one may define an autocorrelation (and thus a diffraction). The framework of [11] is applicable in the more general setting of a locally compact Abelian group. However, if one considers only objects in $\mathbb{R}^d$, these initial examples suggest that consideration of tempered distributions that are the weak-$*$ limits of measures may be a good place to begin.

Acknowledgements This work was supported by the German Research Council (DFG), via the CRC 701, and by the RCM^2, at the University of Bielefeld.

References

1. Baake M, Grimm U (2007) Homometric model sets and window covariograms. Z Kristallogr 222:54–58
2. Baake M, Grimm U (2012) Mathematical diffraction of aperiodic structures. Chem Soc Rev 41:6821–6843
3. Baake M, Grimm U (2013) Theory of aperiodic order: a mathematical invitation. Cambridge University Press, Cambridge, to appear
4. Baake M, Moody RV (2004) Weighted Dirac combs with pure point diffraction. J Reine Angew Math 573:61–94
5. Córdoba A (1989) Dirac combs. Lett Math Phys 17:191–196
6. Gähler F, Klitzing R (1995) The diffraction pattern of self-similar tilings. In: Moody RV (ed) The mathematics of long- range aperiodic order. NATO ASI series C, vol 489. Kluwer, Dordrecht, pp 141–174
7. Grimm U, Baake M (2008) Homometric point sets and inverse problems. Z Kristallogr 223:777–781

8. Grünbaum FA, Moore CC (1995) The use of higher-order invariants in the determination of generalized Patterson cyclotomic sets. Acta Crystallogr, Ser A 51:310–323
9. Hof A (1995) On diffraction by aperiodic structures. Commun Math Phys 169:25–43
10. Hof A (1995) Diffraction by aperiodic structures. In: Moody RV (ed) The mathematics of long-range aperiodic order. NATO ASI series C, vol 489. Kluwer, Dordrecht, pp 239–268
11. Lenz D, Moody RV Stationary processes with pure point diffraction. Preprint. arXiv:1111.3617
12. Terauds V (2013) The inverse problem of pure point diffraction—examples and open questions, in preparation

Chapter 6
Well-Rounded Sublattices and Coincidence Site Lattices

P. Zeiner

Abstract A lattice is called well-rounded, if its lattice vectors of minimal length span the ambient space. We show that there are interesting connections between the existence of well-rounded sublattices and coincidence site lattices (CSLs). Furthermore, we count the number of well-rounded sublattices for several planar lattices and give their asymptotic behaviour.

6.1 Introduction

A lattice in $\mathbb{R}^d$ is called well-rounded, if its (non-zero) lattice vectors of minimal length span $\mathbb{R}^d$. This means that there exist at least $2d$ lattice vectors of minimal positive length, and $\mathbb{R}^d$ has a basis consisting of lattice vectors of minimal length. However, such a basis need not be a primitive lattice basis in dimensions $d \geq 4$.

Well-rounded lattices are important for several reasons. Many important lattices occurring in mathematics and physics are well-rounded. For instance, the hexagonal lattice and the square lattice in $\mathbb{R}^2$ and the cubic lattices in $\mathbb{R}^3$ are well-rounded, as are the hypercubic lattices and the A_4-lattice in $\mathbb{R}^4$, which play an important role in quasicrystallography. Examples in higher dimensions are the Leech lattice, the Barnes–Wall lattices, and the Coxeter–Todd lattice; see [6] for background.

Let us briefly mention two problems of mathematical crystallography where well-rounded lattices occur. They are connected to the question of densest lattice sphere packings, as all extreme lattices (those lattices corresponding to densest lattice sphere packings) are perfect (i.e. the lattice vectors of minimal length determine the Gram matrix uniquely) and are thus well-rounded. They also play an important role in reduction theory, as they are exactly those lattices for which all the successive minima are equal [9].

Here, we want to deal with two specific questions: Has a given lattice well-rounded sublattices, and if so, what are the well-rounded sublattices and how many are there. The first question is answered in Sect. 6.2 for planar lattices and a par-

P. Zeiner (✉)
Faculty of Mathematics, Bielefeld University, 33501 Bielefeld, Germany
e-mail: pzeiner@math.uni-bielefeld.de

S. Schmid et al. (eds.), *Aperiodic Crystals*, DOI 10.1007/978-94-007-6431-6_6,

tial answer is given for $d > 2$. The second question is much more difficult in general. Thus we restrict the discussion to 2 dimensions, and present some results in Sect. 6.3.

6.2 Well-Rounded Lattices and CSLs

Here, we want to deal with the question whether a lattice has a well-rounded sublattice. It turns out that this question is related to the theory of coincidence site lattices (CSLs), so let us review the notion of CSL first. Let Λ be a lattice in $\mathbb{R}^d$ and let $R \in O(d)$ be an isometry. Then $\Lambda(R) = \Lambda \cap R\Lambda$ is called a *coincidence site lattice (CSL)* if $\Lambda(R)$ is a sublattice of full rank in Λ; the corresponding R is called coincidence isometry. The corresponding index of $\Lambda(R)$ in Λ is called coincidence index $\Sigma_\Lambda(R)$, or $\Sigma(R)$ for short. The set of all coincidence isometries forms a group, which we call $OC(\Lambda)$, see [2] for details.

Let us look at the planar case first. Here, any two linearly independent lattice vectors of minimal (non-zero) length form a basis of Λ. Let γ be the angle between them. Now a well-rounded lattice is necessarily a rhombic (centred rectangular) lattice such that $\frac{\pi}{3} < \gamma < \frac{2\pi}{3}, \gamma \neq \frac{\pi}{2}$ or a square (corresponding to $\gamma = \frac{\pi}{2}$) or a hexagonal lattice (corresponding to $\gamma = \frac{\pi}{3}$ or $\gamma = \frac{2\pi}{3}$). Thus, its symmetry group is at least $D_2 = 2$ mm, or in other words, there is at least one reflection symmetry present. As Λ and all of its sublattices have the same group of coincidence isometries [2], we can infer that a lattice possesses a well-rounded sublattice only if it has a coincidence reflection. As the converse holds as well, we have (compare [5])

Theorem 6.1 *A planar lattice $\Lambda \in \mathbb{R}^2$ has a well-rounded sublattice if and only if it has a coincidence reflection.*

An alternative criterion tells us that a planar lattice has a well-rounded sublattice if and only if it has a rhombic or rectangular sublattice [8]. The existence of well-rounded sublattices can also be characterised by the entries of the Gram matrices of Λ, see [8] and [5] for various criteria.

One is tempted to generalise these criteria to d dimensions, by using orthogonal lattices, the d-dimensional analogue of rectangular lattices and orthorhombic lattices in 3 dimensions. However, this does not work since a lattice may be well-rounded without having an orthogonal sublattice. As an example, consider a rhombohedral lattice in $\mathbb{R}^3$, which in general does not have an orthorhombic sublattice. Nevertheless, an orthogonal lattice has well-rounded sublattices, and one even has

Theorem 6.2 *Let G be the symmetry group of an orthogonal lattice, i.e. a lattice that is spanned by an orthogonal basis. Then Λ has a well-rounded sublattice if $G \subseteq OC(\Lambda)$.*

This theorem can be proved by induction. The idea is to show that $G \subseteq OC(\Lambda)$ implies the existence of an orthogonal sublattice, which in turn implies the existence of well-rounded sublattices.

However, note that the intuitive idea of choosing a "body-centred orthogonal" lattice fails in dimensions $d > 4$. For if we construct a lattice as the linear span of the 2^d vectors $\sum_{i=1}^{d} s_i^{(j)} b_i$, where the b_i form an orthogonal basis of $\mathbb{R}^d$ and $s_i^{(j)} \in \{1, -1\}$, then these vectors do not have minimal lengths as at least one of the vectors $2b_i$ is shorter. Nevertheless, a modification of this idea works where we choose a suitable subset of the vectors $\sum_{i=1}^{d} s_i^{(j)} b_i$. In particular, if the basis vectors b_i all have approximately the same length and d is even, we can construct a well-rounded sublattice as the linear span of $\sum_{i=1}^{d} s_i^{(j)} b_i$, where j runs over all possible solutions of $\sum_{i=1}^{d} s^{(j)} \equiv 0 \pmod{d}$.

An immediate consequence of Theorem 6.2 is that every rational lattice has well-rounded sublattices, as $OC(\Lambda)$ contains all reflections generated by a lattice vector [10].

6.3 Well-Rounded Sublattices of Planar Lattices

We now turn to our second question, i.e. we want to find all well-rounded sublattices of a given lattice. We concentrate on some planar lattices here. To begin with, we want to find all well-rounded sublattices of the square lattice. W.l.o.g. we may identify it with $\mathbb{Z}^2 \simeq \mathbb{Z}[\mathrm{i}]$. The idea now is the following. From the previous section, we know that a planar lattice is well-rounded if and only if it is a rhombic lattice with $\frac{\pi}{3} < \gamma < \frac{2\pi}{3}$, a square or a hexagonal lattice. Now a sublattice of a square lattice cannot be hexagonal, so that we can exclude the latter case, i.e. we only have to find all rhombic and square well-rounded sublattices. The latter are just the similar sublattices of the square lattice, which are well known [3, 4]. The Dirichlet series generating function of their counting function reads

$$\Phi_{\square}(s) = \sum_{n\in\mathbb{N}} \frac{s_{\square}(n)}{n^s} = \zeta(2s)\Phi_{\square}^{\mathsf{pr}}(s) = \zeta_{\mathbb{Q}(\mathrm{i})}(s) = L(s, \chi_{-4})\zeta(s) \tag{6.1}$$

where $s_{\square}(n)$ is the number of similar sublattices of the square lattice with index n. Here, $\Phi_{\square}^{\mathsf{pr}}(s)$ is the generating function of the primitive similar sublattices, $\zeta(s)$ is the Riemann zeta function and $\zeta_{\mathbb{Q}(\mathrm{i})}(s)$ is the Dedekind zeta function of the complex number field $\mathbb{Q}(\mathrm{i})$.

Hence it remains to find all rhombic well-rounded sublattices. Now each rhombic sublattice has a rectangular sublattice of index 2, and it is well-rounded if and only if $\frac{a}{\sqrt{3}} \le b \le a\sqrt{3}$ holds, where a and b are the lengths of the orthogonal basis vectors of the corresponding rectangular sublattice. Thus we only need to find all rectangular sublattices satisfying the condition above. In fact, as all square lattices are similar, it is sufficient to find all rectangular sublattices whose symmetry axes are parallel to those of the square lattice, and we finally get [5]

$$\Phi_{\mathsf{wr,even}}(s) = \frac{2}{2^s}\Phi_{\square}^{\mathsf{pr}}(s) \sum_{p\in\mathbb{N}} \sum_{p<q<\sqrt{3}p} \frac{1}{p^s q^s}, \tag{6.2}$$

$$\Phi_{\mathrm{wr,odd}}(s) = \frac{2}{1+2^{-s}} \Phi_{\square}^{\mathrm{pr}}(s) \sum_{k\in\mathbb{N}} \sum_{k<\ell<\sqrt{3}k+\frac{\sqrt{3}-1}{2}} \frac{1}{(2k+1)^s(2\ell+1)^s} \tag{6.3}$$

where $\Phi_{\mathrm{wr,even}}(s)$ and $\Phi_{\mathrm{wr,odd}}(s)$ are the generating functions counting the rhombic well-rounded sublattices of even and odd indices, respectively. Putting everything together, we arrive at the following result [5]

Theorem 6.3 *Let $a_{\square}(n)$ be the number of well-rounded sublattices of the square lattice with index n, and $\Phi_{\square,\mathrm{wr}}(s) = \sum_{n=1}^{\infty} a_{\square}(n) n^{-s}$ the corresponding Dirichlet series generating function. It is given by $\Phi_{\square,\mathrm{wr}}(s) = \Phi_{\square}(s) + \Phi_{\mathrm{wr,even}}(s) + \Phi_{\mathrm{wr,odd}}(s)$ with the functions from Eqs. (6.1), (6.2) and (6.3).*

If $s > 1$, we have the inequality

$$D_{\square}(s) - \Phi_{\square}(s) < \Phi_{\square,\mathrm{wr}}(s) < D_{\square}(s) + \Phi_{\square}(s),$$

with $\Phi_{\square}(s)$ from Eq. (6.1) and the function

$$D_{\square}(s) = \frac{2+2^s}{1+2^s} \frac{1-\sqrt{3}^{1-s}}{s-1} \frac{L(s,\chi_{-4})}{\zeta(2s)} \zeta(s)\zeta(2s-1).$$

As a consequence, the summatory function $A_{\square}(x) = \sum_{n\le x} a_{\square}(n)$ possesses the asymptotic growth behaviour

$$A_{\square}(x) = \frac{\log(3)}{2\pi} x\log(x) + o\big(x\log(x)\big)$$

as $x \to \infty$.

The lower and upper bounds are obtained by approximating the sums in Eqs. (6.2) and (6.3) by integrals via the Euler summation formula, whereas the statement about the asymptotic behaviour of $A_{\square}(x)$ follows from Delange's theorem, which relates the asymptotic behaviour of $A_{\square}(x)$ with the analytic properties of $\Phi_{\square,\mathrm{wr}}(s)$, in particular with its pole at $s = 1$.

In fact, we can get additional information about the asymptotic behaviour of $A_{\square}(x)$ by applying some methods of analytic number theory, including Dirichlet's hyperbola method and the above mentioned Euler summation formula (see, e.g. [1]).

Theorem 6.4 *Let $a_{\square}(n)$ be the number of well-rounded sublattices of the square lattice with index n. Then, the summatory function $A_{\square}(x) = \sum_{n\le x} a_{\square}(n)$ possesses the asymptotic growth behaviour*

$$\begin{aligned} A_{\square}(x) &= \frac{\log(3)}{3} \frac{L(1,\chi_{-4})}{\zeta(2)} x\big(\log(x) - 1\big) + c_{\square} x + \mathcal{O}\big(x^{3/4}\log(x)\big) \\ &= \frac{\log(3)}{2\pi} x\log(x) + \left(c_{\square} - \frac{\log(3)}{2\pi}\right) x + \mathcal{O}\big(x^{3/4}\log(x)\big) \end{aligned}$$

where

$$\begin{aligned}
c_\square :=& \frac{L(1,\chi_{-4})}{\zeta(2)}\Bigg(\zeta(2)+\frac{\log(3)}{3}\bigg(\frac{L'(1,\chi_{-4})}{L(1,\chi_{-4})}+\gamma-2\frac{\zeta'(2)}{\zeta(2)}\bigg)\\
&+\frac{\log(3)}{3}\bigg(2\gamma-\frac{\log(3)}{4}-\frac{\log(2)}{6}\bigg)-\sum_{p=1}^{\infty}\frac{1}{p}\bigg(\frac{\log(3)}{2}-\sum_{p<q<p\sqrt{3}}\frac{1}{q}\bigg)\\
&-\frac{4}{3}\sum_{k=0}^{\infty}\frac{1}{2k+1}\bigg(\frac{1}{4}\log(3)-\sum_{k<\ell<k\sqrt{3}+(\sqrt{3}-1)/2}\frac{1}{2\ell+1}\bigg)\Bigg)\\
\approx& 0.6272237
\end{aligned}$$

is the coefficient of $(s-1)^{-1}$ *in the Laurent series of* $\sum_n \frac{a_\square(n)}{n^s}$ *around* $s=1$. *Here,* γ *is the Euler–Mascheroni constant.*

Similar calculations are also possible for the hexagonal lattice. If $a_\triangle(n)$ is the number of well-rounded sublattices of the triangular lattice with index n, then the corresponding Dirichlet series generating function $\Phi_{\triangle,\mathrm{wr}}(s)=\sum_{n=1}^{\infty}a_\triangle(n)n^{-s}$ is given by

$$\Phi_{\triangle,\mathrm{wr}}(s)=\Phi_\triangle(s)+\Phi_{\triangle,\mathrm{wr,even}}(s)+\Phi_{\triangle,\mathrm{wr,odd}}(s),$$

where

$$\Phi_\triangle(s)=\zeta_{\mathbb{Q}(\rho)}(s)=L(s,\chi_{-3})\zeta(s) \tag{6.4}$$

is the generating function for the similar sublattices of the hexagonal lattice and

$$\Phi_{\triangle,\mathrm{wr,even}}(s)=\frac{3}{4^s(1+3^{-s})}\sum_{p\in\mathbb{N}}\sum_{p<q<3p}\frac{1}{p^s q^s}\Phi_\triangle^{\mathrm{pr}}(s), \tag{6.5}$$

$$\Phi_{\triangle,\mathrm{wr,odd}}(s)=\frac{3}{1+3^{-s}}\sum_{k\in\mathbb{N}}\sum_{k<\ell<3k+1}\frac{1}{(2k+1)^s(2\ell+1)^s}\Phi_\triangle^{\mathrm{pr}}(s) \tag{6.6}$$

are the corresponding Dirichlet series for the number of rhombic well-rounded sublattices with even and odd indices, respectively. For the asymptotic behaviour, we get [5]

Theorem 6.5 *The summatory function* $A_\triangle(x)=\sum_{n\le x}a_\triangle(n)$ *possesses the asymptotic growth behaviour*

$$\begin{aligned}
A_\triangle(x) &= \frac{9\log(3)}{16}\frac{L(1,\chi_{-3})}{\zeta(2)}x\big(\log(x)-1\big)+c_\triangle x+\mathcal{O}\big(x^{3/4}\log(x)\big)\\
&= \frac{3\sqrt{3}\log(3)}{8\pi}x\big(\log(x)-1\big)+c_\triangle x+\mathcal{O}\big(x^{3/4}\log(x)\big)
\end{aligned}$$

where $c_\triangle \approx 0.4915036$ *is the coefficient of* $(s-1)^{-1}$ *in the Laurent series of* $\sum_n \frac{a_\triangle(n)}{n^s}$ *around* $s=1$.

In both examples, we have infinitely many coincidence reflections, which results in a large number of well-rounded sublattices and in an asymptotic growth behaviour of $x\log(x)$. A similar behaviour is to be expected for all rational lattices, but so far only weaker results have been obtained [7].

However, in general we have fewer coincidence reflections, and we want to conclude with this case. In fact, if the lattice is not rational, there are either no or exactly two coincidence reflections [5, 10], and both of them have the same coincidence index. It is remarkable that in the latter case the asymptotic behaviour does not depend on the details of the lattice but only on the coincidence index of its two coincidence reflections. In particular, we have [5]

Theorem 6.6 *Let* Λ *be a planar lattice that has exactly two coincidence reflections. Let* Σ *be their common coincidence index and let* $a_\Lambda(n)$ *denote the number of well-rounded sublattices of* Λ *with index* n. *Then, the summatory function* $A_\Lambda(x) = \sum_{n\le x} a_\Lambda(n)$ *possesses the asymptotic growth behaviour* $A_\Lambda(x) = \frac{\log 3}{4\Sigma}x + \mathcal{O}(\sqrt{x})$.

Acknowledgements The author thanks M. Baake, and R. Scharlau, for fruitful discussions. This work was supported by the German Research Council (DFG) within the CRC 701.

References

1. Apostol TM (1976) Introduction to analytic number theory. Springer, New York
2. Baake M (1997) Solution of the coincidence problem in dimensions $d \le 4$. In: Moody RV (ed) The mathematics of long-range aperiodic order. Kluwer, Dordrecht, pp 9–44. Revised version: arXiv:math.MG/0605222
3. Baake M, Grimm U (2004) Bravais colourings of planar modules with N-fold symmetry. Z Kristallogr 219:72–80. math.CO/0301021
4. Baake M, Scharlau R, Zeiner P (2011) Similar sublattices of planar lattices. Can J Math 63:1220–1237. arXiv:0908.2558
5. Baake M, Scharlau R, Zeiner P (2013) Well-rounded sublattices of planar lattices, in preparation
6. Conway JH, Sloane NJA (1993) Sphere packings, lattices and groups. Springer, New York
7. Fukshansky L (2013) Well-rounded zeta-function of planar arithmetic lattices. Proc Am Math Soc, to appear
8. Kühnlein S (2012) Well-rounded sublattices. Int J Number Theory 8:1133–1144
9. van der Waerden BL (1956) Die Reduktionstheorie der positiven quadratischen Formen. Acta Math 96:265–309
10. Zou YM (2006) Structures of coincidence symmetry groups. Acta Crystallogr, Ser A 62:109–114

Chapter 7
Octagon-Based Quasicrystalline Formations in Islamic Architecture

Rima Al Ajlouni

Abstract The unexpected discovery of ancient Islamic ornaments with quasicrystalline symmetries has triggered significant discussion and a number of debates on the mathematical sophistication of Islamic geometry and its generating principles. Astonishingly, eight centuries before its description in Modern Science, ancient artists had constructed patterns with perfect quasicrystalline formations. Recent studies have provided enough evidence to suggest that ancient designers, by using the most primitive tools (a compass and a straight edge), were able to resolve the complicated long-range principles of quasicrystalline formations. Derived from these principles, a global multi-level structural model is presented that is able to describe the global long-range order of octagon-based quasicrystalline formations in Islamic Architecture. This new method can be used as a general guiding principle for constructing infinite patches of octagon-based quasicrystalline formations, including Ammann–Beenker tiling, without the need for local strategies (matching, scaling, etc.) or complicated mathematics.

7.1 Quasiperiodic Patterns in Islamic Architecture

The discovery of ancient Islamic patterns with similar quasicrystalline symmetries has attracted many scientists to investigate their ancient mathematics and generating principles. To date, three different types of ancient quasiperiodic symmetries were documented; decagonal, octagonal, and dodecagonal. In their attempt to understand these complex and chaotic formations, scientists investigated the relationship between these patterns and abstract tiling models which describes quasicrystals. Decagonal patterns were investigated in relation to Penrose tiling systems [14], and octagonal patterns were often compared to Ammann–Beenker tiling [7]. One of the earliest investigations into decagonal quasiperiodic Islamic patterns was carried out by Makovicky in 1992 [10]. Based on his analyses of the decagonal quasiperiodic cartwheel pattern in western Iran (1197), Makovicky developed new variations of

R. Al Ajlouni (✉)
College of Architecture, Texas Tech University, Lubbock, USA
e-mail: rima.ajlouni@ttu.edu

S. Schmid et al. (eds.), *Aperiodic Crystals*, DOI 10.1007/978-94-007-6431-6_7,

the Penrose aperiodic tiling [10]. In 1996, Makovicky and Hach-Alı, based on their investigation of the octagonal patterns in Alhambra, Granada, and Alcazar, Sevilla, in Spain, concluded that these '*close-to-ideal*' patterns are based on quasiperiodic sequences (quasilattice) with nested '*phantom*' octagons [11]. Two years later, with colleagues, Makovicky investigated ancient patterns with similar decagonal quasiperiodic symmetry in Spain and Morocco [13]. Bonner [4] investigated three types of Islamic geometric ornaments in the fourteenth and fifteenth centuries, focusing on their self-similarity attributes. In 2007, Lu and Steinhardt [9], based on their analyses of the decagonal quasiperiodic formations on the Darb-I Imam shrine (1453) in Isfahan, Iran, concluded that these patterns were constructed by combining a special set of decorated tiles with self-similar transformations. Recently, the first dodecagonal quasiperiodic pattern was discovered by Makovicky and Makovicky [12] in Fez, Morocco, who suggested that the dodecagonal quasiperiodic pattern is based on the Amman quasilattice grid.

Whereas these investigations have greatly contributed to our understanding of some aspects of these complicated formations, none were able to describe their global long-range order or generating principles. It was not until recently that the first global method for generating ancient quasiperiodic patterns was proposed by Al Ajlouni [2]. Based on her investigations of three quasiperiodic patterns found on the external walls of Gunbad-I Kabud tomb tower in Maragha, Iran (1197 C.E.), the walls of the Madrasa of al-'Attarin (1323 C.E.) in Fez, Morocco, and the walls of Darb-i Imam shrine and the Friday Mosque in Isfahan (1453 C.E.) in Isfahan, Iran, she proposed the first global multi-level Hierarchical Framework Model (HFM) for describing the long-range order of five-fold and ten-fold quasiperiodic formations in Islamic Architecture [2], including Penrose tiling systems [1]. By using a compass and a straight edge, ancient designers were able to resolve the complicated long-range principles of ten-fold quasiperiodic formations. Derived from the same principles, this paper presents a structural model for describing the global-long range order of octagon-based quasiperiodic symmetries in Islamic architecture, including Ammann–Beenker tiling.

7.2 The HFM Model

The proposed mathematical model (HFM) conforms to the ancient method of using a compass and straightedge in which all geometric formations are generated by arrangements of overlapping circles [2, 5, 8]. In general, all patterns in Islamic Architecture were generated based on a combination of an underlying basic grid and repeating units [2, 8]. Whereas the decoration of the repeating units dictates the design variation of the pattern, the underlying basic grid controls the type of symmetry. In this process, the underlying basic grid is considered the key in resolving the long-range order of quasiperiodic symmetries [2]. Consequently, a vast variety of octagon-based quasiperiodic patterns can be generated through manipulating a combination of different underlying basic grids with a variety of repeating units. As

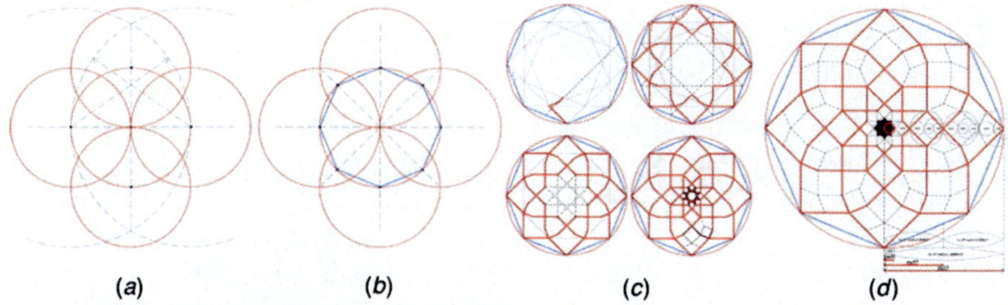

Fig. 7.1 The process of constructing the underlying basic grid

an illustration of the HFM, consider the following examples of octagonal quasiperiodic patterns in Islamic Architecture.

- The quasiperiodic panels from the Patio of Virgins, the Royal Alcazar in Seville (1364–1366) (Figs. 7.2(a) and 7.4(a)).

The Basic Grid From the continuous strip of patterns surrounding the Patio of Virgins in the Alcazar, Seville, two main variations of octagon-based quasiperiodic formations can be recognized. The two panels (Figs. 7.2(a) and 7.4(a)) were also investigated by Makovicky and Fenoll Hach-Alı [11]. An exact copy of both formations can also be found in the Great Mosque of Cordoba.

Based on my analyses of the first pattern (Fig. 7.2(a)), a framework of nested octagrams (Fig. 7.1(d)), which is easily attained by using a compass and a straight edge (Fig. 7.1), serves as the underlying basic grid for generating this quasiperiodic pattern. The framework grows based on the $\sqrt{2}$ sequence. If we denote the radius of the largest octagram by rad_3 and the radius of the smallest octagram by rad_1, then the ratio rad_3/rad_1 is equal to $6 + (4 * \sqrt{2})$. If we denote the radius of the middle octagram by rad_2 and the radius of the smallest octagram by rad_1, then the ratio rad_2/rad_1 is equal to $3 + (2 * \sqrt{2})$. All dimensions within this pattern are related to each other by the $\sqrt{2}$ proportional system.

The First Hierarchy The full sequence of constructing the first hierarchy quasiperiodic cartwheel pattern from the Patio of Virgins in the Royal Alcazar of Seville is demonstrated in Fig. 7.2. According to the HFM model, the octagon-based quasiperiodic empire is generated around one point; the center of the global eight-fold proportional system. The size of the central eight-fold star '*seed*' unit is proportional to the size of the framework and is strictly derived from the diminution sequence of the nested octagrams (Fig. 7.2(d)). In this system, the ratio of the framework to the seed unit is equal to $6 + (4 * \sqrt{2})$. The design of the repeating star '*seed*' unit is also derived from the same geometry (Fig. 7.2(b)).

The positions of all repeating '*seed*' units are determined entirely by the network of the nested octagrams. The dark dots in Fig. 7.2(d) correspond to the center position of all instances of the star units (Fig. 7.2(e)). The connecting formations between the main star units are determined by extending the lines of the repeating units (Figs. 7.2(c) and (f)). As shown in Fig. 7.2(f), some of the repeating star

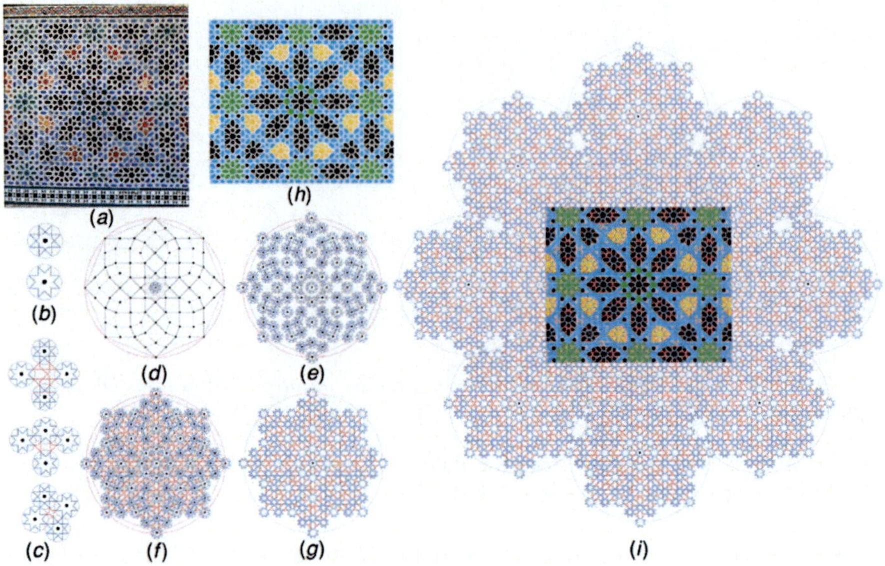

Fig. 7.2 The sequence of constructing the first-level hierarchy of the quasiperiodic cartwheel pattern from the Patio of Virgins in the Royal Alcazar of Seville

units were partially modified to allow different connecting formations. The final line configuration of the quasiperiodic cartwheel pattern is shown in Fig. 7.2(g). Historically, when incorporated into different material, line patterns were often thickened and sometimes broken up to suggest an interlacing pattern [6]. Generating the overall quasiperiodic panel from the Patio of Virgins requires a combination of nine cartwheel patterns and eight connecting formations (Fig. 7.2(i)). These combinations are explained in the next section.

Growing the Empire According to the HFM, constructing the global empire of octagon-based quasiperiodic patterns requires building a progression of multi-level hierarchical formations. In this infinite multi-generation order, the geometric arrangement of the higher-level order is built on the previous order and governed by the same rules of the underlying basic grid. In this order, the final constructed cartwheel pattern of the first-level hierarchy (Fig. 7.3(a)) acts as the '*seed*' unit for the second-level hierarchy. These cartwheel patterns are distributed according to a new generation of nested octagrams (Fig. 7.3(c)). The dark dots in Fig. 7.3(c) correspond to the center position of all instances of the cartwheel unit (Fig. 7.3(d)). This specific distribution of the cartwheels reveals a specific formation of connecting octagons. The internal design of these octagons is shown in Fig. 7.3(b). It is important to note that the connecting formations can take different internal designs without affecting the overall symmetry of the pattern. Fig. 7.3(e) shows the final pattern of the second-level hierarchy of the global quasiperiodic empire.

Building on the same sequence, generating the next higher-level cluster also follows the same process, in which the new higher-generation order is built on the

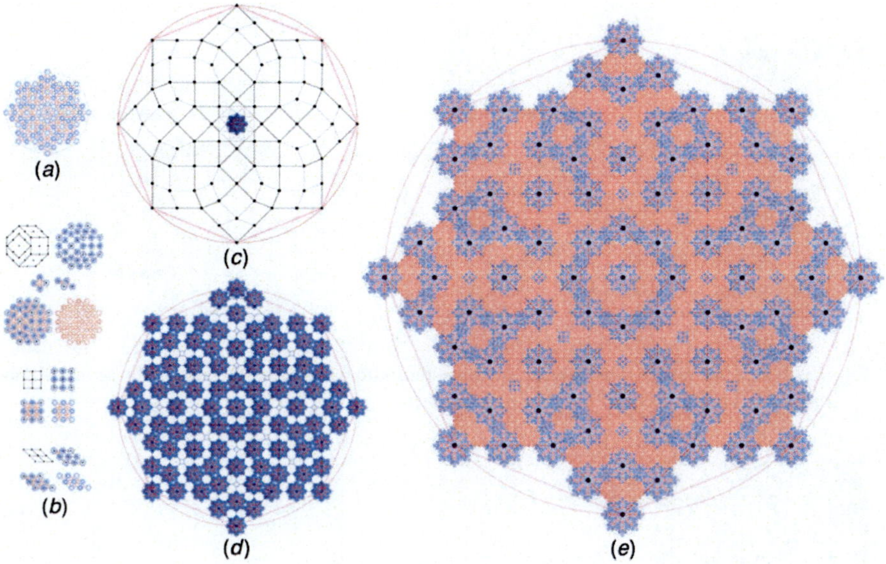

Fig. 7.3 The sequence of constructing the second-level hierarchy of the quasiperiodic cartwheel pattern from the Patio of Virgins in the Royal Alcazar of Seville

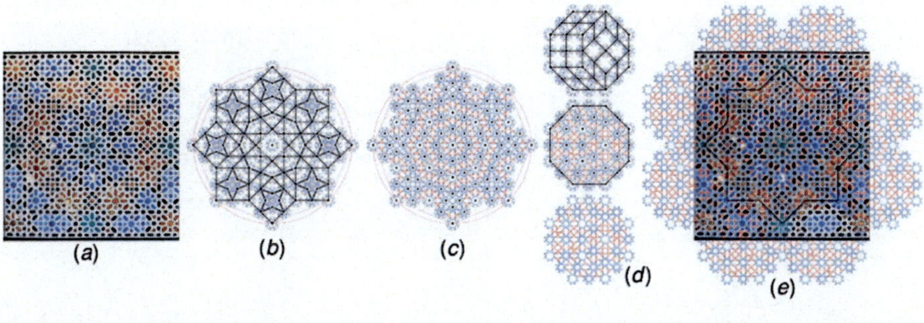

Fig. 7.4 The sequence of constructing the second quasiperiodic panel from the Patio of Virgins in the Royal Alcazar of Seville

previous order. The final generated pattern of the previous order (Fig. 7.3(e)) acts as the '*seed*' unit for the third-level generation order. This process can grow indefinitely to build an infinite empire of quasiperiodic formations.

Constructing the second variation (Fig. 7.4(a)) of the octagon-based quasiperiodic formations surrounding the Patio of Virgins follows the same process explained earlier. However, by manipulating some internal points within the underlying basic grid, ancient designers were able to create many design variations of these complicated formations. Figure 7.4 demonstrates the process of constructing the first-level hierarchy of the second quasiperiodic panel from the Patio of Virgins in Seville. Growing the empire follows the same process demonstrated in Fig. 7.3.

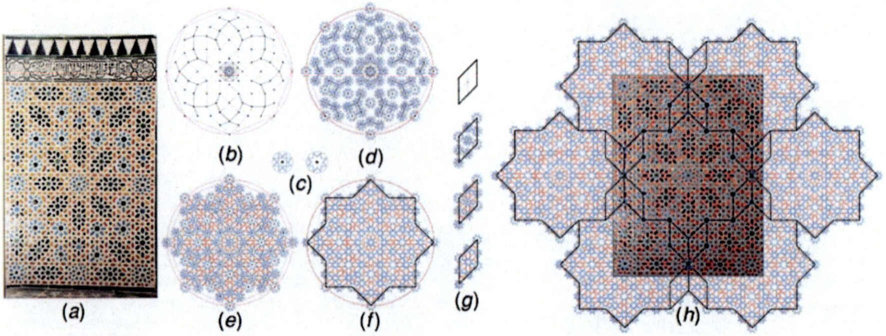

Fig. 7.5 The sequence of constructing the quasiperiodic panel from the side walls of the Mirador de Lindaraja, the Hall of the Two Sisters, Alhambra

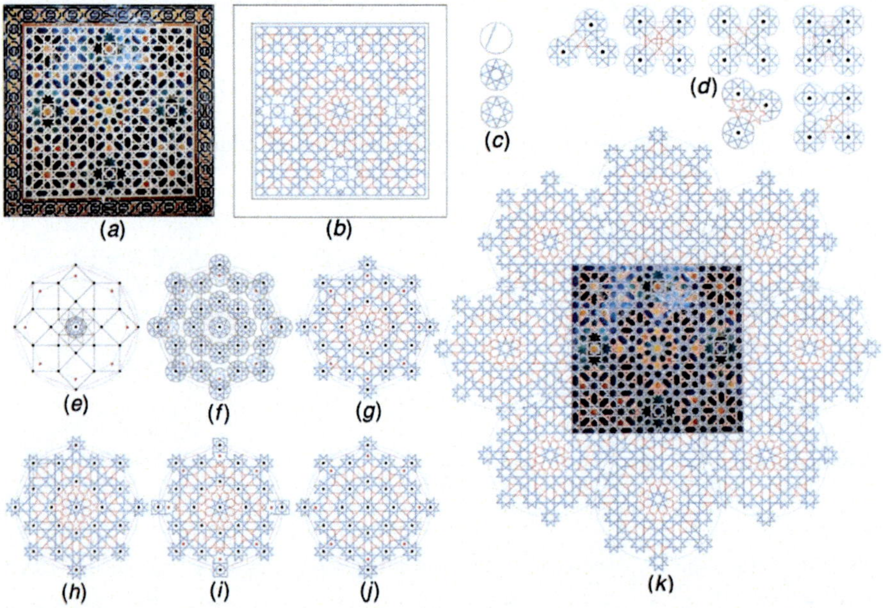

Fig. 7.6 The sequence of constructing the quasiperiodic panel from the Hall of the Ambassadors, Alhambra

- The quasiperiodic panel from the side walls of the Mirador de Lindaraja, the Hall of the Two Sisters, Alhambra (1354–1359).

The same underlying basic grid (Figs. 7.2(d) and 7.3(c)) can be used to construct the cartwheel quasiperiodic pattern from the Mirador de Lindaraja (Fig. 7.5). Growing the quasiperiodic empire can follow the process explained in Fig. 7.3. However, as shown in Fig. 7.5(h), the overall formation of the panel is actually periodic.

- The quasiperiodic panels from the Hall of the Ambassadors, Alhambra (1354).

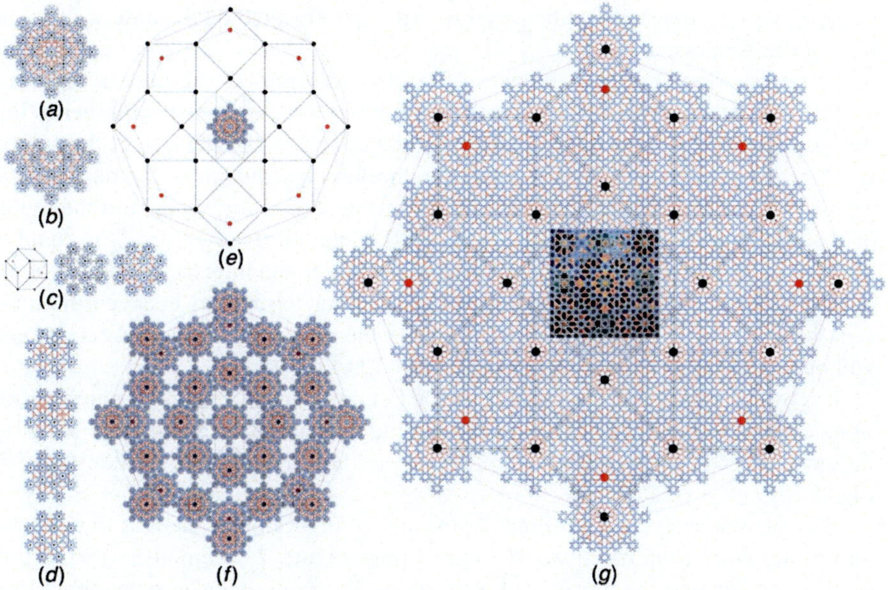

Fig. 7.7 The sequence of constructing the second-level hierarchy of the quasiperiodic panel from the Hall of the Ambassadors, Alhambra

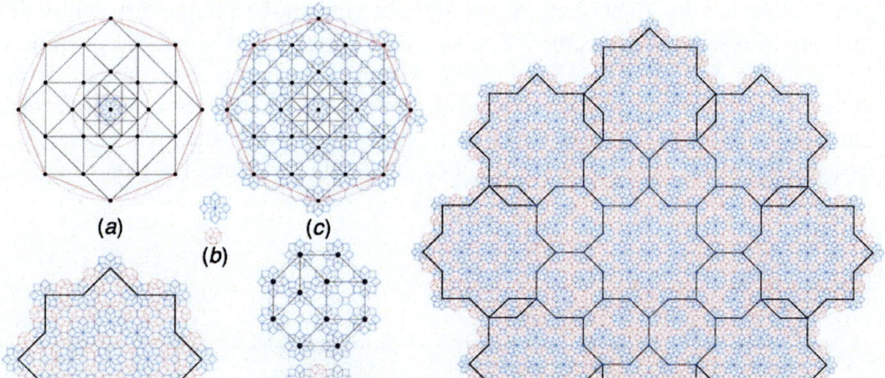

Fig. 7.8 The sequence of constructing the first-level hierarchy of Ammann–Beenker tiling

The Hall of the Ambassadors in Alhambra contains a variety of panels that exhibit similar octagonal quasiperiodic symmetries. As an example of these formations, consider the quasiperiodic panel in Fig. 7.6(a) located on the base wall of the Hall of the Ambassadors.

A framework of nested octagrams (Fig. 7.6(e)) serves as the underlying basic grid. If we denote the radius of the nth octagram by rad_n and the next larger radius by rad_{n+1}, then the ratio rad_{n+1}/rad_n is equal to $(\sqrt{2}+1)$. The size of the central star '*seed*' unit is strictly derived from the diminution sequence of the nested octagrams (Fig. 7.6(e)). If we denote the radius of the *seed* unit by rad_n and the radius of the first hierarchy framework by rad_{n+2}, then the ratio rad_{n+2}/rad_n is equal to $(2\sqrt{2}+3)$. The seed units are distributed according to the intersection points of the basic grid (Fig. 7.6(f)). By changing the connecting formations between the main star '*seed*' units, it is possible to manipulate the design of the cartwheel pattern, without affecting the overall symmetry (Figs. 7.6(g)–(j)).

The overall quasiperiodic panel from the Patio of Virgins requires a combination of nine cartwheel patterns and eight connecting formations (Fig. 7.6(k)). Following the same process, the construction of the second-level hierarchy is demonstrated in Fig. 7.7.

The same underlying grid (Figs. 7.6(e) and 7.7(e)) can also be used to construct an infinite empire of Ammann–Beenker tiling. Figure 7.8 demonstrates the process of constructing the first-level hierarchy of Ammann–Beenker octagonal tiling. Growing the infinite empire would follow the same process demonstrated in Fig. 7.7. An algorithm for constructing the global empire of Ammann–Beenker tiling, based on HFM, can be found in [3].

In conclusion, a global multi-level structural model (HFM) is presented that is able to describe the global long-range order of octagon-based quasicrystalline formations in Islamic Architecture. The suggested model shows that the position of the repeating units, locally and globally, is defined by one global framework, and not based on local rules. This method provides an easy tool for scientists, mathematicians, teachers, designers and artists to generate and study a wide range of octagon-based quasicrystalline formations, including Ammann–Beenker tiling.

References

1. Al Ajlouni R (2011) A long-range hierarchical clustering model for constructing perfect quasicrystalline formations. Philos Mag 91:2728–2738
2. Al Ajlouni R (2012) The global long-range order of quasiperiodic patterns in Islamic architecture. Acta Crystallogr, Ser A 68:235–243
3. Al Ajlouni R (2012) The forbidden symmetries. In: Cabrinha M, Kelly J, Steinfeld K (eds) Synthetic digital ecologies. Acadia, San Francisco, pp 391–400. ISBN: 978-1-62407-267-3
4. Bonner J (2003) Three traditions of self-similarity in fourteenth and fifteenth century Islamic geometric ornament. In: Sarhangi R, Friedman N (eds) Mathematical connections in art, music and science. Proceedings ISAMA/bridges. University of Granada, Granada pp 1–12.
5. El-Said E (1993) Islamic art and architecture: the system of geometric design. Garnet, Reading
6. Gonzalez V (2001) Beauty and Islam: aesthetic in Islamic art and architecture. Islamic Publications, London
7. Grünbaum B, Shephard GC (1986) Tilings and patterns. Freeman, New York

8. Kritchlow K (1976) Islamic patterns: an analytical and cosmological approach. Thames & Hudson, New York
9. Lu P, Steinhardt P (2007) Decagonal and quasicrystalline tilings in Medieval Islamic architecture. Science 315:1106–1110
10. Makovicky E (1992) 800-year-old pentagonal tiling from Maragha, Iran, and the new varieties of aperiodic tiling it inspired. In: Hargittai I (ed) Fivefold symmetry. World Scientific, Singapore, pp 67–86
11. Makovicky E, Fenoll Hach-Ali P (1996) Mirador de Lindaraja: Islamic ornamental patterns based on quasiperiodic octagonal lattices in Alhambra, Granada, and Alcazar, Sevilla, Spain. Bol Soc Esp Mineral 19:1–26
12. Makovicky E, Makovicky N (2011) The first find of dodecagonal quasiperiodic tiling in historical Islamic architecture. J Appl Crystallogr 44:569–573
13. Makovicky E, Rull Pérez F, Fenoll Hach-Alí P (1998) Decagonal patterns in the Islamic ornamental art of Spain and Morocco. Bol Soc Esp Mineral 21:107–127
14. Penrose R (1974) The role of aesthetics in pure and applied mathematical research. Bull Inst Math Appl 10:266–271

Chapter 8
The Ammann–Beenker Tilings Revisited

Nicolas Bédaride and Thomas Fernique

Abstract This paper introduces two tiles whose tilings form a one-parameter family of tilings which can all be seen as digitization of two-dimensional planes in the four-dimensional Euclidean space. This family contains the Ammann–Beenker tilings as the solution of a simple optimization problem.

8.1 Introduction

Having decided to retile your bathroom this weekend, you go to your favorite retailer of construction products. There, you see an unusual special offer on two strange notched tiles (Fig. 8.1): "Pay the squares cash, get the rhombi for free!"

Fearing that this might be a scam, you try to figure out how your bathroom could be tiled at little cost. After careful consideration, you see that the possible tilings are exactly those where any two rhombi adjacent or connected by lined up squares have different orientations (see Fig. 8.2). In particular, rhombi only do not tile, so you would have to buy at least some squares. You could, of course, tile with squares only (on a grid), but this would mean missing this special offer!

We will show that the cheapest (if not the simplest) way to tile your bathroom is to form a non-periodic tiling, namely an *Ammann–Beenker tiling*. Furthermore, we will show that the set of all possible tilings form a one-parameter family of tilings which can all be seen as digitization of two-dimensional planes in the four-dimensional Euclidean space. Figure 8.3 depicts some possible tilings, with the rightmost one being an Ammann–Beenker tiling.

This is, of course, not only of interest to tile bathrooms, but it could provide a new insight into the theory of quasicrystals. Indeed, digitizations of irrational planes in higher dimensional spaces (also called *projection tilings*) are a common model

N. Bédaride
LATP, Univ. Aix-Marseille, Marseille, France
e-mail: nicolas.bedaride@latp.univ-mrs.fr

T. Fernique (✉)
LIPN, CNRS & Univ. Paris 13, Paris, France
e-mail: thomas.fernique@lipn.univ-paris13.fr

S. Schmid et al. (eds.), *Aperiodic Crystals*, DOI 10.1007/978-94-007-6431-6_8,

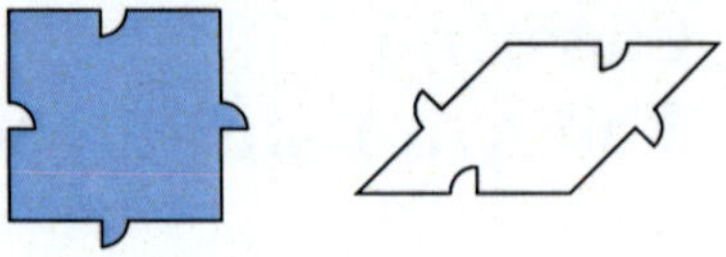

Fig. 8.1 Two notched tiles

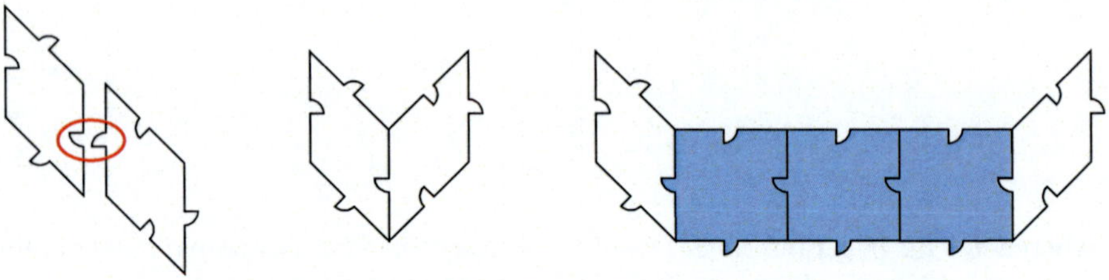

Fig. 8.2 Two rhombi match only if they have different orientations. This still holds with lined up squares between them, since those just carry the notching

Fig. 8.3 Three different possible tilings (notching are not depicted)

of quasicrystals, and the above results give an example of how very simple local constraints can enforce long range order, with the non-periodicity simply coming from tile proportions. In particular, slight variations of tile proportions around those of a non-periodic tiling can lead to close periodic tilings, reminding approximants of quasicrystals.

The rest of the paper is organized as follows. Section 8.2 briefly recalls the history of Ammann–Beenker tilings. Sections 8.3 and 8.4 introduce the main notions, Sect. 8.5 makes a simple but powerful connection with classic results of algebraic geometry, and the technical part of our proof is exposed in Sect. 8.6. We conclude in Sect. 8.7 by formally stating our main result (Theorem 8.1).

8.2 Ammann–Beenker Tilings

Ammann–Beenker tilings are non-periodic tilings of the plane by a square and a rhombus with a 45° angle. Enjoying a (local) 8-fold symmetry, they became a popular model of the 8-fold quasicrystals [10]. They were introduced by Ammann in the 1970s and Beenker in 1982, independently and from different viewpoints.

On the one hand, Ammann defined these tilings as the ones that can be formed by two specific notched tiles and a "key" tile, with the non-periodicity deriving from the hierarchical structure enforced by the notching. This can be compared to the first (and concomitant) definition of Penrose tilings [9].

On the other hand, following the algebraic approach of de Bruijn for Penrose tilings [3], Beenker defined these tilings as digitizations of parallel planes in $\mathbb{R}^4$, with the non-periodicity deriving from the irrationality of the slope of these planes, calling them *Grid-Rhombus* tilings [2]. Unfortunately, Beenker was unaware of the work of the amateur mathematician Ammann, published only some years later [1, 6], and he was unable to find notched tiles which can form only these tilings. Instead, he introduced the notching of Fig. 8.1, called *Arrowed-Rhombus* tilings, the tilings which can be formed, and proved that they strictly contain the Grid-Rhombus tilings.

To conclude this short review, let us mention that Ammann–Beenker tilings cannot be characterized by their local patterns, that is, for any $r \geq 0$, there exists a tiling whose patterns of radius r all appear in an Ammann–Beenker tiling but which is not itself an Ammann–Beenker tiling [4]. Suitable notchings of tiles must thus carry some information over arbitrarily long distances!

8.3 Octagonal Tilings and Planarity

Let $\mathbf{v}_1, \ldots, \mathbf{v}_4$ be pairwise non-collinear unitary vectors of the Euclidean plane. We define the six rhombi $\{\lambda \mathbf{v}_i + \mu \mathbf{v}_j \,|\, 0 \leq \lambda, \mu \leq 1\}$, for $1 \leq i < j \leq 4$, and we call *octagonal tiling* any covering of the Euclidean plane by translated rhombi, where rhombi can intersect only on a vertex or along a complete edge (Fig. 8.3).

Let $\mathbf{e}_1, \ldots, \mathbf{e}_4$ be the canonical basis of $\mathbb{R}^4$. A *lift* of an octagonal tiling is obtained by mapping its rhombi onto faces of unit hypercubes $\mathbb{Z}^4$ so that any two rhombi adjacent along $\mathbf{v}_k$ are mapped onto unit faces adjacent along $\mathbf{e}_k$. This is a two-dimensional surface of $\mathbb{R}^4$ which is uniquely defined up to translation.

An octagonal tiling is said to be *planar* if there are a two-dimensional plane $E \subset \mathbb{R}^4$ and $t \geq 1$ such that it can be lifted into the "slice" $E + [0, t]^4$. The plane E is called its *slope* and the smallest suitable t its *thickness* (both are unique). A planar octagonal tiling can be seen as a digitization of its slope.

For example, the Ammann–Beenker tilings are the planar octagonal tilings of thickness one whose slope is generated by $(\cos \frac{k\pi}{4})_{0 \leq k < 4}$ and $(\sin \frac{k\pi}{4})_{0 \leq k < 4}$.

Planar octagonal tilings form a subclass of the so-called *projection tilings*. Those of thickness one are periodic for a rational slope, *quasiperiodic* otherwise, i.e. any pattern of radius r which appears somewhere in a tiling reappears in this tiling at a distance uniformly bounded in r. This perfect order weakens when the thickness increases, but the long range order nevertheless persists.

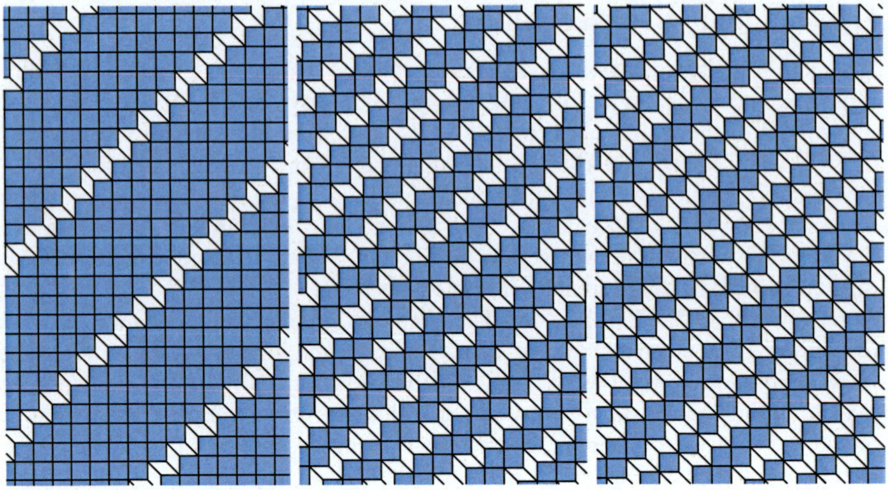

Fig. 8.4 Shadows of the tilings depicted in Fig. 8.3

8.4 Shadows and Subperiods

The *kth shadow* of an octagonal tiling is the orthogonal projection of its lift along $\mathbf{e}_k$. Formally, a kth shadow is a lift of an octagonal tiling, i.e. a two-dimensional surface of $\mathbb{R}^4$, but since it does not contain unit faces with the edge $\mathbf{e}_k$, it can be convenient to see it as a two-dimensional surface of $\mathbb{R}^3$.

A *period* of a shadow is a translation vector leaving invariant the shadow. The *subperiods* of an octagonal tilings are the periods of its shadows.

Figure 8.4 depicts the fourth shadows of the tilings of Fig. 8.3: they are periodic. Actually, the alternation of rhombus orientations in these tilings, discussed in the introduction, precisely enforces a period for each shadow. Formally, one checks that with $\mathbf{v}_k = e^{i\frac{k\pi}{4}}$ (complex notation) for $1 \leq k \leq 4$, the kth shadow of any such tiling admits the period $\mathbf{p}_k$ defined by

$$\mathbf{p}_1 = \mathbf{e}_2 - \mathbf{e}_4, \qquad \mathbf{p}_2 = \mathbf{e}_1 + \mathbf{e}_3, \qquad \mathbf{p}_3 = \mathbf{e}_2 + \mathbf{e}_4, \qquad \mathbf{p}_4 = \mathbf{e}_1 - \mathbf{e}_3.$$

8.5 Grassmann Coordinates and Plücker Relations

First, recall (see, e.g. [7, Chap. 7]) that a two-dimensional plane E of $\mathbb{R}^4$ generated by (u_1, u_2, u_3, u_4) and (v_1, v_2, v_3, v_4) has for *Grassmann coordinates* the numbers $G_{ij} = u_i v_j - u_j v_i$, $1 \leq i < j \leq 4$. Theses coordinates are unique up to a common multiplicative constant; one writes $E = (G_{12}, G_{13}, G_{14}, G_{23}, G_{24}, G_{34})$. Conversely, any G_{ij}'s not all equal to zero are the Grassmann coordinates of some two-dimensional plane of $\mathbb{R}^4$ if and only if they satisfy the *Plücker relation*

$$G_{12}G_{34} = G_{13}G_{24} - G_{14}G_{23}.$$

Then, it is not hard to see that if the lth shadow of a planar octagonal tiling of slope E admits a period (p,q,r), then the Grassmann coordinates satisfy

$$pG_{jk} - qG_{ik} + rG_{ij} = 0,$$

where $l \notin \{i,j,k\}$. Indeed, if E is generated by (u_1,u_2,u_3,u_4) and (v_1,v_2,v_3,v_4), then the lth shadow can be seen as a digitization of the plane of $\mathbb{R}^3$ generated by (u_i,u_j,u_k) and (v_i,v_j,v_k). If (p,q,r) is a period of this plane, it belongs to this plane and thus has a zero dot product with the normal vector $(G_{jk},-G_{ik},G_{ij})$.

One can also use shadows to show that in any planar octagonal tiling of slope E, the ratio between the proportions of tiles with edges $\mathbf{v}_i$ and $\mathbf{v}_j$ and those with edges $\mathbf{v}_k$ and $\mathbf{v}_l$ is $|G_{ij}/G_{kl}|$.

Now, consider a tiling by tiles of Fig. 8.1: it is octagonal up to the notching. If we assume that it is planar, then its subperiods yield

$$G_{23} = G_{34}, \qquad G_{14} = G_{34}, \qquad G_{12} = G_{14}, \qquad G_{12} = G_{23},$$

and plugging this into the Plücker relation, a short computation shows that the slope must be one of the planes

$$E_0 := (0,0,0,0,1,0), \qquad E_{t\neq 0} := (1,t,1,1,2/t,1), \qquad E_\infty := (0,1,0,0,0,0).$$

Conversely, any planar octagonal tiling with one of these slopes and thickness one satisfies the alternation of rhombi orientations (two rhombi with the same orientation would not fit into the slice), thus can be tiled by the tiles of Fig. 8.1.

For example, the tilings of Fig. 8.3 have respective slope $E_{1/4}$, E_1 and $E_{\sqrt{2}}$. In the latter case, which is an Ammann–Beenker tiling, there is thus $\sqrt{2}$ rhombi for each square (since the square area is $\sqrt{2}$ times the rhombus area, each tile covers exactly half of the plane). Tilings by squares only have slope E_0 or E_∞.

However, nothing yet ensures that tilings by Fig. 8.1 tiles are indeed planar!

8.6 Planarity

Lemma 8.1 *Figure 8.1 tiles form only planar tilings of uniformly bounded thickness.*

Proof Let $E := E_{\sqrt{2}}$. One checks that the orthogonal projection of the $\mathbf{e}_i$'s onto E are pairwise non-collinear vectors. Let us identify E with the two-dimensional Euclidean plane and the above projections (up to rescaling) with the $\mathbf{v}_i$'s which define the tiles, so that the orthogonal projection onto E is a homeomorphism from any lift of any tiling of the Euclidean plane by these tiles onto E. Let $\mathcal{T}$ be such a tiling and $\mathcal{S}$ be a lift of it. Define

$$\mathbf{q}_1 = \mathbf{p}_1 + \sqrt{2}\mathbf{e}_1, \qquad \mathbf{q}_2 = \mathbf{p}_2 + \sqrt{2}\mathbf{e}_2,$$
$$\mathbf{q}_3 = \mathbf{p}_3 + \sqrt{2}\mathbf{e}_3. \qquad \mathbf{q}_4 = \mathbf{p}_4 - \sqrt{2}\mathbf{e}_4.$$

Those are pairwise non-collinear vectors of E. Let also $\mathbf{r}_i$ be obtained by changing $\sqrt{2}$ in $-\sqrt{2}$ in $\mathbf{q}_i$. The $\mathbf{r}_i$'s are pairwise non-collinear vectors of $E' := E_{-\sqrt{2}}$. One checks that E and E' are orthogonal planes, so that there exist two real functions z_1 and z_2 defined on E such that the lift $\mathcal{S}$ is the image of E under

$$\rho : \mathbf{x} \mapsto \mathbf{x} + z_1(\mathbf{x})\mathbf{r}_1 + z_2(\mathbf{x})\mathbf{r}_2.$$

Let π_i denote the orthogonal projection along $\mathbf{e}_i$. One has $\pi_i(\mathbf{q}_i) = \pi_i(\mathbf{r}_i) = \mathbf{p}_i$. For any $\mathbf{x} \in E$, the plane $\pi_i(\mathbf{x} + E')$ intersects the shadow $\pi_i(\mathcal{S})$ along the curve

$$\mathcal{C}_i(\mathbf{x}) = \{\pi_i(\mathbf{x}) + z_1(\mathbf{x} + \lambda\mathbf{q}_i)\pi_i(\mathbf{r}_1) + z_2(\mathbf{x} + \lambda\mathbf{q}_i)\pi_i(\mathbf{r}_2) \mid \lambda \in \mathbb{R}\}.$$

Since both $\pi_i(\mathcal{S})$ and $\pi_i(\mathbf{x} + E')$ are $\mathbf{p}_i$-periodic, so is $\mathcal{C}_i(\mathbf{x})$. In particular, it stays at bounded distance from some line directed by $\mathbf{p}_i$. For $i = 1$, since $\pi_1(\mathbf{r}_1) = \mathbf{p}_1$, this ensures that $\lambda \mapsto z_2(\mathbf{x} + \lambda\mathbf{q}_1)$ is uniformly bounded. In other words, z_2 has bounded fluctuations in the direction $\mathbf{q}_1$. Similarly, for $i = 2$, $\pi_2(\mathbf{r}_2) = \mathbf{p}_2$ yields that z_1 has bounded fluctuations in the direction $\mathbf{q}_2$. For $i = 3$, one computes $\mathbf{p}_3 = -\pi_3(\mathbf{r}_1) - \sqrt{2}\pi_3(\mathbf{r}_2)$, what yields bounded fluctuations for $z_2 - \sqrt{2}z_1$ in the direction $\mathbf{q}_3$. Since $\mathbf{q}_1$ and $\mathbf{q}_2$ form a basis of E, let $z_i(\lambda, \mu)$ stand for $z_i(\lambda\mathbf{q}_1 + \mu\mathbf{q}_2)$, $i \in \{1, 2\}$, and write $f \equiv g$ if the difference of two functions f and g is uniformly bounded. The bounded fluctuations of z_1 and z_2 in the directions $\mathbf{q}_1$ and $\mathbf{q}_2$ yield the existence of real functions f and g such that $z_2(\lambda, \mu) \equiv f(\mu)$ and $z_1(\lambda, \mu) \equiv g(\lambda)$. Further, since $\mathbf{q}_3 = \sqrt{2}\mathbf{q}_2 - \mathbf{q}_1$, the bounded fluctuations of $z_2 - \sqrt{2}z_1$ in the direction $\mathbf{q}_3$ yield the existence of a real function h such that $(z_2 - \sqrt{2}z_1)(\lambda, \mu) \equiv h(\sqrt{2}\mu - \lambda)$. Thus

$$f(\mu) - \sqrt{2}g(\lambda) \equiv h(\sqrt{2}\mu - \lambda).$$

Fix $\lambda = 0$ to get $f(\mu) \equiv h(\sqrt{2}\mu)$. Fix $\mu = 0$ to get $-\sqrt{2}g(\lambda) \equiv h(-\lambda)$. Hence

$$h(\sqrt{2}\mu) + h(-\lambda) \equiv h(\sqrt{2}\mu - \lambda).$$

From this easily follows that h, hence f, g, z_1, z_2 and ρ, are linear (up to bounded fluctuations). The tiling $\mathcal{T}$ is thus planar. The thickness is uniformly bounded because the lifts are Lipschitz surfaces with a constant which depends only on E. □

8.7 Conclusion

The following theorem summarizes the results obtained in Sects. 8.5 and 8.6:

Theorem 8.1 *Figure 8.1 tiles can form only planar tilings with slope in* $\{E_t\}_{t \in \mathbb{R} \cup \{\infty\}}$ *and uniformly bounded thickness, and they form at least those of thickness one.*

Moreover, the Ammann–Beenker tilings have the slope which maximizes the area covered by rhombi: they provide the cheapest way to tile your bathroom! Let us also note that among the tilings by Fig. 8.1 tiles, Ammann–Beenker tilings are exactly (up to the thickness) those whose slope satisfies the relation $G_{13} = G_{24}$, i.e.

where the squares appear with the same frequency in their two possible orientations. The above-mentioned result of [4] shows that this relation, although simple, cannot be enforced by local patterns: when t tends towards $\sqrt{2}$, the tilings of slope E_t and $E_{\sqrt{2}}$ (and thickness one) become locally indistinguishable.

Comments

At the time we wrote this paper, we unfortunately were unaware of Ref. [5]. There, A. Katz already obtained Theorem 8.1, and moreover showed that the uniform bound on the thickness of the tilings that can be formed is actually one. We, however, think that our proof deserve to be published. Indeed, the proof in [5] relies on purely geometric considerations in the four-dimensional space, which can be hard to follow by the reader (as acknowledged by the author himself). Alternatively, the notions of shadows and subperiods we rely on reduce much of the problem to the more usual three-dimensional space, while the use of Grassmann coordinates points the way towards a purely algebraic way to solve a wide range of similar tiling problems.

Acknowledgements We thank T.Q.T. Le for sending us the preprint [8] which inspired the proof of Lemma 8.1, and the referee, notably for pointing us the highly relevant reference [5] (see below).

References

1. Ammann R, Grünbaum B, Shephard GC (1992) Aperiodic tiles. Discrete Comput Geom 8:1–25
2. Beenker FPM (1982) Algebraic theory of non periodic tilings of the plane by two simple building blocks: a square and a rhombus. TH Report 82-WSK-04, Technische Hogeschool, Eindhoven
3. de Bruijn NG (1981) Algebraic theory of Penrose's nonperiodic tilings of the plane. Indag Math 43:39–66
4. Burkov SE (1988) Absence of weak local rules for the planar quasicrystalline tiling with the 8-fold rotational symmetry. Commun Math Phys 119:667–675
5. Katz A (1995) Matching rules and quasiperiodicity: the octagonal tilings. In: Axel F, Gratias D (eds) Beyond quasicrystals, pp 141–189
6. Grünbaum B, Shephard GC (1986) Tilings and patterns. Freemann, New York
7. Hodge WVD, Pedoe D (1984) Methods of algebraic geometry, vol 1. Cambridge University Press, Cambridge
8. Le TQT (1992) Necessary conditions for the existence of local rules for quasicrystals. Preprint
9. Penrose R (1974) The Role of aesthetics in pure and applied research. Bull Inst Maths Appl 10
10. Wang N, Chen H, Kuo K (1987) Two-dimensional quasicrystal with eightfold rotational symmetry. Phys Rev Lett 59:1010–1013

Chapter 9
Substitution Rules and Topological Properties of the Robinson Tilings

Franz Gähler

Abstract A relatively simple substitution for the Robinson tilings is presented, which requires only 56 tiles up to translation. In this substitution, due to Joan M. Taylor, neighbouring tiles are substituted by partially overlapping patches of tiles. We show that this overlapping substitution gives rise to a normal primitive substitution as well, implying that the Robinson tilings form a model set and thus have pure point diffraction. This substitution is used to compute the Čech cohomology of the hull of the Robinson tilings via the Anderson–Putnam method, and also the dynamical zeta function of the substitution action on the hull. The dynamical zeta function is then used to obtain a detailed description of the structure of the hull, relating it to features of the cohomology groups.

9.1 Introduction

Robinson's aperiodic set of tiles [8] was the first reasonably small such set which could tile the plane only aperiodically. The local matching rules enforce a hierarchical structure into the tilings, which is used to prove that only aperiodic tilings are admitted. Despite this hierarchical structure, for a long time it was not known whether the Robinson tilings can be generated also by a substitution, which would have enormous advantages for a more detailed study. Only very recently, a substitution for the Robinson tilings could be constructed explicitly [4], albeit a rather complicated one. The Robinson tilings therefore remain an interesting example, not only for historical reasons. In this paper, we present a much simpler substitution, derived from an overlapping substitution due to Joan M. Taylor, which we then use to analyse the structure of the hull of the Robinson tilings in more detail, and relate it to some of the topological invariants of the hull.

F. Gähler (✉)
Faculty of Mathematics, Bielefeld University, 33615 Bielefeld, Germany
e-mail: gaehler@math.uni-bielefeld.de

S. Schmid et al. (eds.), *Aperiodic Crystals*, DOI 10.1007/978-94-007-6431-6_9,

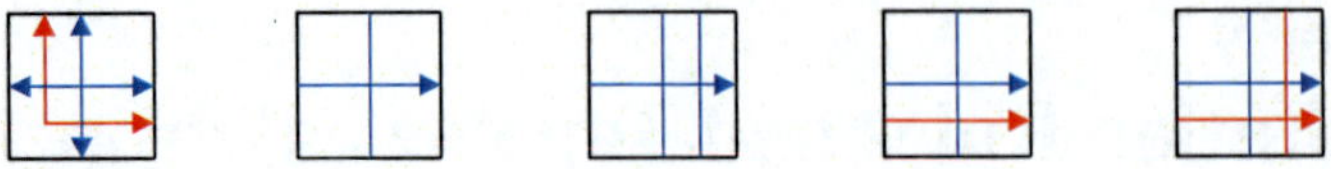

Fig. 9.1 The Robinson tiles. The one on the left is called a cross and plays a special role. All tiles can also be rotated and reflected

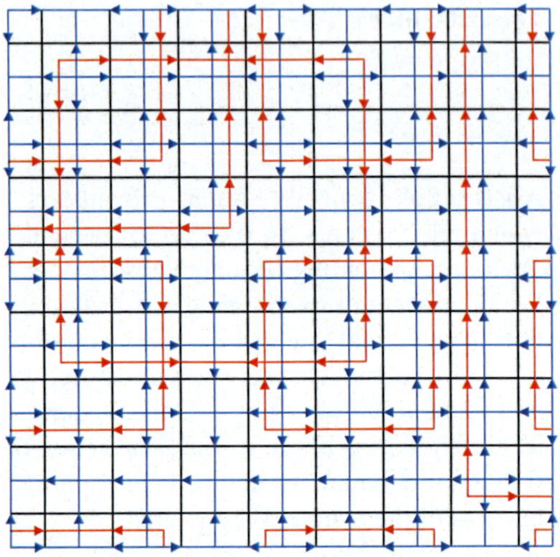

Fig. 9.2 A patch of a Robinson tiling. Note the *red square* frames with corners at cross tiles, which occur at all sizes of the form 2^n, proving the aperiodicity of the Robinson tilings. The corners of the smallest square frames are at crosses forming the odd/odd sublattice of tiles

9.2 A Simple Substitution for the Robinson Tilings

Robinson tilings consist of the five square tiles shown in Fig. 9.1. As the tiles are allowed to be rotated and reflected, there are 28 tiles up to translation. In a legal Robinson tiling, the tiles must obey some local rules. Firstly, the decoration lines must continue across edges, with exactly one arrow head at each line join. Secondly, there must be a square sublattice of index 4 whose tiles are all cross tiles. Apart from this lattice of cross tiles, there may be other crosses as well. We assume in the following, that this sublattice of cross tiles is a the odd/odd position. All tilings satisfying the two rules (which are both local) are called Robinson tilings. In any Robinson tiling, the decoration lines form a hierarchy of square frames of all sizes 2^n (see Fig. 9.2), which proves that Robinson tilings cannot be periodic.

The local rules given above admit also some tilings with defect lines, which are not repetitive (for details, see [5, 8]). As we are heading at primitive substitution rules, by which we can reach only repetitive tilings, we want to discard these defective tilings. We therefore confine ourselves to the minimal subspace of repetitive tilings which is closed and invariant under translations and substitutions. The tilings which we discard form a set of measure zero. In particular, their exclusion does not change any spectral properties.

The hierarchy of square frames of all sizes (Fig. 9.2) suggests a hierarchical structure in the tilings, and it would only be natural if the Robinson tilings could be constructed also by a substitution rule. The construction of such a substitution rule was achieved only recently [4], and the substitution proved to be rather complicated, with 208 tiles up to translation. The reason is that the self-similarity inherent in the Robinson tilings scales around the tile centres, not the vertices. For the substitution, one therefore had to dissect and reassemble the original tiles to new ones, having their vertices at the original tile centres, which results in the rather large number of tiles.

Here, we want to follow a different route, starting from a proposal of Joan M. Taylor (private communication). Recall that the self-similarity scales about tile centres. The idea now is to replace a tile by a 3×3-patch of tiles under the substitution. This patch is larger than the original tile inflated by a factor of 2, so that there are consistency conditions to be obeyed: the substitutions of neighbouring tiles have an overlap, on which they must agree. A relatively simple solution is obtained if we pass to new tiles which are larger by a factor 2. These new tiles have their centres at the tiles at even/even positions (recall that the tiles at odd/odd positions are all crosses). If we add to those even/even tiles a layer of thickness one half, all the remaining tiles are consumed, and we end up with new square tiles of edge length 2 at even/even positions. It turns out that the 28 translation classes of tiles at even/even positions split up into two classes each, so that we now have 56 tile types up to translation. Moreover, these tiles admit a well-defined overlapping substitution, as shown in Fig. 9.3.

The overlapping substitution of Fig. 9.3 is considerably simpler than the one found previously [4]. The set of translation classes of tiles has been cut to a mere 56, from 208 previously. For certain applications, however, such as the computation of the cohomology via the Anderson–Putnam method [1], an overlapping substitution is not suitable. To avoid this problem, we observe that we can always pass to a normal substitution by replacing a tile not by a full 3×3-patch, but by the 2×2-subpatch at the upper right corner, say. Note that we always have to take the subpatch at the same corner, also for the rotated tiles, so that each tile is assigned to a unique supertile. As a result, this assignment breaks the rotation/reflection covariance of the substitution rules, but this is a small price to pay.

Having derived our substitution from an overlapping substitution has yet another advantage. Since the 3×3-patches cover more than the inflated tiles, the overlapping substitution obviously forces the border [6], a property which is inherited also by the normal substitution derived from it. This allows to avoid the use of collared tiles [1] in the Anderson–Putnam method, which is a tremendous advantage, as it also helps to keep the number of tile types small.

9.3 The Structure of the Hull

Due to the repetitivity, the translation group acts minimally on the space of all repetitive Robinson tilings: every translation orbit is dense. The tiling space is therefore the hull of any of its member tilings. Having a substitution, the hull can now be constructed as an inverse limit space [1], and having a *simple* substitution which

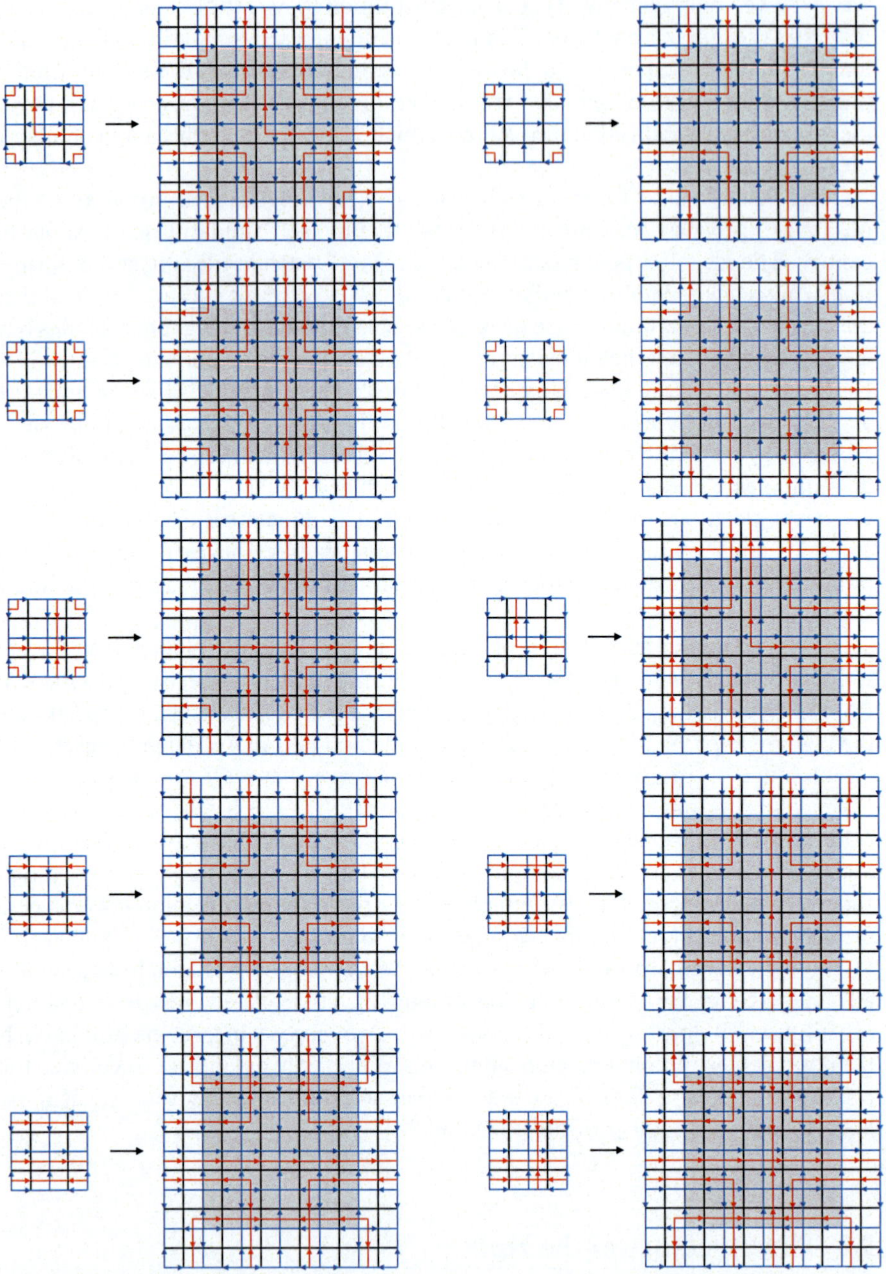

Fig. 9.3 Overlapping substitution for the Robinson tiling. Each tile is replaced by a 3 × 3-patch of tiles. Rotated/reflected tiles are substituted by the corresponding rotated/reflected patches. The inflated tiles cover only the area shaded in *gray*. The substitutions of neighbouring tiles have thus an overlap, on which they agree

forces the border and requires only 56 tiles up to translation simplifies the task considerably. The mere fact of having a lattice substitution tiling has some immediate consequences. Since the crosses at odd/odd positions form a lattice-periodic subset of tiles (with period 4 in each direction), results of Lee and Moody [7] allow concluding that the Robinson tilings form a model set and are thus pure-point diffractive. Since the defective Robinson tilings are a subset of measure zero, the pure-point diffractiveness extends even to all Robinson tilings.

As a limit-periodic model set, the space of Robinson tilings must project 1-to-1 almost everywhere to an underlying 2d, 2-adic solenoid $\mathbb{S}_2^2$ via the torus parametrisation [2]. In the following, we will analyse the structure of the set where this projection fails to be 1-to-1, and try to connect it to the Čech cohomology of the hull. The latter was obtained in [4] via the Anderson–Putnam method [1] as

$$H^2 = \mathbb{Z}\left[\frac{1}{4}\right] \oplus \mathbb{Z}\left[\frac{1}{2}\right]^{10} \oplus \mathbb{Z}^8 \oplus \mathbb{Z}_4, \qquad H^1 = \mathbb{Z}\left[\frac{1}{2}\right]^2 \oplus \mathbb{Z}, \qquad H^0 = \mathbb{Z}, \quad (9.1)$$

which is confirmed using our new, simpler substitution. There is a natural substitution action on the hull, whose Artin–Mazur zeta function is defined as

$$\zeta(z) = \exp\left(\sum_{m=1}^{\infty} \frac{a_m}{m} z^m\right) \quad (9.2)$$

where a_m is the number of points in the hull that are invariant under an m-fold substitution. Note that if the hull consists of two components for which the periodic points can be counted separately, $a_m = a'_m + a''_m$, the corresponding partial zeta functions have to be multiplied: $\zeta(z) = \zeta'(z) \cdot \zeta''(z)$.

Anderson and Putnam have given a different way to compute the dynamical zeta function, as a by-product of computing the Čech cohomology [1]. Recall that the hull is obtained as the inverse limit of the substitution acting on an approximant cell complex. As a consequence, the cohomology of the hull is the direct limit of the substitution action on the cohomology of that cell complex. Suppose $A^{(m)}$ is the matrix of the substitution action on the mth cohomology group (with rational coefficients) of the hull of a substitution tiling. The dynamical zeta function is then given by [1]

$$\zeta(z) = \frac{\prod_{k\ \text{odd}} \det(1 - zA^{d-k})}{\prod_{k\ \text{even}} \det(1 - zA^{d-k})} = \frac{\prod_{k\ \text{odd}} \prod_i (1 - z\lambda_i^{d-k})}{\prod_{k\ \text{even}} \prod_i (1 - z\lambda_i^{d-k})} \quad (9.3)$$

where the latter equality holds if the matrices $A^{(m)}$ diagonalizable, and the $\lambda_i^{(m)}$ are their eigenvalues. Note that Anderson and Putnam have used the matrices of the substitution action on the cochain groups of the approximant complex, rather than the cohomology, but the additional terms in their formula cancel between numerator and denominator.

If we apply this to the Robinson tilings, and take into account the eigenvalues of the substitution action on the cohomology, we obtain for the zeta function

$$\zeta(z) = \frac{(1-2z)^2(1-z)}{(1-z)(1-4z)(1-2z)^{10}(1-z)^8} \tag{9.4}$$

$$= \frac{(1-2z)^2}{(1-z)(1-4z)} \cdot \left(\frac{1-z}{1-2z}\right)^{10} \cdot \frac{1}{(1-z)^{17}}, \tag{9.5}$$

where in the second line we have written the zeta function as the product of the zeta functions of one 2d solenoid $\mathbb{S}_2^2$, ten 1d solenoids $\mathbb{S}_2$, and 17 extra fixed points.

How can this be interpreted? A Robinson tiling generically consists of a single, infinite order supertile. Such tilings project 1-to-1 to the solenoid $\mathbb{S}_2^2$. However, a Robinson tiling can consist also of two infinite order supertiles, which are separated by a horizontal or vertical row of tiles without any crosses. These are the tilings where the projection to $\mathbb{S}_2^2$ is not 1-to-1. A separating row of tiles can be decorated with a single blue line, or a double line with the second line (red) on either side of the middle blue line, and all three cases can be combined with arrows in one or the other direction. All six possibilities, everything else being the same, project to the same point on $\mathbb{S}_2^2$. Moreover, if we take the translation orbit along the defect line, we obtain a whole 1d sub-solenoid $\mathbb{S}_2$ of such 6-tuples. So, in addition to the 2d solenoid $\mathbb{S}_2^2$, the hull contains 5 extra 1d solenoids $\mathbb{S}_2$ in horizontal and 5 in vertical direction. Further, there are 28 fixed points of the substitution, consisting of 4 infinite order supertiles, which all project to the origin of $\mathbb{S}_2^2$. The 2d solenoid and the 10 extra 1d solenoids contain one such fixed point each, so that in addition to those there must be 17 further ones, which all show up in the zeta function (9.5). We finally note that the structure of the hull is in line with the interpretation of [3], were terms $\mathbb{Z}[\frac{1}{2}]$ in H^2 are associated with extra 1d sub-solenoids $\mathbb{S}_2$ in the hull.

Acknowledgements The author would like to thank J.M. Taylor for sharing her ideas on the overlapping substitution rules for the Robinson tilings.

References

1. Anderson JE, Putnam IF (1998) Topological invariants for substitution tilings and their associated C^*-algebras. Ergod Theory Dyn Syst 18:509–537
2. Baake M, Lenz D, Moody RV (2007) Characterization of model sets by dynamical systems. Ergod Theory Dyn Syst 27:341–382
3. Barge M, Sadun L (2011) Quotient cohomology for tiling spaces. NY J Math 17:579–599
4. Gähler F, Julien A, Savinien J (2012) Combinatorics and topology of the Robinson tiling. C R Acad Sci Paris, Ser I 350:627–631
5. Johnson A, Maddden K (1997) Putting the pieces together: understanding Robinson's nonperiodic tilings. Coll Math J 28:172–181

6. Kellendonk J (1995) Noncommutative geometry of tilings and gap labelling. Rev Math Phys 07:1133–1180
7. Lee J-Y, Moody RV (2001) Lattice substitution systems and model sets. Discrete Comput Geom 25:173–201
8. Robinson RM (1971) Undecidability and nonperiodicity for tilings of the plane. Invent Math 12:177–209

Chapter 10
Short-Range Spin Fluctuation in the Zn–Mg–Tb Quasicrystal and Its Relation to the Boson Peak

I. Kanazawa, M. Saito, and T. Sasaki

Abstract We introduce the mechanism of localized collective fluctuation of short-range ordered spin in a dodecahedral spin cluster in Zn–Mg–Tb icosahedral quasicrystals. In addition, we shall discuss the relation to the boson peak in topological glasses.

10.1 Introduction

Since the discovery of quasicrystals, much research has been directed at their unique structure and physical properties, especially quasiperiodicity [1, 2]. Ordering and excitations of quasiperiodically arranged magnetic moments (spins) remain fundamental open problems, despite the intensive efforts continuously made since the discovery of quasicrystals. The Zn–Mg–R (R = rare-earth) icosahedral quasicrystals [3] are the most extensively studied magnetic quasicrystals because of the following experimental advantages. These quasicrystals have well-localized, mostly isotropic and sizable 4f magnetic moments. Their atomic structure is relatively well known. Sato et al. [4] have investigated the low-temperature spin dynamics in the face-centered-icosahedral Zn–Mg–Tb quasicrystal around its spin-glass-like freezing temperature by inelastic neutron scattering. They observed the broad inelastic peak, which can be interpreted as localized collective fluctuations of short-range-ordered spins in a dodecahedral spin cluster. Furthermore, they gave a very important suggestion. That is, they indicated a possible close relation between the broad inelastic spin-excitation peak and the so-called boson peak in topological glasses [5, 6]. The boson peak is a broad inelastic excitation peak universally observed in vibrational spectra of topological glasses at a Q-independent excitation energy of a few meV. Its intensity shows a Q-dependence similar to the static structure factor, whereas its temperature dependence is given by the Bose temperature factor. The boson peak is given by the Bose temperature factor. The boson peak is believed to be related to collective atomic vibrations in a small structural unit.

I. Kanazawa (✉) · M. Saito · T. Sasaki
Department of Physics, Tokyo Gakugei University, Koganei-shi, Tokyo 184-8501, Japan
e-mail: kanazawa@u-gakugei.ac.jp

S. Schmid et al. (eds.), *Aperiodic Crystals*, DOI 10.1007/978-94-007-6431-6_10,

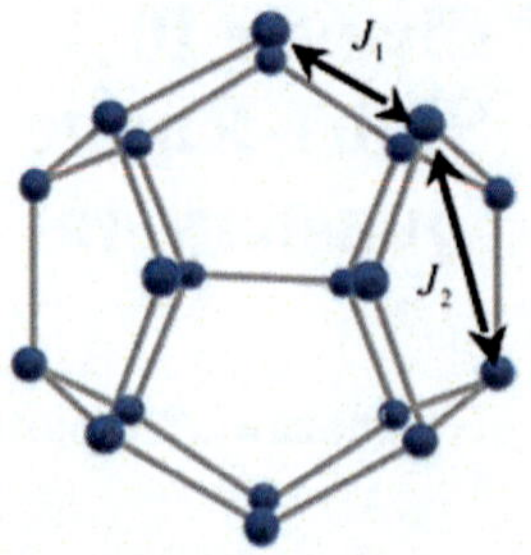

Fig. 10.1 Dodecahedral spin cluster model. A typical ground-state spin configuration is shown by the thick *solid lines*. The nearest-neighbor (J_1) and next nearest-neighbor (J_2) interactions are also indicated by *arrows*

These characteristics are surprisingly similar to those of the observed inelastic spin-excitation peak, and thus, at least phenomenologically, we may regard the inelastic peak as a spin analogue of the boson peak.

One of the present authors (I.K.) [7–11] has introduced a generalized view of the physical origin of the boson peak in the gauge-invariant formula. Especially the localized modes (massive gauge modes), which correspond to the boson peak, are required naturally through the Higgs mechanism. In this study, we shall propose the mechanism of localized collective fluctuations of short-range-ordered spin in a dodecahedral spin cluster (Fig. 10.1) and discuss the relation to the boson peak in topological glasses, by using the theoretical formula [8–11].

10.2 Localized Collective Spin-Fluctuation and the Boson Peak

It has been proposed that the parameter $\phi_a(t, r, u)$ ($a = 1$–4), whose t, r, and u are the time axis, the spatial axes and the perpendicular axes, in the quasicrystal is specified by the rotation, which is related to the gauge fields SO(4) of A^a_μ [11, 12], where $a = 1$–4 and $\mu = 1, 2, 3$ correspond to the physical space axes, and $\mu = 4, 5, 6$ correspond to the perpendicular space axes. To represent the dodecahedral cluster, we set the symmetry breaking $\langle 0|\phi_a|0\rangle = (0, 0, 0, \nu)$ of the Bose parameter field ϕ_a in the Lagrange density as follows:

$$L = \frac{1}{2}(\partial_i S^i - g_1 \varepsilon_{ijk} A^a_i S^k)^2 + \frac{1}{2}(\partial_\mu \phi_a - g_4 \varepsilon_{abc} A^b_\mu \phi_c)^2$$
$$- \frac{1}{4}(\partial_\nu A^a_\mu - \partial_\mu A^a_\nu + g_3 \varepsilon_{abc} A^b_\mu A^c_\nu)^2 - \lambda^2(\phi_a \phi_a - \nu^2)^2. \quad (10.1)$$

After the symmetry breaking $\langle 0|\phi_a|0\rangle = (0, 0, 0, \nu)$, we can obtain the effective Lagrange density as follows:

$$L_{\text{eff}} = \frac{1}{2}(\partial_i S^j - g_1 \varepsilon_{ijk} A^a_i S^k)^2 - \frac{1}{4}(\partial_\nu A^a_\mu - \partial_\mu A^a_\nu + g_1 \epsilon_{abc} A^b_\mu A^c_\nu)^2$$
$$+ \frac{1}{2}(\partial_\mu \rho^\beta - g \epsilon_{\beta\alpha\gamma} A^\alpha_\mu \rho^\gamma)^2 + \frac{m_1^2}{2}[(A^1_\mu)^2 + (A^2_\mu)^2 + (A^3_\mu)^2]$$
$$+ m_1(A^1_\mu \partial_\mu \rho^2 - A^2_\mu \partial_\mu \rho^1) + m_1(A^2_\mu \partial_\mu \rho^3 - A^3_\mu \partial_\mu \rho^2)$$

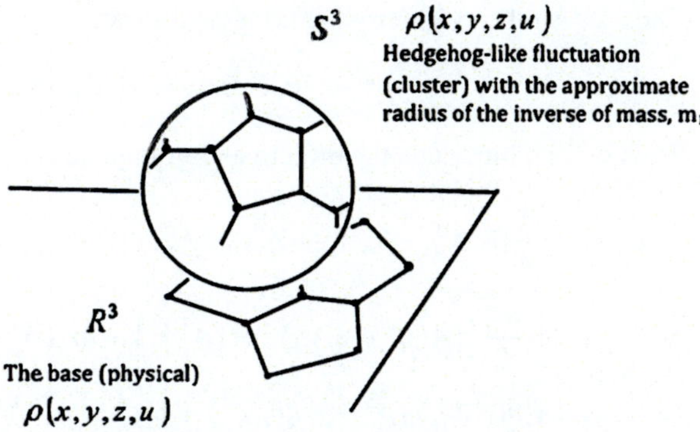

Fig. 10.2 Hedgehog-like fluctuation (cluster), which is similar to the three-dimensional sphere S^3, with the approximate radius of the inverse of mass, m^1, introduced by the Higgs mechanism

$$+m_1(A^3_\mu \partial_\mu \rho^1 - A^1_\mu \partial_\mu \rho^3) + gm_1\{\rho^4[(A^1_\mu)^2 + (A^2_\mu)^2 + (A^3_\mu)^2] - A^4_\mu[\rho^1 A^1_\mu + \rho^2 A^2_\mu + \rho^3 A^3_\mu]\} - \frac{m_1^2}{2}(\rho^4)^2 - \frac{m_2 g}{2m_1}\rho^4(\rho^a)^2 - \frac{m_2^2 g^2}{8m_1^2}(\rho^a\rho^a)^2 \tag{10.2}$$

where S^j is the spin of Tb and $m_1 = \nu \cdot g_4$, $m_2 = 2(2)^{\frac{1}{2}}\lambda \cdot \nu$. The effective Lagrange density describes three massive gauge fields. A^1_μ, A^2_μ and A^3_μ are created through the Anderson–Higgs mechanism by introducing the dodecahedral cluster, the fields A^1_μ, A^2_μ, and A^3_μ are then localized around the dodecahedral cluster within the length of $\sim 1/|m_1| = (\nu.g_4)^{-1} \equiv R_c.R_c$ approximately corresponds to the radius of the dodecahedral cluster.

From the first term in Eq. (10.2), it is shown that the massive gauge fields A^1_μ, A^2_μ, and A^3_μ induce localized collective fluctuation of spins in a dodecahedral spin cluster, taking into account the short-range spin fluctuation mechanism by massive gauge fields [15].

Now we shall consider the relation to the so-called boson peak in topological glasses. It is preferable that we think of the anomalous density fluctuations in three-dimensional liquids (glasses) as the hedgehog-like clusters, taking account of the curvature, as shown intuitively in Fig. 10.2.

We adopt the parameter $\rho(r,u) \equiv \rho^a (a = 1,2,3,4)$, which is similar to that in the Sachdev and Nelson model [12]. The SO(4) quadruplet fields A^a_μ are spontaneously broken through the fluid host around the hedgehog-like fluctuation (cluster) [13, 14]. When the hedgehog-like cluster (soliton) is created, we set the symmetry breaking of the quadruplet fields, $0|\rho|0\rangle$, equal to $\langle 0,0,0,\nu_4\rangle$.

Now we approximately introduce the Lagrange density as follows:

$$L = -\frac{1}{4}(\partial_\nu A^a_\mu - \partial_\mu A^a_\nu + g_1\epsilon_{abc}A^b_\mu A^c_\nu)^2 + \frac{1}{2}(\partial_\mu \rho^B - g\epsilon_{\beta\alpha\gamma}A^\alpha_\mu \rho^\gamma)^2 + c^2\rho^a\rho^a - \lambda(\rho^a\rho^a)^2 - \lambda^2(\rho^a\rho^a - \nu^2)^2. \tag{10.3}$$

Then we set the symmetry breaking as follows:

$$\rho^a \to (0,0,0,\nu_4) + (\rho^1, \rho^2, \rho^3, \rho^4).$$

Thus we can introduce the effective Lagrange density

$$\begin{aligned} L_{\text{eff}} = & -\frac{1}{4}(\partial_\nu A_\mu^a - \partial_\mu A_\nu^a + g_1\epsilon_{abc}A_\mu^b A_\nu^c)^2 + \frac{1}{2}(\partial_\mu \rho^\beta - g\epsilon_{\beta\alpha\gamma}A_\mu^\alpha \rho^\gamma)^2 \\ & + \frac{m_1^2}{2}\left[(A_\mu^1)^2 + (A_\mu^2)^2 + (A_\mu^3)^2\right] + m_1(A_\mu^1\partial_\mu\rho^2 - A_\mu^2\partial_\mu\rho^1) \\ & + m_1(A_\mu^2\partial_\mu\rho^3 - A_\mu^3\partial_\mu\rho^2) + m_1(A_\mu^3\partial_\mu\rho^1 - A_\mu^1\partial_\mu\rho^3) \\ & + gm_1\left\{\rho^4\left[(A_\mu^1)^2 + (A_\mu^2)^2 + (A_\mu^3)^2\right] - A_\mu^4\left[\rho^1 A_\mu^1 + \rho^2 A_\mu^2 + \rho^3 A_\mu^3\right]\right\} \\ & - \frac{m_1^2}{2}(\rho^4)^2 - \frac{m_2 g}{2m_1}\rho^4(\rho^a)^2 - \frac{m_2^2 g^2}{8m_1^2}(\rho^a\rho^a)^2. \end{aligned} \tag{10.4}$$

Here m_1 is $\nu_4 \cdot g$ and m_2 is $2\sqrt{2\lambda \cdot \nu_4}$.

The effective Lagrange density, L_{eff}, represents three massive vector fields A_μ^1, A_μ^2, and A_μ^3, and the masses are created through the Higgs mechanism by introducing the hedgehog-like clusters (solitons), the gauge fields A_μ^1, A_μ^2, and A_μ^3 are only present around clusters. The inverse, $1/|m|$, of the mass of A_μ^1, A_μ^2, and A_μ^3 reveals approximately the radius of the clusters.

Since the gauge field A_μ^4 is massless, it is thought that the gauge field A_μ^4 mediates the long-range interaction between two excited clusters (the hedgehog-like solitons).

In glasses and amorphous materials, the broad maximum of Raman spectra and neutron scattering is due to excess vibrational density of states. It is the so-called boson peak because its intensity changes with T in accordance with the Bose–Einstein factor. It is thought that the vibrational states responsible for the boson peak contribute also to the thermal conductivity plateau because the energy range spanned by the plateau covers that of the boson peak spectra, indicating that acoustic excitations must cease to propagate when their wavelength λ reaches the nm range. That is, acoustic modes may become strongly localized modes, satisfying the Ioffe–Regel condition. By a computer simulation of a soft sphere glass, it is found that there are (quasi)localized modes with effective masses ranging from 10 atomic masses upwards, which are related to the boson peak. In the present theoretical formulation, the effective Lagrangian represents three massive vector fields A_μ^1, A_μ^2, and A_μ^3 which are localized within a radius, $1/|m|$, around the hedgehog-like clusters [13]. Thus, it is suggested that the localized gauge fields A_μ^1, A_μ^2, and A_μ^3 around the hedgehog-like clusters (solitons) are related to the (quasi)localized modes of the boson peak. Expanding the present formula, we can introduce a more generalized view of the origin of the boson peak. We adopt the generalized parameter,

$$\rho(r,u) \equiv \rho^a \quad (a = 1, 2, 3, 4, 5, \ldots, N).$$

When the locally favored cluster is created, we set the symmetry breaking of $0|\rho^b|0 = 0$, in which b represents components within the components from $a = m$ to N. As a result, $m - 1$ massive gauge modes (the localized modes) are introduced around the locally favored cluster through the Higgs mechanism.

It should be noticed that the massive gauge fields A^1_μ, A^2_μ, and A^3_μ in Eq. (10.2) are certainly similar to the ones in Eq. (10.4). Thus, the massive gauge fields A^1_μ, A^2_μ, and A^3_μ, which correspond to the boson peak, induce localized collective fluctuations of spins in a dodecahedral spin cluster. This suggests that we may regard the inelastic peak as a spin analogy of the boson peak.

10.3 Conclusion

We have proposed the mechanism of localized collective spin-fluctuation in a dodecahedral spin cluster. Massive gauge fields, which correspond to the boson peak, induce localized collective spin-fluctuation in a dodecahedral spin clusters.

References

1. Shechtman D, Blech I, Gratias D, Cahn JW (1984) Phys Rev Lett 53:1951
2. Yamamoto A (1996) Acta Crystallogr A, Found Crystallogr 52:509
3. Niikura A, Tsai AP, Inoue A, Matsumoto T (1994) Philos Mag Lett 69:351
4. Sato TJ, Takahara H, Tsai AP, Shibata K (2006) Phys Rev B 73:054417
5. Frick B, Richter D (1995) Science 267:1939
6. Kanaya T, Kaji K (2001) Adv Polym Sci 154:87
7. Kanazawa I (1997) Prog Theor Phys Suppl 126:393
8. Kanazawa I (2001) J Non-Cryst Solids 293–295:615
9. Kanazawa I (2002) J Non-Cryst Solids 312–314:608
10. Suzuki H, Kanazawa I (2010) Intermetallics 18:1809
11. Sethna JP (1983) Phys Rev Lett 51:2198
12. Sachdev S, Nelson DR (1984) Phys Rev Lett 53:1947
13. Kanazawa I (1996) J Radioanal Nucl Chem 210:451
14. Kanazawa I (2000) Radiat Phys Chem 58:456
15. Kanazawa I (2003) J Phys A 36:9371

Chapter 11
Anomalous Properties and the Electronic Glass-Like State in Al-Based Stable Quasicrystals

Kohei Yamada, Tomoaki Sasaki, and Ikuzo Kanazawa

Abstract We discuss the slow dynamics mechanism of the excited carriers in the Al-based quasicrystal-like system. This glassy relaxation mechanism is closely related to the long recombination time of the excited carriers in Al–Pd–Re quasicrystals.

11.1 Introduction

Stable quasicrystals have been found in numerous metallic systems, mainly Al-based systems. Although they consist of metallic atoms, these icosahedral quasicrystals have anomalous electrical conductivities, such as the semiconductor-like properties [1, 2]. That is, their electrical conductivity is extremely low compared with that of crystalline and amorphous phases, and the conductivity decreases with decreasing temperature. It has been suggested that the semiconductor-like properties are due to the existence of a pseudogap in the electron density of states at the Fermi level and the localization tendency of electrons near the Fermi level [3].

In order to obtain detailed information of the density of states (DOS) spectrum, modulated photocurrent measurements for Al–Pd–Re icosahedral quasicrystals have been performed [4, 6].

The amplitude and phase shift of modulated photocurrent for Al–Pd–Re quasicrystals can be well explained by the model in which the two processes, carrier

K. Yamada (✉) · T. Sasaki · I. Kanazawa
Department of Physics, Tokyo Gakugei University, Nukuikita-machi 4-1-1, Koganei-Shi, Tokyo 184-8501, Japan
e-mail: f077151p@st.u-gakugei.ac.jp
url: http://www.springer.com/lncs

K. Yamada
e-mail: m111814k@st.u-gakugei.ac.jp

T. Sasaki
e-mail: f107143f@st.u-gakugei.ac.jp

I. Kanazawa
e-mail: kanazawa@u-gakugei.ac.jp

S. Schmid et al. (eds.), *Aperiodic Crystals*, DOI 10.1007/978-94-007-6431-6_11,

generation and recombination, are involved. It is interesting that the recombination time is by about six orders of magnitude larger than those reported for conventional semiconductors. This suggests a slow dynamics such as glassy relaxation.

The present authors [7–12] have considered the transport property in the randomly distributed system of the correlated configurations (the aggregation) such as the prolate and oblate rhombohedral, in which the nearest distance between each configuration is $\sim 2\pi/2k_F$ (the quasicrystal-like system).

The quasicrystal-like system is regarded as the system composed of the Gaussian correlated distribution of the icosahedral cluster such as the Bergman type and the Mackay type, which includes $2k_F$-phase shift scattering.

Taking into account the short mean free path of 15–20 Å of the electrons in quasicrystals, it looks like that the anomalous transport properties of quasicrystals are not directly related to quasicrystal structures in the longer range than the mean free path of 15–20 Å.

In addition, it has been suggested that there may exist the electron-glassy properties in the quasicrystal-like system [12].

In this study, we shall consider the slow dynamic mechanism of the excited carrier in Al-based quasicrystal such as Al–Pd–Re.

11.2 Quasicrystal-Like System and the Slow Relaxation

The $2k_F$-phase shift scattering induces strongly the density wave of *sp* electrons with a wavelength of $\sim 2\pi/2k_F$. When the high density region of the standing wave occupies a transition metal atom, the *sp* electrons hybridize more strongly the *d* wave-function of the transition metal atoms located in the configuration. Thus the $2k_F$ phase shift scattering and the *sp–d* hybridization are more correlated to each other. In this case, $\gamma \propto n_i N |V_{d,sp}|^2$ is a large value, where the matrix element $V_{d,sp}$ represents the *sp–d* hybridization, and n_i is the density of the aggregation, which is composed of N connected configurations with distance $\sim 2\pi/2k_F$. The aggregation might be identified with the icosahedral cluster such as the Bergman type and the Mackay type. Intuitively, the electron state of a quasicrystal-like system is regarded as the aggregation of incoherent standing waves due to $2k_F$-phase shift scattering. This is the main difference compared to typical amorphous states.

When the system is amorphous or when many defects are introduced, the aggregation, which is composed of N correlated connected configurations with the distance $\sim 2\pi/2k_F$, is broken. As a result, the value of γ will decrease remarkably. This is consistent with the experimental results [20].

In a quasicrystal-like system, we have assumed that the temperature dependence of the conductivity $\Delta\sigma_{1+2}$ is given as $\Delta\sigma_{1+2} \propto T^{\frac{3}{4}}$ in the region of low temperature [11, 12].

Figure 11.1 shows the conductivities of the $Al_{63}Cu_{25}Fe_{12}$, $Al_{65}Cu_{20}Ru_{15}$, and $Al_{70}Cu_{15}Ru_{15}$ quasicrystals. The solid lines show the fits. The solid lines from 20 to 70 K show the temperature curves described by power laws, $\sigma(T) = a + bT^{\frac{3}{4}}$.

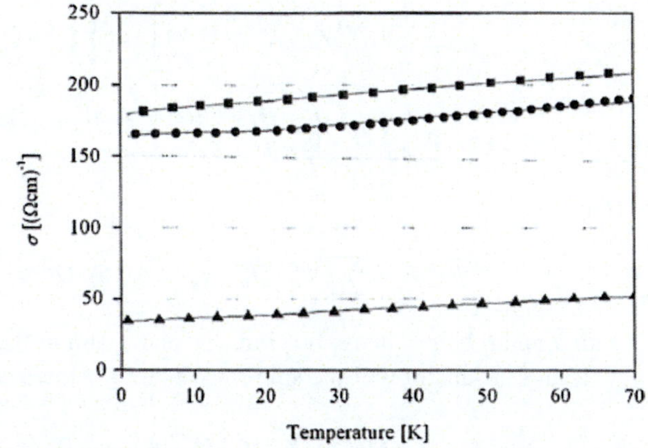

Fig. 11.1 Conductivities in the temperature region below 70 K for the $Al_{63}Cu_{25}Fe_{12}$, $Al_{65}Cu_{20}Ru_{15}$, and $Al_{70}Cu_{15}Ru_{15}$ quasicrystals. The *solid circles*, the *solid squares*, and *triangles* are data points for $Al_{63}Cu_{25}Fe_{12}$ [21], $Al_{65}Cu_{20}Ru_{15}$, and $Al_{70}Cu_{15}Ru_{15}$ [22], respectively. The *solid lines* show the fits [12]

The solid lines below 20 K show the temperature dependence curves given by the $T^{\frac{1}{2}}$ power law.

The coefficient constants of the residual conductivity are coincident within ±5 %.

The $T^{\frac{1}{2}}$ power law below 20 K is attributed to the electron–electron interaction contribution [15, 16]. To average over the randomly distributed aggregation, which is composed of N connected configurations with distance $\sim 2\pi/2k_F$, we use the replica method [17]. We calculate the partition function Z_n of n replicas of the system and average Z_n over the aggregations such the icosahedral clusters. Performing the averaging over the aggregation such as the icosahedral cluster, the potential will be regarded as a random quantity with a Gaussian δ-correlated distribution

$$\langle V(r)V(r')\rangle = \frac{n_i N|V_{d,sp}|^2}{2\pi\nu}\delta(r-r'), \tag{11.1}$$

where $V(r)$ is the effective potential of the clusters, and ν is the state density per spin at the Fermi energy at temperature T. Averaging Z_n over the distribution (1), we obtain

$$\langle\langle Z_n\rangle\rangle = \int \exp(S)\prod_a d\bar{\psi}^a d\psi^a,$$
$$S = \int dr\left[\sum_{n,a}\left[\bar{\psi}_n^a\left(i\varepsilon_n + \frac{1}{2m}\Delta + \mu\right)\right]\psi_n^a + \frac{n_i N|V_{d,sp}|^2}{4\pi\nu}(\bar{\psi}\psi)^2\right], \tag{11.2}$$

where

$$\exp\left[\frac{n_i N|V_{d,sp}|^2}{4\pi\nu}\cdot\frac{1}{\Omega}\sum_{P_1,P_2,k}(\bar{\psi}_{P_1}\psi_{P_2})(\bar{\psi}_{P_2+k+2k_F}\psi_{P_1+k+2k_F})\right]$$
$$= \int \exp\left\{-\int\left[\frac{\pi\nu n_i N|V_{d,sp}|^2}{4}\,\mathrm{Sp}\,Q^2\right.\right.$$

$$-\frac{i}{2}n_i N|V_{d,sp}|^2(\bar{\psi}Q\psi)\Big]\cdot\mathrm{dr}\Big\}\prod \mathrm{d}Q$$

$$\times\Big[\int \exp\Big(-\int\frac{\pi\nu n_i N|V_{d,sp}|^2}{4}\operatorname{Sp}Q^2\mathrm{dr}\Big)\prod\mathrm{d}Q\Big]^{-1}. \tag{11.3}$$

Here,

$$\bar{\psi}Q\psi=\sum\psi_n^{a\dagger}Q_{nm}^{ab}\psi_m^b,\qquad \operatorname{Sp}Q^2=\sum Q_{nm}^{ab}Q_{mn}^{ba},$$

with a and b being the replica indices, and n and m the energy states.

Using the analogy of the spin by Wegner [18], we can show the relation

$$\begin{aligned}\bar{\psi}Q\psi&=\sum_{\substack{ab\\nm}}\psi_n^{a\dagger}Q_{nm}^{ab}\psi_m^b\\&=\sum_{\substack{ab\\nm}}\psi_n^{a\dagger}\psi_n^a\sigma_a\psi_n^{a\dagger}\psi_m^b\sigma_b\psi_m^{\dagger b}\psi_m^b\\&=\sum_{\substack{ab\\nm}}|\psi_n^a|^2q_{nm}^{ab}|\psi_m^b|^2=\sum_{\substack{ab\\nm}}P_n^a q_{nm}^{ab}P_m^b.\end{aligned}\tag{11.4}$$

Here σ_a (σ_b) is the set of Pauli matrices, $P_n^a=|\psi_n^a|^2$, $P_m^b=|\psi_m^b|^2$, and $q_{nm}^{ab}\equiv\sigma_a\psi_n^{a\dagger}\psi_m^b\sigma_b$. P_n^a means the probability that the state (the replica index, a, and the energy level, n) is one of different valleys. The overlap function, $P(q)$, and $\bar{q}$ are defined as follows:

$$P(q)=\Big\langle\sum_{\substack{ab\\nm}}P_n^aP_m^b\delta\big(q-q_{nm}^{ab}\big)\Big\rangle_{\mathrm{av}},$$

$$\bar{q}\equiv\Big\langle\sum_{\substack{ab\\nm}}P_n^aq_{nm}^{ab}P_m^b\Big\rangle_{\mathrm{av}},$$

where $\langle\cdot\rangle_{\mathrm{av}}$ represents the bond averaged quantity. By using the overlap function $P(q)$, we can introduce the relation $\bar{q}=\int_0^{q_{\max}}\mathrm{d}q\,qP(q)$.

An almost insulating quasicrystal-like system might have the broadly distributed overlap function, $P(q)$, which corresponds to the replica symmetry breaking states. The *sp* electron energies, in one icosahedral cluster, depend strongly on *sp*–*d* hybridization and the correlation between the state a of *sp* electron wavefunctions in its icosahedral cluster and the state b from the other icosahedral clusters around its icosahedral cluster.

After the bonding averaging, the distribution of the correlation energy E_c is $\propto-(\sum_{nm}P_n^aq_{nm}^{ab}P_n^b)N|V_{sp-d}|^2\propto P(q)$.

That is, the distribution of energy E_c is $\rho(E_c)\propto P(q)$.

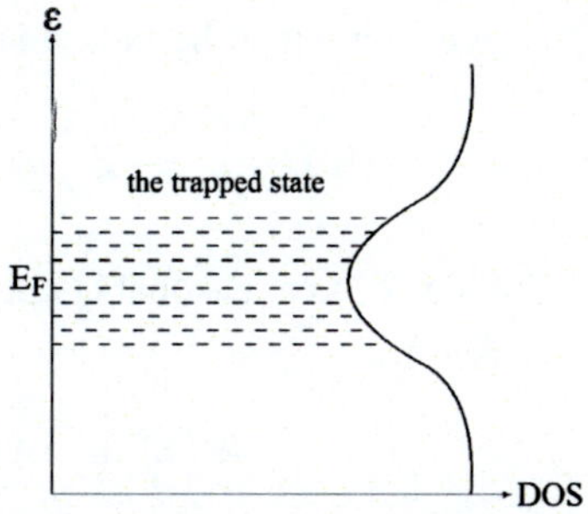

Fig. 11.2 A model of the trapped states in the quasicrystal

A trapped state of the correlation energy E_c is defined by a distribution $\rho(E_c)$ of the trap energy (the energy levels of the correlation energy E_c) as shown in Fig. 11.2.

The primary dynamical quantity is then $P_0(E_c, t)$, the distribution of the energy in a trap of energy E_c at time t, where the subscript 0 indicates that for now we are considering the dynamics without perturbing fields. The evolution of P_0 is given by the master equation

$$\frac{\partial}{\partial t} P_0(E_c, t) = -\Gamma_0(E_c) P_0(E_c, t) + \rho(E_c) \int dE'_c\, W_0(E_c \leftarrow E'_c) P_0(E'_c, t), \tag{11.5}$$

where $W_0(E_c \leftarrow E'_c)$ is the rate of transfer between traps of energy E'_c and E_c,

$$\Gamma_0(E_c) = \int dE'_c \rho(E'_c)\, W_0(E'_c \leftarrow E_c), \tag{11.6}$$

$W_0(E'_c \leftarrow E_c)$ is represented as follows [19]:

$$W_0(E'_c \leftarrow E_c) = \frac{1}{1 + \exp[\beta(E'_c - E_c)]}.$$

Here $\beta = 1/T$ as usual.

The two-time correlation of the excited carrier density $n(t)$ is then

$$\begin{aligned} C(t, t_w) &= \langle n(t) n(t_w) \rangle \\ &= \int dE_c\, dn\, dE'_c\, dn'\, nn'\, P_0(E_c, n | E'_c, n', t - t_w) \\ &\quad \times \rho(n'/E'_c) P_0(E'_c, t_w), \end{aligned}$$

so that $C(t, t_w)$ is approximately represented in the simple form

$$C(t, t_w) = \int dE_c\, \Delta^2(E_c) e^{-\Gamma_0(E_c)(t - t_w)} P_0(E_c, t_w). \tag{11.7}$$

Here $\Delta^2(E_c) = \int dn\, n^2 \rho(n|E_c)$.

Assuming that the field h shifts all energy, the rate is then

$$W(E_c', n' \leftarrow E_c, n) = \frac{1}{1+\exp\{\beta[(E_c' - hn') - (E_c - hn)]\}}. \tag{11.8}$$

For low T, as a reasonable approximation [23] of multiplicatively perturbed rates is given by

$$W(E_c', n' \leftarrow E_c, n) = \mathrm{e}^{h(\gamma n' - \mu n)} W_0(E_c' \leftarrow E_c). \tag{11.9}$$

Then the relation between response is as follows:

$$R(t, t_w) = -\mu \frac{\partial C}{\partial t} + \gamma \frac{\partial C}{\partial t_w}. \tag{11.10}$$

In the case of t and $t_w \gg 1$, $C(t, t_w) \sim t_w/t$ is introduced approximately [19]. For the integrated response function, we get

$$\begin{aligned}\chi(t, t_w) \equiv \chi(C) &= \int_{t_w}^{t} \mathrm{d}t \, R(t, t') \\ &= \gamma(1 - C) + \frac{\mu}{2}(1 - C^2) \\ &\sim \gamma\left(1 - \frac{t_w}{t}\right) + \frac{\mu}{2}\left(1 - \frac{t_w^2}{t^2}\right).\end{aligned}$$

This represents the glassy relaxation such as the slow dynamics of the excited carrier in the quasicrystal-like system. This glassy relaxation mechanism might be closely related to the long recombination time in Al–Pd–Re quasicrystals [5, 6].

11.3 Conclusion

We have proposed the slow dynamics mechanism of the excited carriers in a stable Al-based quasicrystal-like system. It is suggested that this mechanism is closely related to the long recombination time of the excited carriers in Al–Pd–Re quasicrystals.

References

1. Kimura K, Iwahashi H, Hashimoto T, Takeuchi S, Mizutani U, Ohashi S, Itoh O (1989) J Phys Soc Jpn 58:2472
2. Pierce FS, Guo Q, Poon SJ (1994) Phys Rev Lett 73:2220
3. Berger C (1994) In: Hippert F, Gratias D (eds) Lectures on quasicrystals. Les Editions de Physiquo Les Ulis, Paris, pp 463–504

4. Takeda M, Tamura R, Sakairi Y, Kimura K (1997) J Phys Soc Jpn 66:1924
5. Sakairi Y, Takeda M, Tamura R, Edagawa K, Kimura K (1999) In: Dubois JM, Thiel PA, Tsai AP, Urban K (eds) Proceeding of the MRS symposium, vol 553, p 385
6. Sakairi Y, Takeda M, Tamura R, Edogawa E, Kimura K (2000) Mater Sci Eng A 294–296:519
7. Kanazawa I (2002) J Alloys Compd 342:381
8. Kanazawa I (2003) Physica E 18:306
9. Kanazawa I (2003) Mod Phys Lett B 17:841
10. Kanazawa I (2003) Physica B 328:111
11. Akiyama T, Takagiwa Y, Kanazawa I (2006) Philos Mag 86:747
12. Kanazawa I, Kitahata H, Sekiyama Y (2008) Z Kristallogr 223:813
13. Janot C, Boissien M (1994) Phys Rev Lett 72:1674
14. Katz A, Gratias D (1993) J Non-Cryst Solids 153–154:187
15. Aptshuler BL, Aronov A (1979) Sov Phys JETP 50:968
16. Fukuyama H (1985) Prog Theor Phys Suppl 84:47
17. Edwards SF (1975) J Phys C 8:1660
18. Wegner F (1979) Z Phys C 35:207
19. Barnat A, Mezard M (1995) J Phys I:941
20. Rapp O, Karkin AE, Goshchitskii BN, Voronin VI, Srinivas V, Poon SJ (2008) J Phys Condens Matter 20:114120
21. Klein T, Rakoto H, Berger C, Foorcandor G, Gyrot-Lockmann F (1992) Phys Rev B 45:2046
22. Biggs BD, Poon SJ, Manirathnam NR (1990) Phys Rev Lett 65:2700
23. Ritort F (2003) J Phys A 36:10791

Chapter 12
Quantum Diffusion in Separable d-Dimensional Quasiperiodic Tilings

Stefanie Thiem and Michael Schreiber

Abstract We study the electronic transport in quasiperiodic separable tight-binding models in one, two, and three dimensions. First, we investigate a one-dimensional quasiperiodic chain, in which the atoms are coupled by weak and strong bonds aligned according to the Fibonacci chain. The associated d-dimensional quasiperiodic tilings are constructed from the product of d such chains, which yields either the square/cubic Fibonacci tiling or the labyrinth tiling. We study the scaling behavior of the mean square displacement and the return probability of wave packets with respect to time. We also discuss results of renormalization group approaches and lower bounds for the scaling exponent of the width of the wave packet.

12.1 Introduction

Understanding the relations between the atomic structure and the physical properties of materials remains one of the elementary questions of condensed-matter physics further emphasized by the discovery of quasicrystals [1]. Quasicrystals are characterized by a perfect long range order without having a three-dimensional translational periodicity. The former is manifested by the occurrence of sharp spots in the diffraction pattern and the latter in the occurrence of rotational symmetries forbidden for conventional crystals. Already in the 1970s, works by Penrose and Ammann showed that the Euclidean space can be filled gapless and non-overlapping by two or more tiles which are arranged in a nonperiodic way according to matching rules. It turned out that these tilings are suitable to describe the structure of quasicrystals.

Experimental studies revealed rather exotic physical properties. For instance, quasicrystalline surfaces are anti-adhesive in combination with a high level of hardness making them suitable for the production of coatings for medical equipment, engines, cookware, etc. Further, they possess a low thermal and electrical conduc-

S. Thiem (✉) · M. Schreiber
Institut für Physik, Technische Universität Chemnitz, 09107 Chemnitz, Germany
e-mail: stefanie.thiem@physik.tu-chemnitz.de

M. Schreiber
e-mail: schreiber@physik.tu-chemnitz.de

S. Schmid et al. (eds.), *Aperiodic Crystals*, DOI 10.1007/978-94-007-6431-6_12,

tance although they contain a high amount of well-conducting elements. For one-dimensional quasicrystals, many numerical studies helped to establish a better understanding of the physical properties [2–4]. Further, several exact theoretical results are known in one dimension today [5]. However, the characteristics in two or three dimensions have been clarified to a much lesser degree because numerical studies are usually restricted to small approximants.

To address this problem, we study models of d-dimensional quasicrystals with a separable Hamiltonian in a tight-binding approach [6]. This method is based on the Fibonacci sequence, which describes the weak and strong couplings of atoms in a quasiperiodic chain. After a iterations of the inflation rule $\mathcal{P} = \{s \longrightarrow w, w \longrightarrow ws\}$, we obtain the ath order approximant $\mathcal{C}_a$ of the Fibonacci chain. The length f_a of an approximant $\mathcal{C}_a$ is given by the recursive rule $f_a = f_{a-1} + f_{a-2}$ with $f_0 = f_1 = 1$. Further, the ratio of the lengths of two successive approximants approaches the golden mean τ for $a \to \infty$. Solving the time-independent Schrödinger equation

$$\mathbf{H}|\Psi^i\rangle = E^i|\Psi^i\rangle \quad \Leftrightarrow \quad t_{l-1,l}\Psi^i_{l-1} + t_{l,l+1}\Psi^i_{l+1} = E^i\Psi^i_l \tag{12.1}$$

for the Fibonacci chain, we obtain discrete energy values E^i and wave functions $|\Psi^i\rangle = \sum_{l=1}^{f_a+1} \Psi^i_l |l\rangle$ represented in the orthogonal basis states $|l\rangle$ associated to a vertex l. The hopping strength t in the Schrödinger equation is given by the Fibonacci sequence $\mathcal{C}_a$ with $t_s = s$ for a strong bond and $t_w = w$ for a weak bond $(0 < w \le s)$.

The d-dimensional separable quasiperiodic tilings are then constructed from the product of d quasiperiodic chains which are perpendicular to each other. For this setup two special cases are known for which the systems becomes separable [6]:

- *Hypercubic Tiling* $\mathcal{H}_a^{\mathrm{dd}}$—This tiling corresponds to the usual Euclidean product of d linear quasiperiodic chains, i.e., only vertices connected by vertical and horizontal bonds interact as shown in Fig. 12.1. The electronic structure and the transport properties are understood quite well for these systems [3, 7].
- *Labyrinth Tiling* $\mathcal{L}_a^{\mathrm{dd}}$—Only coupling terms to neighbors along the diagonal bonds are considered (cf. Fig. 12.1), where the bond strengths of this tiling equal the products of the corresponding bond strengths of the one-dimensional chains.

The eigenstates of these tilings in d dimensions can be constructed from the eigenstates of d one-dimensional chains. In both cases, the wave functions of the higher-dimensional tiling are constructed as the products of the one-dimensional wave functions, i.e., $\Phi^{\mathbf{s}}_{\mathbf{r}} = \Phi^{i,j,\ldots,k}_{l,m,\ldots,n} \propto \Psi^{1i}_l \Psi^{2j}_m \cdots \Psi^{dk}_n$. The superscripts $\mathbf{s} = (i, j, \ldots, k)$ enumerate the eigenvalues E, and $\mathbf{r} = (l, m, \ldots, n)$ denotes the coordinates of the vertices in the tiling. The energies are treated in a different way. In d dimensions, they are given by $E^{\mathbf{s}} = E^{i,j,\ldots,k} = E^{1i} + E^{2j} + \cdots + E^{dk}$ for the hypercubic tiling and by $E^{\mathbf{s}} = E^{i,j,\ldots,k} = E^{1i}E^{2j}\cdots E^{dk}$ for the labyrinth tiling. This approach allows us to study very large systems in higher dimensions with up to 10^{10} sites.

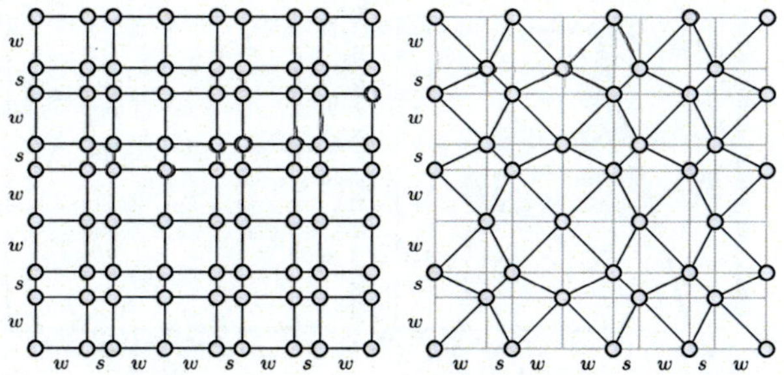

Fig. 12.1 Two-dimensional square tiling $\mathcal{H}_5^{2\mathrm{d}}$ (*left*) and labyrinth tiling $\mathcal{L}_5^{2\mathrm{d}}$ (*right*) constructed from two Fibonacci chains C_5

12.2 Quantum Diffusion

To obtain a deeper understanding of the connections of the electronic transport and the quasiperiodic structure of a system, we study the time evolution of wave packets $|\Upsilon(\mathbf{r}_0,t)\rangle = \sum_{\mathbf{r}\in\mathcal{L}} \Upsilon_\mathbf{r}(\mathbf{r}_0,t)|\mathbf{r}\rangle$ initially localized at a position $\mathbf{r}_0$ which are constructed from the solutions $\Upsilon_\mathbf{r}(\mathbf{r}_0,t) = \sum_\mathbf{s} \Phi^\mathbf{s}_{\mathbf{r}_0}\Phi^\mathbf{s}_\mathbf{r} e^{-\mathrm{i}E^\mathbf{s}t}$ of the time-dependent Schrödinger equation. The temporal autocorrelation function of a wave packet equals the integrated probability to be at the position $\mathbf{r}_0$ up to time $t>0$ [8], i.e., $C(\mathbf{r}_0,t) = \frac{1}{t}\int_0^t |\Upsilon_{\mathbf{r}_0}(\mathbf{r}_0,t')|^2\,\mathrm{d}t'$. We denote the integrand as return probability

$$P(\mathbf{r}_0,t) = \left|\Upsilon_{\mathbf{r}_0}(\mathbf{r}_0,t)\right|^2. \tag{12.2}$$

Another quantity often considered for the description of the electronic transport properties is the mean square displacement of the wave packet (also called width)

$$d(\mathbf{r}_0,t) = \sqrt{\sum\nolimits_{\mathbf{r}\in\mathcal{L}} |\mathbf{r}-\mathbf{r}_0|^2 \left|\Upsilon_\mathbf{r}(\mathbf{r}_0,t)\right|^2}. \tag{12.3}$$

The wave-packet dynamics reveal anomalous diffusion with $d(\mathbf{r}_0,t)\propto t^{\beta(\mathbf{r}_0)}$, and the electronic transport is governed by the wave-packet dynamics averaged over different initial positions $\mathbf{r}_0$, i.e., $d(t) = \langle d(\mathbf{r}_0,t)\rangle \propto t^\beta$. The exponent β is related to the conductivity σ via the generalized Drude formula, where $\beta = 0$ corresponds to no diffusion, $\beta = 1/2$ to classical diffusion, and $\beta = 1$ to ballistic spreading.

The autocorrelation function is expected to decay with $C(\mathbf{r}_0,t)\propto t^{-\delta(\mathbf{r}_0)}$ [8]. The exponent $\delta(\mathbf{r}_0)$ is equivalent to the correlation dimension D_2^μ of the local density of states (LDOS) $\rho(\mathbf{r}_0,E)$ of a system [8]. In one dimension, $\delta\to 1$ corresponds to the ballistic motion of an electron. More information about the transport can be obtained by studying the return probability which shows a power-law behavior according to $P(t)\propto t^{-\delta'}$. The integration is only used to smooth the results. However, this leads to some disadvantages: For $\delta' = 1$ one obtains an additional logarithmic contribution $C(t)\propto \ln(t)/t$, leading to δ significantly lower than 1 [3]. Further, in higher dimensions $\delta' > 1$ is possible for large coupling parameters w. This leads to the convergence of the integral in $C(t)$ as shown in Figs. 12.2 and 12.3 [3].

Some typical results for the mean square displacement $d(t)$ and the temporal autocorrelation function $C(t)$ are shown in Fig. 12.2. We average the results over different initial positions of the wave packet, i.e., we study the scaling behavior

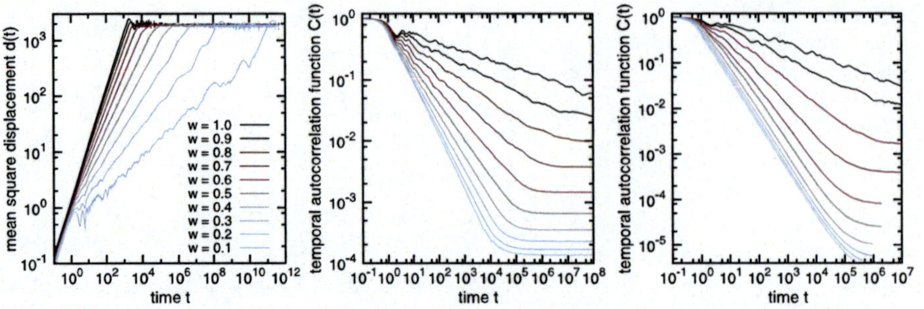

Fig. 12.2 Averaged mean square displacement $d(t)$ for the Fibonacci chain $\mathcal{C}_{19}$ (*left*), averaged temporal autocorrelation function $C(t)$ for the Fibonacci chain $\mathcal{C}_{20}$ (*center*) and the 2D labyrinth tiling $\mathcal{L}_{16}^{2d}$ (*right*) for $s = 1$

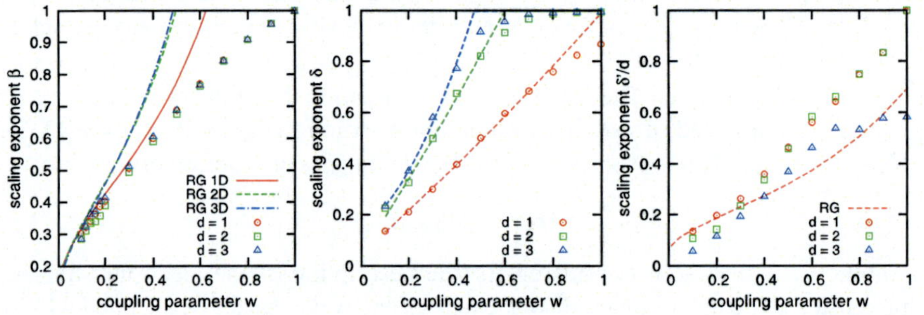

Fig. 12.3 Scaling exponents β (*left*), δ including the limit behavior (*lines*) for infinite systems (*center*), and δ'/d (*right*) for the Fibonacci chain and the labyrinth tilings in 2D and 3D for $s = 1$. Renormalization group (RG) results are shown by lines in *left* and *right panel*

of $d(t) = \langle d(\mathbf{r}_0, t)\rangle \propto t^{-\beta}$ and $C(t) = \langle C(\mathbf{r}_0, t)\rangle \propto t^{-\delta}$. The power-law behavior can be observed over several orders of magnitude in time, before $d(t)$ and $C(t)$ approach a constant due to finite size effects. The corresponding scaling exponents β and δ obtained by least squares fits are compiled in Fig. 12.3. We find $\delta' > 1$ for $w > 0.55s$ in two dimensions and $w > 0.43s$ in three dimensions. Consequently, the scaling exponent δ slowly approaches 1 for this threshold w as indicated by the dashed lines [9]. This is in reasonable agreement with the known transition to an absolute continuous energy spectrum for $w_{\text{th}}^{2d} \approx 0.6s$ and $w_{\text{th}}^{3d} \approx 0.46s$ [7, 10]. The exponent β indicates anomalous transport for the hypercubic and the labyrinth tiling for all $w < s$. Only for $w = s$ ballistic transport is found. Further, we observe that the exponent δ'_{2d} is about twice as large as the one-dimensional exponent δ'_{1d}. However, the results for the three-dimensional labyrinth tiling do not entirely fit into this scheme. While for $w \leq 0.6s$ the expression δ'_{3d}/d is only slightly smaller than the exponent δ'_{1d}, the behavior changes completely for large values of w. In the latter case, the scaling exponent δ'_{3d} becomes almost constant and approaches $\delta'_{3d}(w = s) \approx 1.75$.

For the hypercubic tiling the relation $\delta'_{1d} = \delta'_{dd}/d$ holds, which also characterizes the transition to an absolutely continuous energy spectrum according to $d\delta_{1d}(w_{th}^{dd}) = 1$ [3]. While the numerical results for the 2D labyrinth tiling seem to satisfy these relations as well, there are significant differences in the wave-packet dynamics in three dimensions, which are probably related to the different grid structure. In three dimensions for $w \to s$, the labyrinth tiling approaches a *body centered cubic* lattice while the cubic tiling approaches a *simple cubic* lattice.

12.3 RG Approach and Lower Bound for the Scaling Exponent β

For the hypercubic tiling the scaling exponents of the width fulfill the relation $\beta_{dd} = \beta_{1d}$ due to the separability of the time-evolution operator. Numerical results suggest that this relation is also valid for the labyrinth. For weak couplings ($w \ll s$) it is possible to find a connection between the quasiperiodic structure of the Fibonacci chain [11] or the labyrinth tiling [12, 13] with the wave-packet dynamics by an RG approach proposed by Niu and Nori [14]. The results show that the scaling exponents β for different dimensions approach each other for $w \to 0$ but are not identical [12, 13] (cf. Fig. 12.3). The hierarchic structure of the RG enables us to describe the properties of the wave functions and energy spectrum of the Fibonacci chain for $w \ll s$ [14, 15]. This can be used to derive analytical relations, too, for the mean square displacement [15] and the return probability (cf. Fig. 12.3) [13, 16].

The computation of the scaling exponent β is computationally very expensive. However, one can make use of two general lower bounds for β. As a rule of thumb, the wave-packet propagation is faster for smoother spectral measures. This was proved by Guarneri et al. with the lower bound $\beta \geq D_1^\mu/d$ based on the information dimension D_1^μ of the spectral measure of the LDOS [17]. Ketzmerick et al. showed that the spreading of a wave packet in a space of reduced dimension D_2 (i.e., the correlation dimension of the wave functions) is described by the bound $\beta \geq D_2^\mu/D_2$ as long as $D_2^\mu < 1$, i.e., for singular continuous energy spectra [18]. This result is based on the normalization condition and the known decay of the center of the wave packet according to $t^{-D_2^\mu}$. A comparison with the numerical results in Fig. 12.4 shows that all inequalities are clearly satisfied. However, in the regime of an absolutely continuous energy spectrum, the spectral dimensions fulfill $D_q^\mu = 1$, and the two inequalities are no longer good lower bounds. We like to point out that the decay of the center of the wave packet is only described by the correlation dimension D_2^μ of the LDOS until the integral in $C(t)$ converges [3], i.e., as long as the exponent δ', which describes the decay of the center of the wave packet, is smaller than 1. Hence, for d-dimensional systems the bound by Ketzmerick et al. should be replaced by $\beta \geq \delta'/D_2$ in the absolutely continuous regime. For the hypercubic tiling this bound is as good as for the Fibonacci chain. This bound should hold also in general because it solely makes use of the normalization condition of the wave packet. We have checked in Fig. 12.4 whether this relation is satisfied for the labyrinth tiling and found that this is a significantly better lower bound for $w > w_{th}$.

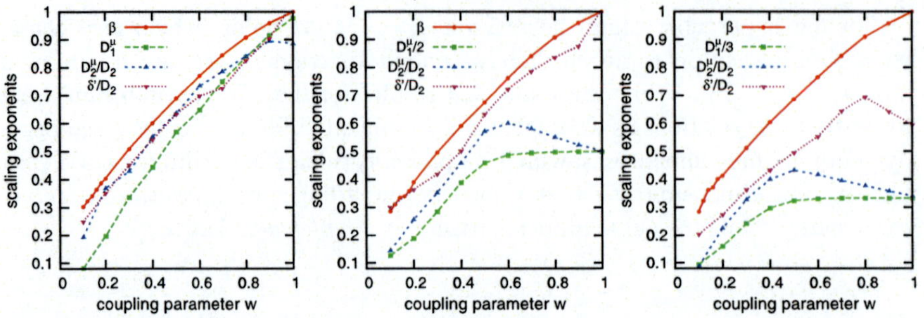

Fig. 12.4 Comparison of β with the lower bounds D_1^{μ}/d, D_2^{μ}/D_2, and δ'/D_2 for the Fibonacci chain (*left*) and the corresponding labyrinth tilings in two (*center*) and three dimensions (*right*) for $s = 1$

12.4 Conclusion

We have studied the wave-packet dynamics for separable quasiperiodic tilings. In higher dimensions, $\delta \to 1$ indicates the existence of an absolutely continuous part in the energy spectrum rather than the occurrence of ballistic transport. To understand the diffusive properties of a system, it is also necessary to compute the width $d(t)$ of the wave packet and the return probability $P(t)$ because in three dimensions the scaling exponents β hardly differ for the considered models although the scaling exponents δ' are rather different for large coupling parameters w. We also found good agreement between the scaling exponents δ' and β with the analytical expressions derived by an RG approach in the regime of strong quasiperiodic modulation. Further, the exponent δ' can be used to define a better lower bound for the scaling exponent β of the wave-packet width in the absolute continuous regime.

References

1. Shechtman D, Blech I, Gratias D, Cahn JW (1984) Phys Rev Lett 53(20):1951
2. Kohmoto M, Sutherland B, Tang C (1987) Phys Rev B 35(3):1020
3. Zhong J, Mosseri R (1995) J Phys Condens Matter 7(44):8383
4. Sánchez V, Wang C (2004) Phys Rev B 70:144207
5. Damanik D, Gorodetski A (2011) Commun Math Phys 205(1):221
6. Sire C (1989) Europhys Lett 10(5):483
7. Even-Dar Mandel S, Lifshitz R (2008) Philos Mag 88(13–15):2261
8. Ketzmerick R, Petschel G, Geisel T (1992) Phys Rev Lett 69(5):695
9. Cerovski VZ, Schreiber M, Grimm U (2005) Phys Rev B 72(5):054203
10. Yuan HQ, Grimm U, Repetowicz P, Schreiber M (2000) Phys Rev B 62(23):15569
11. Abe S, Hiramoto H (1987) Phys Rev A 36(11):5349
12. Thiem S, Schreiber M (2012) Phys Rev B 85(22):224205
13. Thiem S (2012) PhD thesis, Chemnitz University of Technology
14. Niu Q, Nori F (1986) Phys Rev Lett 57(16):2057
15. Piéchon F (1996) Phys Rev Lett 76(23):4372
16. Thiem S, Schreiber M (2013) J Phys Condens Matter 25(7):75503
17. Guarneri I (1993) Europhys Lett 21(7):729
18. Ketzmerick R, Kruse K, Kraut S, Geisel T (1997) Phys Rev Lett 79(11):1959

Chapter 13
Hume–Rothery Stabilization Mechanism of Be-Based Complex Alloys

H. Sato, M. Inukai, E.S. Zijlstra, and U. Mizutani

Abstract We performed first-principles FLAPW (Full potential Linearized Augmented Plane Wave) band calculations for $Be_{13}Mg$ and $Be_{13}Sb$. Furthermore, we calculated the Hume–Rothery plot and $\mathbf{e/a}$ with the tetrahedron method from the case.output1 file generated from WIEN2k. These complex alloys belong to fcc structures with almost the same atom density as hcp Be. From the FLAPW-Fourier spectrum, we could point out that, in both alloys, the pseudogap is formed by Fs–Bz interactions with the spheres just coinciding to reciprocal lattice vectors, $|\mathbf{G}| = 32, 35, 36$ and 40.

13.1 Introduction

Both the Fermi surface–Brillouin zone (Fs–Bz) interactions and orbital hybridizations have been considered to be responsible for the formation of a pseudogap across the Fermi level in structurally complex metallic alloys (CMAs) [1]. Research along this line had been initiated in the framework of the nearly free electron (NFE) model and later the linear muffin-tin orbital-atomic sphere approximation [2]. The NFE model has a serious drawback since it cannot properly handle transition metals (TM) bearing localized d-band near the Fermi level. To overcome this difficulty, Mizutani and co-workers employed first-principles FLAPW band calculations and established a powerful technique named the FLAPW-Fourier method to extract Fs–Bz interac-

H. Sato (✉)
Aichi University of Education, Kariya-shi, Aichi 448-8542, Japan
e-mail: hsato@auecc.aichi-edu.ac.jp

M. Inukai
Toyota Technological Institute, Hisakata, Tempaku-ku, Nagoya 468-8511, Japan

E.S. Zijlstra
Theoretical Physics, University of Kassel, 34132 Kassel, Germany

U. Mizutani
Nagoya Industrial Science Research Institute, 1-13 Yotsuya-dori, Chikusa-ku, Nagoya 464-0819, Japan

S. Schmid et al. (eds.), *Aperiodic Crystals*, DOI 10.1007/978-94-007-6431-6_13,

tions even in strongly hybridizing CMAs [3–5]. The square of the effective Fermi diameter $(2k_F)^2$ and the number of itinerant electrons per atom $\mathbf{e/a}$ can be determined by extracting major plane waves of itinerant electrons outside the muffin-tin spheres and subsequently averaging the square of the wave vectors at the Fermi level. The method above has been specifically named the Hume–Rothery plot (hereafter abbreviated as the HR plot), since it can determine the $\mathbf{e/a}$ value serving as a key parameter in the Hume–Rothery electron concentration rule. Using the HR plot method, they could determine $\mathbf{e/a}$ values even for TM metals and their alloys [6], which had been a longstanding subject in the electron theory of metals since the 1930s [7, 8].

In the present work, attention is directed to the $Be_{13}X$ (X = Mg, Ca, Zr, Sb and La) compounds containing 112 atoms per unit cell (cF112) with space group $Fm\bar{3}c$. The structure information of $Be_{13}X$ compounds is available in the literature [9]. They are characterized by a deep pseudogap across the Fermi level. Our objective is to discuss the origin of the pseudogap from the viewpoint of Fs–Bz interactions by applying the FLAPW-Fourier analysis to $Be_{13}Mg$ and $Be_{13}Sb$, both of which are composed of only sp-bands.

We have recently established the tetrahedron method to enhance the accuracy in the HR plot method. Its principles will be described in Sect. 13.2. We have performed the HR plot analysis for not only fcc-$Be_{13}Mg$ and $Be_{13}Sb$ but also for hcp-Be, hcp-Mg, trigonal-Sb as references.

13.2 Electronic Structure Calculations

FLAPW band calculations have been executed with INTEL version Linux personal computers by using the WIEN2k program package [10]. It provides us the list of the jth eigenvalue $E^j_{\mathbf{k}}$ and the corresponding Fourier coefficient $C^j_{\mathbf{k+G}}$ of an allowed reciprocal lattice vector $\mathbf{G}$ for the wave vector $\mathbf{k}$ in the irreducible wedge of the first Brillouin zone.

The tetrahedron method is newly developed for extracting the set of LAPW states having the largest Fourier coefficient $|C^j_{\mathbf{k+G}}|^2_{\max}$. The values of $E^j_{\mathbf{k}}$ and $C^j_{\mathbf{k+G}}$ for any $\mathbf{k}$ point in the Brillouin zone of the parallelepiped are replaced by those for the equivalent $\mathbf{k}$ point in the irreducible wedge of the zone, by means of symmetry operations in a given crystal structure. The wedge is further divided into the assembly of tetrahedra. The LAPW state $\{2|\mathbf{k}+\mathbf{G}|\}^2_E$ having $|C^j_{\mathbf{k+G}}|^2_{\max}$ is calculated at the centre of gravity of each cross-sectional area $S_\ell(E)$ formed by cutting the ℓth tetrahedron through a given energy surface E, using a linear interpolation approximation [11], and is averaged over all the tetrahedra in the Brillouin zone:

$$\left\langle \{2|\mathbf{k}+\mathbf{G}|\}^2 \right\rangle_E = \frac{\sum_{\ell \text{ in BZ}} \frac{\{2|\mathbf{k}^\ell_{\mathrm{cg}}+\mathbf{G}|\}^2 S_\ell(E)}{|\nabla E|_\ell}}{\sum_{\ell \text{ in BZ}} \frac{S_\ell(E)}{|\nabla E|_\ell}} \tag{13.1}$$

where $\mathbf{k}_{cg}^{\ell}$ specifies the wave vector at the centre of gravity (cg) of $S_\ell(E)$. The number of itinerant electrons per atom **e/a** is calculated from

$$\left(\frac{e}{a}\right)_{\text{local}} = \frac{\pi}{3N_{\text{atom}}}\left[\frac{V_{\text{uc}}^{3/2}}{4\pi^2}\left\langle\{2|\mathbf{k}+\mathbf{G}|\}^2\right\rangle_{E_F}\right]^{3/2}, \tag{13.2}$$

where N_{atom} and V_{uc} are the number of atoms per unit cell and the volume of unit cell, respectively.

13.3 Results and Discussions

The HR plot, i.e. the energy dependence of $\frac{V_{\text{uc}}^{3/2}}{4\pi^2}\langle\{2|\mathbf{k}+\mathbf{G}|\}^2\rangle_E$ along with the total and partial DOSs for hcp-Be and hcp-Mg are shown in Figs. 13.1 and 13.2, respectively. A DOS pseudogap of about 5 eV in width is observed across the Fermi level in Be while a free electron-like DOS with small van-Hove singularities is present in Mg. As can be seen in Fig. 13.1, the pseudogap in Be can be ascribed to orbital hybridizations mainly due to the p states in Be. The direct reading of the ordinate at the Fermi level in Be and Mg yield $(2k_F)^2$ of 2.64 and 2.46 in units of $(2\pi/a)^2 \times \sqrt[3]{3a^2/4c^2}$, respectively. The effective $(\mathbf{e/a})_{\text{local}}$ value is now immediately calculated from Eq. (13.2) to be 2.24 and 2.02, respectively. The HR data points in Mg fall on a straight line from the bottom of the valence band up to +10 eV. The resulting electronic parameters $(2k_F)^2$ and $(\mathbf{e/a})_{\text{local}}$ are in perfect agreement with the free electron value.

In the case of Be, we obtained an almost +10 % deviation from the nominal valence of two. The enhancement in **e/a** is explained in terms of the existence of strong orbital hybridization effects in Be. Note that almost 70 % of valence electrons reside inside the MT-sphere of 2 Å in radius while the remaining 30 % as plane waves in intermediate regions. The p-states in the neighbouring Be atoms are hybridized via plane waves in intermediate regions strongly enough to split them into bonding and anti-bonding states across the Fermi level. This effect is reflected in the HR plot as an upward deviation from otherwise the free electron behaviour, as is drawn with red in colour in Fig. 13.1. Similarly, the HR plot for trigonal Sb known as a semimetal provided the value of $(\mathbf{e/a})_{\text{local}}$ which is 0.33 larger than its nominal valence of five due to strong hybridization effects between the p-states in the same way as in Be.

Now we are ready to discuss Fs–Bz interactions in $Be_{13}Mg$ and $Be_{13}Sb$. As shown in Fig. 13.3, we found the DOS pseudogap in Be to remain essentially unchanged except for the growth of new states immediately above the Fermi level in fcc-$Be_{13}Mg$. A horizontal line with small open circles at both ends in Fig. 13.3 refers to bonding and anti-bonding states caused by the zone splitting associated with $\{2|\mathbf{k}+\mathbf{G}|\}^2 = G^2$ (see Fig. 13.3). It is clear that electronic states over the energy range, where both the pseudogap and the new states are involved, are dominated by those specified by $G^2 = 35$, 36 and 40 in units of $(2\pi/a)^2$.

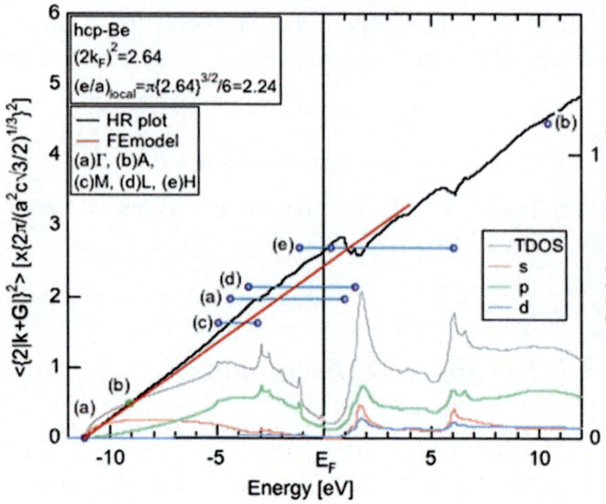

Fig. 13.1 Hume–Rothery plot, and total and partial densities of states for hcp Be. *Small open circles* denote the electronic state $|\mathbf{G}|^2$ versus the energy eigen value at given symmetry points, at which the square of the Fourier coefficient $|C^j_{\mathbf{G}}|^2$ is the largest in the wave function outside the MT sphere

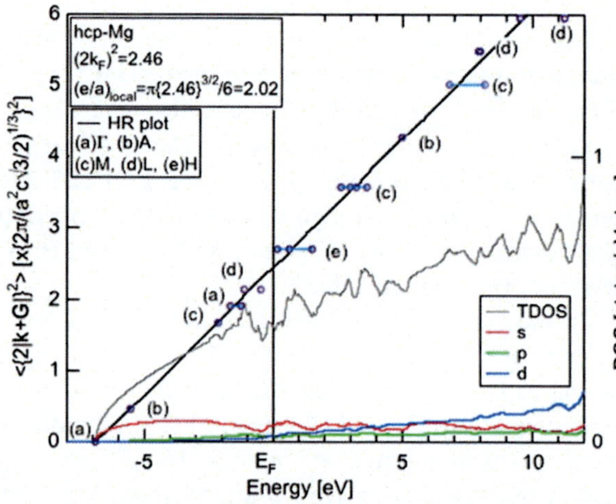

Fig. 13.2 Hume–Rothery plot, and total and partial densities of states for hcp Mg. *Small open circles* denote the electronic state $|\mathbf{G}|^2$ versus the energy eigen value at given symmetry points, at which the square of the Fourier coefficient $|C^j_{\mathbf{G}}|^2$ is the largest in the wave function outside the MT sphere

The HR plot data deviate from otherwise the free electron-like straight line (see the red line in Fig. 13.3) from −3 up to −1 eV but resume the free electron behaviour across the Fermi level. The square of the Fermi diameter $(2k_F)^2 = \langle\{2|\mathbf{k}+\mathbf{G}|\}\rangle^2_{E_F}$ can be read off from the ordinate and turned out to be 35.92. This means that the Fermi surface lies in contact with the zones formed by $|\mathbf{G}|^2 = 35$ and 36, indicating the fulfillment of the matching condition $(2k_F)^2 = |\mathbf{G}|^2$ in the compound $Be_{13}Mg$. The $(\mathbf{e/a})_{\text{local}}$ value is calculated to be 2.01, which is very close to its nominal value of two: $(\mathbf{e/a})_{\text{Be}} = (\mathbf{e/a})_{\text{Mg}} = 2.0$. We consider this to be brought about by the restoration of the free electron behaviour at the Fermi level, thanks to the growth of new states arising from orbital hybridizations between Mg-p and Be-p states.

The HR plot data along with the total DOS for $Be_{13}Sb$ are shown in Fig. 13.4. The total DOS is again found to be similar to that in Be except for the growth of new states across the Fermi level. The growth of the new states must be attributed

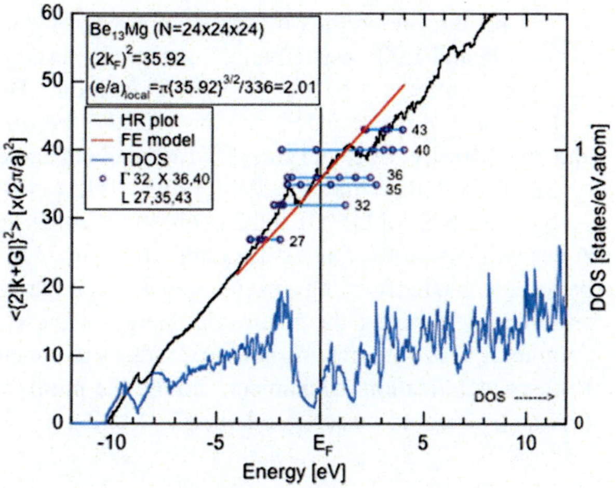

Fig. 13.3 Hume–Rothery plot, and total density of states for $Be_{13}Mg$. *Small open circles* denote the electronic state $|\mathbf{G}|^2$ versus the energy eigen value at given symmetry points, at which the square of the Fourier coefficient $|C^j_{\mathbf{G}}|^2$ is the largest in the wave function outside the MT sphere

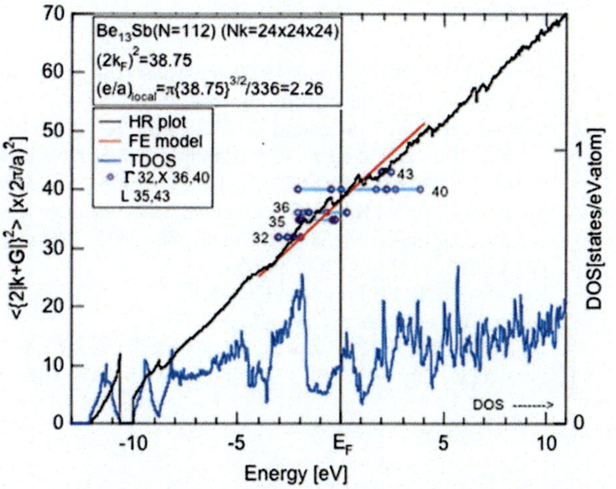

Fig. 13.4 Hume–Rothery plot, and total density of states for $Be_{13}Sb$. *Small open circles* denote the electronic state $|\mathbf{G}|^2$ versus the energy eigen value at given symmetry points, at which the square of the Fourier coefficient $|C^j_{\mathbf{G}}|^2$ is the largest in the wave function outside the MT sphere

to orbital hybridization effect between Be-p and Sb-p states. One can see that the free electron-like behaviour, as guided by the red line in Fig. 13.4, is resumed across the Fermi level owing to the growth of the new states. The value of $(2k_F)^2$ is immediately deduced to be 38.75 from the HR plot. The resulting $(\mathbf{e/a})_{\text{local}} = 2.26$ is very close to its nominal valence of 2.21 (=31/14), lending strong support to the restoration of the free electron behaviour at the Fermi level.

Guided by the same symbols as those in Fig. 13.3, we are convinced to say that electronic states over energies from −2 to +2 eV are heavily perturbed by zone effects associated with $G^2 = 35$, 36 and 40, thereby resulting in not only a wide pseudogap but also the new states near the Fermi level inside the pseudogap. In other words, Fs–Bz interactions involving multi-zones of $G^2 = 35$, 36 and 40 are to produce a wide pseudogap as a result of interference of electrons with relevant zone planes in the fcc Brillouin zone.

The Fs–Bz interactions in hcp-Be and Mg essentially involve a single zone of $G^2 = 1.98$ and 1.90, respectively. Instead, the participation of multi-zones is essential upon forming a pseudogap in CMAs including $Be_{13}Mg$ and $Be_{13}Sb$.

In summary, we revealed that the effective Fermi spheres with $(2k_F)^2 = 35.92$ and 38.75 for $Be_{13}Mg$ and $Be_{13}Sb$, respectively, are embedded in the net of almost spherical Brillouin zones consisting of 12-fold {440} with $G^2 = 32$, 48-fold {531} with $G^2 = 35$, 8-fold {600} and 24-fold {442} zones with $G^2 = 36$ and 24-fold {620} zones with $G^2 = 40$. The Fs–Bz interactions involving the multi-zones above must be responsible for forming a pseudogap and new states as well across the Fermi level and thereby lowering the electronic energy of the system. The involvement of the common Fs–Bz interactions in them leads us to conclude that they obey the Hume–Rothery stabilization mechanism, though the multi-zone effect causes the resulting $(\mathbf{e/a})_{\mathrm{local}}$ values to be scattered over 2.01 to 2.26.

References

1. Mizutani U (2010) Hume–Rothery rules for structurally complex alloy phases. CRC Press/Taylor & Francis Group, Boca Raton
2. Sato H, Takeuchi T, Mizutani U (2004) Phys Rev B 70:094207
3. Asahi R, Sato H, Takeuchi T, Mizutani U (2005) Phys Rev B 71:165103
4. Asahi R, Sato H, Takeuchi T, Mizutani U (2005) Phys Rev B 72:125102
5. Mizutani U, Asahi R, Sato H, Takeuchi T (2006) Phys Rev B 74:235119
6. Mizutani U, Inukai M, Sato H, Zijlstra ES (2012) Philos Mag 92:1691
7. Pauling L (1938) Phys Rev 54:899
8. Raynor GV (1949) Prog Met Phys 1:1
9. Beker TW (1962) Acta Crystallogr 15:175
10. Blaha P, Schwarz K, Madsen G, Kvasnicka D, Luitz J. http://www.wien2k.at/
11. Lehman G, Taut M (1972) Phys Status Solidi (b) 54:469

Chapter 14
Hume–Rothery Stabilization Mechanism in Tsai-Type Cd_6Ca Approximant and e/a Determination of Ca and Cd Elements in the Periodic Table

U. Mizutani, M. Inukai, H. Sato, K. Nozawa, and E.S. Zijlstra

Abstract We performed FLAPW electronic structure calculations with subsequent FLAPW-Fourier analysis for Tsai-type Cd_6Ca containing 168 atoms per unit cell with space group $Pm\bar{3}$. The square of the Fermi diameter $(2k_F)^2$, **e/a** and *critical* reciprocal lattice vector $|\mathbf{G}|^2$s were determined. The origin of the pseudogap across the Fermi level was interpreted in terms of the Hume–Rothery stabilization mechanism based on Fermi surface–Brillouin zone interactions (Fs–Bz) involved. The intuitively expected value of $\mathbf{e/a} = 2.0$ was confirmed. By extending our work to intermetallic compounds existing in the Cd–Ca binary alloy system, we determined the **e/a** values for Ca embedded in the polyvalent matrix Cd. The effective **e/a** for Ca was deduced to be two.

14.1 Introduction

The origin of a pseudogap at the Fermi level in MI-type approximants like Al–Mn and Al–Cu–Fe had been largely discussed in terms of orbital hybridizations due mainly to Al-sp and TM-d (TM = Mn, Fe, etc.) states involved. Unfortunately, the electron concentration-dependent phase stabilization or the Hume–Rothery stabilization mechanism can be hardly approached along this line. Obviously, the Hume–

U. Mizutani (✉)
Nagoya Industrial Science Research Institute, 1-13 Yotsuya-dori, Chikusa-ku, Nagoya 464-0819, Japan
e-mail: uichiro@sky.sannet.ne.jp

M. Inukai
Toyota Technological Institute, Hisakata, Tempaku-ku, Nagoya 468-8511, Japan

H. Sato
Aichi University of Education, Kariya-shi, Aichi 448-8542, Japan

K. Nozawa
Department of Physics, Chuo University, 1-13-27 Kasuga, Tokyo 112-8551, Japan

E.S. Zijlstra
Theoretical Physics, University of Kassel, 34132 Kassel, Germany

S. Schmid et al. (eds.), *Aperiodic Crystals*, DOI 10.1007/978-94-007-6431-6_14,

Rothery stabilization mechanism can be most efficiently studied in the context of the Fermi surface–Brillouin zone (Fs–Bz) interactions in reciprocal space [1]. Mizutani and his coworkers [2, 3] have recently made first-principles WIEN2k-FLAPW electronic structure calculations with subsequent FLAPW-Fourier analysis for both RT- (Rhombic Triacontahedron) and MI- (Mackay Icosahedron) type 1/1–1/1–1/1 approximants and revealed that the Hume–Rothery stabilization mechanism works equally for both families of approximants. They confirmed the existence of common Fs–Bz interactions and could attribute the origin of a pseudogap to the interference phenomenon of electrons across the Fermi level with reciprocal lattice vector $|\mathbf{G}|^2$s centered at 50 in units of $(2\pi/a)^2$, where a is the lattice constant. The square of the Fermi diameter $(2k_F)^2$ was deduced to be equally close to 50. Therefore, the matching condition $(2k_F)^2 = |\mathbf{G}|^2$ involving $|\mathbf{G}|^2 = 50$ as major plane waves holds universally well for both RT- and MI-type approximants.

It is now time for us to examine if the matching condition $(2k_F)^2 = |\mathbf{G}|^2$ with $|\mathbf{G}|^2 = 50$ also holds for the family of Tsai-type 1/1–1/1–1/1 approximants. Tsai et al. [4] reported in 2000 that a stable binary icosahedral quasicrystal can be formed at the composition $Cd_{5.7}Yb$ in the Cd–Yb alloy system and belongs to a new class of the quasicrystal consisting of 66-atom icosahedral clusters. Soon after that, both compounds Cd_6Yb and Cd_6Ca were identified as 1/1–1/1–1/1 approximants with space group $Im\bar{3}$ to the icosahedral quasicrystal in both Cd–Yb and Cd–Ca alloy systems [5, 6].

The Cd_6M (M = Ca and Yb) approximant to the $Cd_{5.7}M$ quasicrystal was found to involve geometrical disorder in the tetrahedral atom cluster in the first shell. At room temperature, the orientation of the cluster is so random that the crystal can be treated on average with space group $Im\bar{3}$. To the best of our knowledge, the atomic structure of the low temperature phase Cd_6Ca has not been experimentally determined. Nozawa and Ishii [7] carried out first-principles structural relaxation to relax short interatomic distances involved in the experimentally derived structure due to Gómez and Lidin [6]. They pointed out that the structure model proposed by Lin and Corbett [8], in which four atoms in the first shell are oriented towards the center of the pentagonal faces of the dodecahedral second shell, is most stable at the lattice constant 15.3 Å, and that the Ishimasa model [9] becomes more stable when the lattice constant is lowered to 15.1 Å. It was noted that the Ishimasa model above is best suited to describe the low temperature ordered phase of Zn_6Sc [7].

We performed first-principles FLAPW band calculations with subsequent FLAPW-Fourier analysis for Cd_6Ca 1/1–1/1–1/1 approximant by using the model structure constructed as an extension of [7]. The aim of the present work is to interpret the origin of a pseudogap at the Fermi level in terms of Fs–Bz interactions involved and to discuss if the Hume–Rothery stabilization mechanism universally holds in all 1/1–1/1–1/1 approximants.

14.2 Electronic Structure Calculations

The model structure of the Cd_6Ca 1/1–1/1–1/1 approximant was obtained by introducing the cubeoctahedron model proposed by Lin and Corbett [8] for the tetrahedral cluster in the first shell into the structure refined by Nozawa and Ishii [7]. Here the relaxation was executed using the generalized gradient approximation (GGA) for the exchange-correlation energy in place of the local density approximation (LDA) employed in [7]. After relaxing both lattice constant and atom positions, we could obtain atomic coordinates for a total of 168 atoms in a cubic unit cell with the lattice constant $a = 15.97$ Å in better agreement with the experimentally derived value of 15.702 Å at 293 K [6].

The FLAPW band calculations have been performed for Cd_6Ca by employing the commercially available WIEN2k-FLAPW program package [10] with INTEL version Linux personal computers. Our in-house Fortran90 program was employed to carry out the FLAPW-Fourier analysis by using "case.output1" file generated by running WIEN2k [1–3]. Briefly, the FLAPW-Fourier spectrum was first constructed to plot the energy dependence of plane waves specified by the square of reciprocal lattice vector $|\mathbf{G}|^2$ in units of $(2\pi/a)^2$ at selected symmetry points of the Brillouin zone. This allows us to extract electronic states dominating at the Fermi level. We also constructed the energy dispersion relation for the LAPW state $\{2|\mathbf{k}_i + \mathbf{G}|\}^2_{E_j}$ having the largest Fourier coefficient for a given energy E_j and wave vector $\mathbf{k}_i$ produced by partitioning the Brillouin zone into $N_\mathbf{k}$ meshes. This is done in an energy interval $E_j \le E < E_j + \Delta E$ for all $\mathbf{k}_i$ values in the Brillouin zone with subsequent averaging of $\{2|\mathbf{k}_i + \mathbf{G}|\}^2_E$ over the Brillouin zone. It provides the energy dispersion relation of the LAPW states having the largest Fourier coefficient. We specifically call this the Hume–Rothery plot since it allows us to determine the square of the effective Fermi diameter $(2k_F)^2$ and $\mathbf{e/a}$ values for any intermetallic compound studied [1–3].

The WIEN2k was run by using the cut-off parameter $R_{\mathrm{MT}}K_{\max} = 6.0$, which determines the number of basis functions or size of the matrices, and $N_\mathbf{k} = 125$ for Cd_6Ca.

14.3 Results and Discussions

The energy dispersion relations and the total DOS for Cd_6Ca are shown in Fig. 14.1. A deep pseudogap is obtained across the Fermi level. The Cd- and Ca-partial DOSs in Cd_6Ca are shown in Figs. 14.2(a) and (b), respectively. The pseudogap at the Fermi level may well be explained in terms of orbital hybridization effects mainly due to Cd-sp and Ca-d states. However, in the present work, we discuss the origin of the pseudogap in terms of Fs–Bz interactions by analyzing both FLAPW-Fourier spectrum and the Hume–Rothery plot.

Figures 14.3(a) and (b) show the FLAPW-Fourier spectra for selected $|\mathbf{G}|^2$s at symmetry points M and point G for Cd_6Ca, respectively. One can immediately find

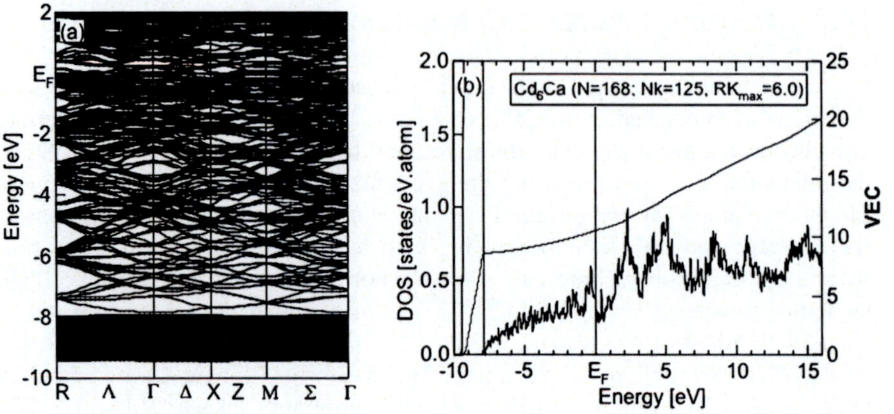

Fig. 14.1 (**a**) Dispersion relations and (**b**) total density of states (DOS) for Cd_6Ca 1/1–1/1–1/1 approximant. **VEC** indicates the integrated DOS or the number of electrons accommodated in the valence band

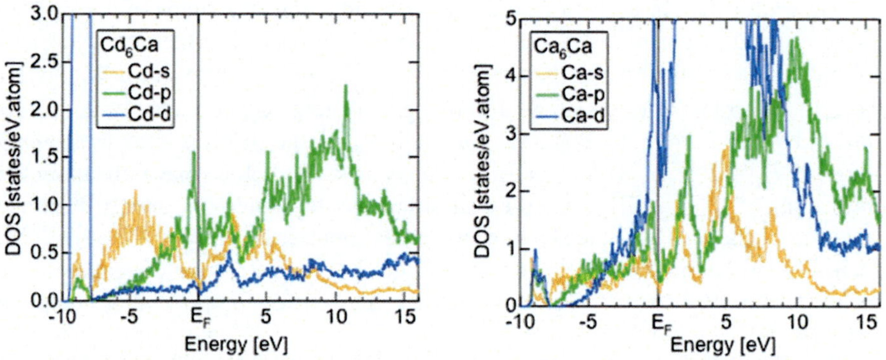

Fig. 14.2 (**a**) Cd-s, Cd-p and Cd-d and (**b**) Ca-s, Ca-p and Ca-d partial DOS of Cd_6Ca

that LAPW waves of $|\mathbf{G}|^2 = 50$ at M and 48 at Γ are most evenly distributed across the Fermi level. By studying the FLAPW-Fourier spectrum at symmetry points X as well, we conclude that electronic states at the Fermi level are dominated over $|\mathbf{G}|^2$s from 44 to 54 and that $|\mathbf{G}|^2 = 48$ and 50 are *critical*. Figure 14.4(a) shows the energy dependence of $\langle \sum_{\mathbf{k+G}} |C_{\mathbf{k+G}}|^2_{\max} \rangle_E$, or briefly $\langle |C|^2_{\max} \rangle_E$, which represents the square of the maximum Fourier coefficient extracted from wave function outside the MT spheres on a given energy surface $E(\mathbf{k}) = E$. Here the summation is carried out over equivalent zone planes. Its value at the Fermi level, $\langle |C|^2_{\max} \rangle_{E_F}$, turns out to be 0.21 for Cd_6Ca, indicating that electrons at the Fermi level are well itinerant [11].

The Hume–Rothery plot, along with its non-dimensional standard deviation $F(E)$ [11] for Cd_6Ca, is shown in Fig. 14.4(b). The data points fall on a straight line from −8 eV above the Cd-4d band to about +2 eV above the Fermi level. A small anomaly found over +3 to +6 eV is due to Ca-3d states. The square of the

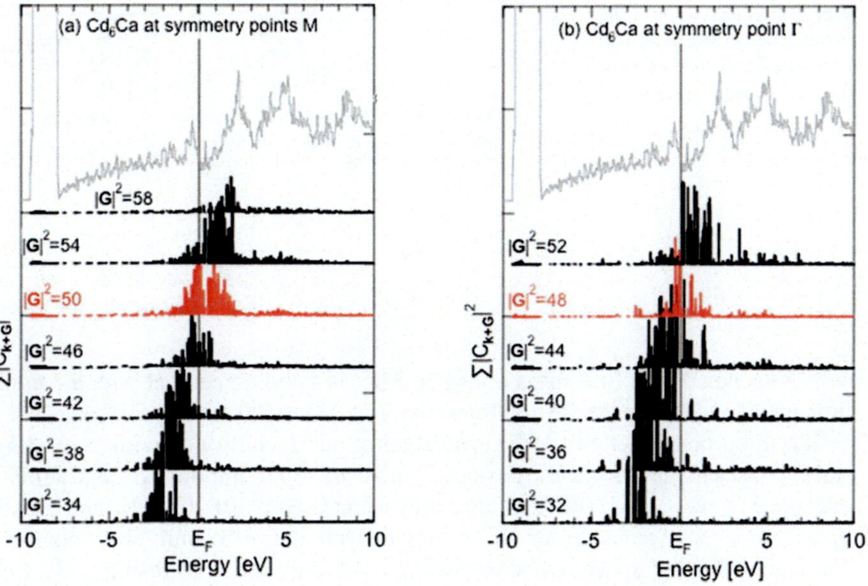

Fig. 14.3 FLAPW-Fourier spectrum at symmetry points (**a**) M and (**b**) G with $|\mathbf{G}|^2$ ranging over 32 to 58 for Cd_6Ca. The total DOS is superimposed

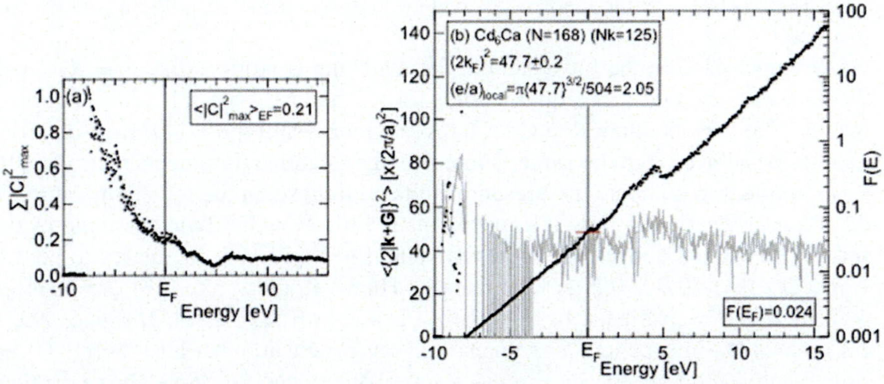

Fig. 14.4 (**a**) Energy dependence of $\langle |C|^2_{\max} \rangle$ and (**b**) Hume–Rothery plot for Cd_6Ca

Fermi diameter, $(2k_F)^2$ in units of $(2\pi/a)^2$, is immediately deduced to be 47.7 ± 0.2 at the intercept with the Fermi level. We confirmed that the matching condition $(2k_F)^2 = |\mathbf{G}|^2$, which plays a key role in the formation of a pseudogap at the Fermi level [1], is well satisfied, since the value of $(2k_F)^2$ thus obtained agrees well with *critical* $|\mathbf{G}|^2$ s $= 48$ and 50 mentioned above.

The effective $\mathbf{e/a}$ value can be immediately calculated by inserting $(2k_F)^2$ obtained above into $\mathbf{e/a} = [\pi\{(2k_F)^2\}^{3/2}]/3N$, where the number of atoms per unit

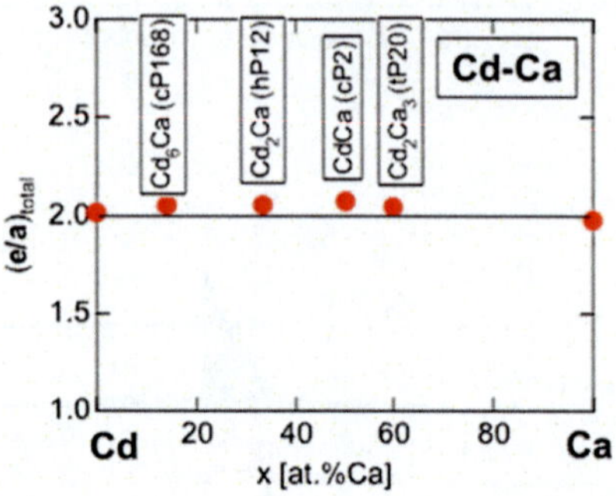

Fig. 14.5 Ca concentration dependence of $(\mathbf{e/a})_{\mathrm{total}}$ for intermetallic compounds in the Cd–Ca alloy system

cell, N, is equal to 168. It turns out to be 2.05, in good agreement with a composition average of 2.0 under the condition that $(\mathbf{e/a})_{\mathrm{Cd}} = 2.0$ and $(\mathbf{e/a})_{\mathrm{Ca}} = 2.0$.

It is important to determine the **e/a** values of intermetallic compounds existing as stable phases in the Cd–Ca alloy system. There are four compounds: Cd_6Ca (cP168) discussed above, Cd_2Ca (hP12), CdCa (cP2), and Cd_2Ca_3 (tP20) [12]. The FLAPW-Fourier analysis was made for all of them as well as for the pure elements Cd and Ca. Due to limited space, we show in Fig. 14.5 only the final results, i.e., the Ca concentration dependences of the **e/a** values deduced from the Hume–Rothery plot for intermetallic compounds in the Cd–Ca alloy system. The data points are found to fall on straight lines connecting $(\mathbf{e/a})_{\mathrm{Cd}} = 2.0$ and $(\mathbf{e/a})_{\mathrm{Ca}} = 2.0$, irrespective of sizes of the unit cell and crystal structures. This confirms that the assignment of $\mathbf{e/a} = 2.0$ to Ca embedded in polyvalent elements like Cd is theoretically well justified.

As emphasized in the Introduction, the matching condition $(2k_F)^2 = |\mathbf{G}|^2$ with the most *critical* $|\mathbf{G}|^2 = 50$ equally holds for RT- and MI-type 1/1–1/1–1/1 approximants. The present studies on Cd_6Ca revealed the Tsai-type 1/1–1/1–1/1 approximant to be subjected to the same Hume–Rothery stabilization mechanism. Finally, it is important to compare the present results with those on the low temperature ordered phase Zn_6Sc 1/1–1/1–1/1 approximant with $N = 336$ twice as large as that for Cd_6Ca studied above. The total **e/a** value for Zn_6Sc was calculated to be 2.18 from $(2k_F)^2 = 79.0 \pm 0.2$ derived from its Hume–Rothery plot [13]. An insertion of 2.18 and $N = 168$ back to the relation $\mathbf{e/a} = [\pi\{(2k_F)^2\}^{3/2}]/3N$ gives rise to $(2k_F)^2 = 49.6$, in excellent agreement with our conclusion that all 1/1–1/1–1/1 approximants including Zn_6Sc forming a superlattice obey the same Hume–Rothery stabilization mechanism specified by $(2k_F)^2 = |\mathbf{G}|^2 = 50$.

Acknowledgements One of the authors (UM) is grateful for the financial support of the Grant-in-Aid for Scientific Research (Contract No. 23560793) from the Japan Society for the Promotion of Science.

References

1. Mizutani U (2010) Hume–Rothery rules for structurally complex alloy phases. CRC Press/Taylor & Francis Group, Boca Raton

2. Inukai M, Zijlstra ES, Sato H, Mizutani U (2011) Philos Mag 91:4247
3. Mizutani U, Inukai M, Sato H, Zijlstra ES (2012) Philos Mag 92:1691
4. Tsai AP, Guo JQ, Abe E, Takakura H, Sato TJ (2000) Nature (London) 408:537
5. Takakura H, Guo J, Tsai AP (2001) Philos Mag Lett 81:411
6. Gómez CP, Lidin S (2003) Phys Rev B 68:024203
7. Nozawa K, Ishii Y (2008) J Phys Condens Matter 31:315206
8. Lin Q, Corbett JD (2004) Inorg Chem 43:1912
9. Ishimasa T, Kasano Y, Tachibana A, Kashimoto S, Osaka K (2007) Philos Mag 87:2887
10. Blaha P, Schwarz K, Madsen G, Kvasnicka D, Luitz J (2012) WIEN2k. http://www.wien2k.at/, last accessed on August 12, 2012
11. Mizutani U, Inukai M, Sato H, Zijlstra ES (2012) Chem Soc Rev 41:6799
12. Villars P (1997) Pearson's handbook, crystallographic data. ASM, Materials Park
13. Mizutani U, Inukai M, Sato H, Zijlstra ES (2012) In: International conference on aperiodic

Chapter 15
Hume–Rothery Stabilization Mechanism in Low-Temperature Phase Zn_6Sc Approximant and e/a Determination of Sc and Y in M–Sc and M–Y (M = Zn, Cd and Al) Alloy Systems

U. Mizutani, M. Inukai, H. Sato, and E.S. Zijlstra

Abstract We have performed FLAPW electronic structure calculations with subsequent FLAPW-Fourier analysis for the low temperature phase Zn_6Sc containing 336 atoms per unit cell with space group B2/b. The square of the Fermi diameter $(2k_F)^2$, electrons per atom ratio **e/a** and *critical* reciprocal lattice vector $|\mathbf{G}|^2$s were determined. The origin of its pseudogap at the Fermi level was interpreted as arising from interference of electrons with $(2k_F)^2 = 79.0 \pm 0.2$ with sets of lattice planes with $|\mathbf{G}|^2$ ranging over 72 to 96. The work was extended to intermetallic compounds existing in M–Sc and M–Y (M = Zn, Cd and Al) binary alloy systems. The effective **e/a** values for Sc and Y were deduced to be 3.0 and 3.1, respectively.

15.1 Introduction

Guided by the empirical Hume–Rothery electron concentration rule, Tsai and his coworkers discovered a series of thermally stable Al–Cu–TM (TM = Fe, Ru and Os) and Al–Pd–TM (TM = Mn, Re) icosahedral quasicrystals in the early 1990s [1, 2]. Here negative valences proposed by Raynor [3] for transition metal (TM) elements were used. Tsai et al. [4] reported in 2000 that a stable binary icosahedral quasicrystal can be formed at the composition $Cd_{5.7}Yb$ and belongs to a new class of packing 66-atom icosahedral clusters. They also revealed that the compound Cd_6Yb lying next to the quasicrystal above in the phase diagram corresponds to its

U. Mizutani (✉)
Nagoya Industrial Science Research Institute, 1-13 Yotsuya-dori, Chikusa-ku, Nagoya 464-0819, Japan
e-mail: uichiro@sky.sannet.ne.jp

M. Inukai
Toyota Technological Institute, Hisakata, Tempaku-ku, Nagoya 468-8511, Japan

H. Sato
Aichi University of Education, Kariya-shi, Aichi 448-8542, Japan

E.S. Zijlstra
Theoretical Physics, University of Kassel, 34132 Kassel, Germany

S. Schmid et al. (eds.), *Aperiodic Crystals*, DOI 10.1007/978-94-007-6431-6_15,

1/1–1/1–1/1 approximant with space group Im$\bar{3}$. More recently, Ishimasa and his associates [5] have succeeded in synthesizing the Zn–Mg–Sc quasicrystal guided by the work of Andrusyak et al. [6]. Later, Ishimasa et al. [7] studied the structures in more details in both quasicrystals and their 1/1–1/1–1/1 approximants in Zn–Mg–Sc, Cu–Ga–Mg–Sc and Zn–Mg–Ti alloys and revealed the presence of a few Zn atoms in the first shell in the 1/1–1/1–1/1 approximants and claimed them to be isostructural to the family of Cd_6M (M = Yb and Ca) 1/1–1/1–1/1 approximants. They also discussed the formation range for the Tsai-type icosahedral quasicrystals in terms of electrons per atom ratio **e/a** by taking a composition average of **(e/a)** values of constituent elements: $(\mathbf{e/a})_{Cu} = (\mathbf{e/a})_{Ag} = 1.0$, $(\mathbf{e/a})_{Mg} = (\mathbf{e/a})_{Ca} = (\mathbf{e/a})_{Zn} = (\mathbf{e/a})_{Cd} = (\mathbf{e/a})_{Yb} = 2.0$, $(\mathbf{e/a})_{Sc} = (\mathbf{e/a})_{Y} = 3.0$ and $(\mathbf{e/a})_{Ti} = 4.0$. As a result, they are commonly stabilized at **e/a** ranging over 2.00 to 2.15. It is worthwhile noting, at this stage, that tri-valence for Sc and Y in Group 3 and quadri-valence for Ti and Zr in Group 4 in the periodic table were originally proposed by Pauling [8].

There is a clear difference in the approach between Pauling and Raynor: Pauling treated the valence band of the TM element itself to define its metallic valence, while Raynor considered the effective **e/a** of the TM element embedded in the host metal Al. Obviously, a substantial difference in the electron concentration parameter emerged between them. In the past, experimentalists have employed either of these two models upon discussing the **e/a**-dependent alloy phase stability or the Hume–Rothery electron concentration rule, though both models were constructed without any rigorous justification based on first-principles electronic structure calculations. To overcome this difficulty, Mizutani and coworkers [9–11] employed first-principles FLAPW (Full potential Linearized Augmented Plane Wave) electronic structure calculations and established a powerful technique to extract Fs–Bz (Fermi surface–Brillouin zone) interactions involved and to determine the **e/a** value for elements and intermetallic compounds, regardless of whether or not the TM element is involved. They have made full use of its formalism, in which the wave functions outside the muffin-tin (MT) spheres are expanded into plane waves over allowed reciprocal lattice vectors **G**.

The Zn_6Sc compound is known to be the 1/1–1/1–1/1 approximant to its quasicrystal and the structure of its low temperature phase has been recently determined by Ishimasa et al. [12]. The geometrically disordered atom cluster in the first shell undergoes an orientational ordering and the structure was described as a perfectly ordered phase containing a total of 336 atoms per a monoclinic unit cell with space group $C2/c$. In the present work, we have performed the WIEN2k-FLAPW band calculations with subsequent FLAPW-Fourier analysis for Zn_6Sc with equivalent space group $B2/b$. The calculations above were extended to existing intermetallic compounds in M–Sc and M–Y (M = Al, Zn and Cd) alloy systems to determine theoretically the **e/a** values of the TM elements Sc and Y.

15.2 Electronic Structure Calculations

FLAPW band calculations have been performed by employing the commercially available WIEN2k-FLAPW program package [13] with INTEL version Linux personal computers. Our in-house Fortran90 Program has been devised to carry out the FLAPW-Fourier analysis by using "case.output1" file generated by running WIEN2k [9–11].

The FLAPW-Fourier spectrum was first constructed to plot the energy dependence of plane wave components specified by the square of reciprocal lattice vector $|\mathbf{G}|^2$ in units of $(2\pi/a)^2$ at selected symmetry points of the Brillouin zone. This allows us to extract electronic states dominating at the Fermi level. The reciprocal lattice vectors thus extracted are called *critical*. As a next step, we construct the energy dispersion relation for the LAPW state $\{2|\mathbf{k}_i + \mathbf{G}|\}^2_{E_j}$ having the largest Fourier coefficient for a given energy E_j and wave vector $\mathbf{k}_i$ produced by partitioning the Brillouin zone into $N_{\mathbf{k}}$ meshes. This is done in an energy interval $E_j < E < E_j + \Box E$ for all $\mathbf{k}_i$ values in the Brillouin zone with subsequent averaging of $\{2|\mathbf{k}_i + \mathbf{G}|\}^2_E$ over the Brillouin zone. It provides the energy dispersion relation reflecting the LAPW states having the largest Fourier coefficient. We specifically call this the Hume–Rothery plot since it allows us to determine the square of the effective Fermi diameter $(2k_F)^2$ and the $\mathbf{e/a}$ values for each intermetallic compound studied [9–11].

The WIEN2k was run by using the cut-off parameter $R_{\mathrm{MT}}K_{\max} = 6.0$, which determines the number of basis functions or size of the matrices, and $N_{\mathbf{k}} = 400$ for Zn_6Sc.

15.3 Results and Discussions

The energy dispersion relations and the total DOS for Zn_6Sc are shown in Fig. 15.1. A deep DOS pseudogap is formed at the Fermi level. The Zn- and Sc-partial DOSs in Zn_6Sc are shown in Figs. 15.2(a) and (b), respectively. Orbital hybridization effects mainly due to Zn-sp and Sc-d states are apparently responsible for opening a pseudogap across the Fermi level. We can alternatively discuss the origin of a pseudogap at the Fermi level in terms of Fs–Bz interactions by analyzing both FLAPW-Fourier and the Hume–Rothery plot.

Figure 15.3 shows the FLAPW-Fourier spectrum at symmetry point Γ for selected $|\mathbf{G}|^2$s for Zn_6Sc. One can immediately find that LAPW waves of $|\mathbf{G}|^2 = 76$ and 80 are most densely distributed across the Fermi level. The electronic states thus extracted are called *critical*. By studying the FLAPW-Fourier spectra at symmetry points M and X as well, we conclude that electronic states at the Fermi level are dominated over $|\mathbf{G}|^2$s from 72 to 96 and that $|\mathbf{G}|^2 = 76$ and 80 are the most *critical*. Figure 15.4(a) shows the energy dependence of $\langle\sum_{\mathbf{k+G}} |C_{\mathbf{k+G}}|^2_{\max}\rangle_E$, or briefly $\langle|C|^2_{\max}\rangle_E$, which represents the square of the maximum Fourier coefficient extracted from wave function outside the MT spheres on a given energy surface $E(\mathbf{k}) = E$. The summation is carried out over equivalent zone planes. Its value at

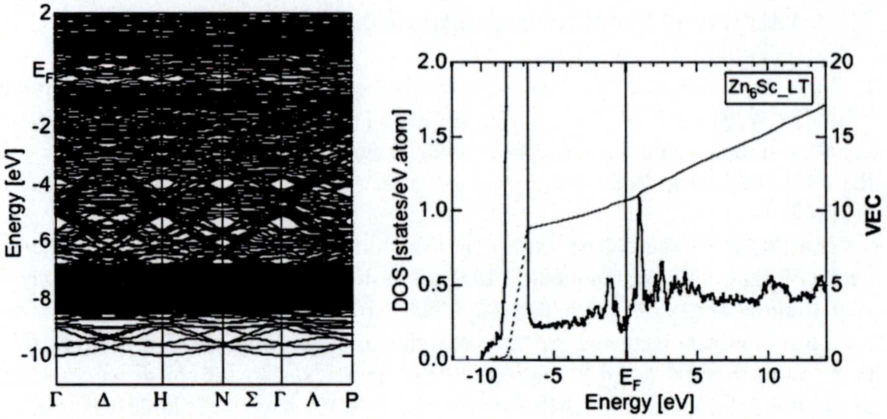

Fig. 15.1 (**a**) Dispersion relations and (**b**) total density of states (DOS) for low temperature phase Zn_6Sc. **VEC** indicates the integrated DOS or the number of electrons accommodated in the valence band

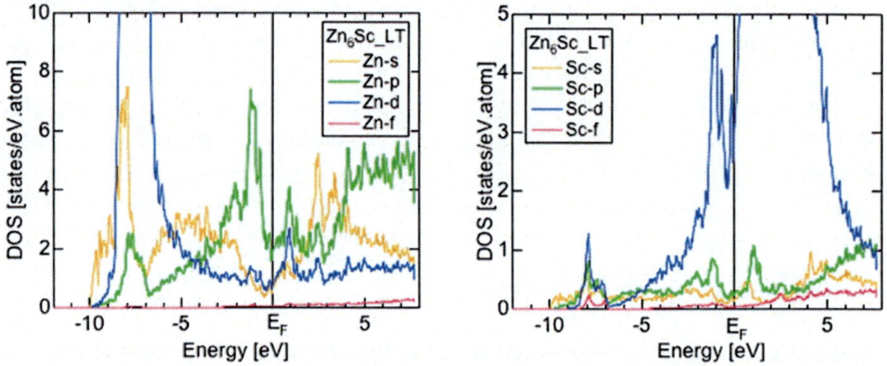

Fig. 15.2 (**a**) Zn-s, Zn-p, Zn-d and Zn-f and (**b**) Sc-s, Sc-p, Sc-d and Sc-f partial DOS of Zn_6Sc

the Fermi level, $\langle |C|^2_{\max}\rangle_{E_F}$, may be used as a measure to judge the itinerancy of electrons at the Fermi level. An electron at the Fermi level is regarded as being itinerant if $\langle |C|^2_{\max}\rangle_{E_F} > 0.1$ [14]. The value for Zn_6Sc turns out to be 0.067, indicating that electrons at the Fermi level are localized.

The Hume–Rothery plot and its non-dimensional standard deviation $F(E)$ [14] for Zn_6Sc are shown in Fig. 15.4(b). The data points fall on a straight line, provided that the free electron model holds well. A triangle shaped anomaly at about -8 eV reflects the highly localized Zn-3d band. One can also see that anomalies due to Sc-d states occur across the Fermi level but are well suppressed. A straight line can be drawn through the Fermi level, as indicated in Fig. 15.4(b). The square of the Fermi diameter, $(2k_F)^2$ in units of $\{2\pi/(abc)^{1/3}\}^2$, is deduced to be 79.0 ± 0.2. It is clear that the matching condition $(2k_F)^2 = |\mathbf{G}|^2$, which plays a key role in the formation of a pseudogap at the Fermi level, is well satisfied, since $(2k_F)^2$ thus obtained agrees

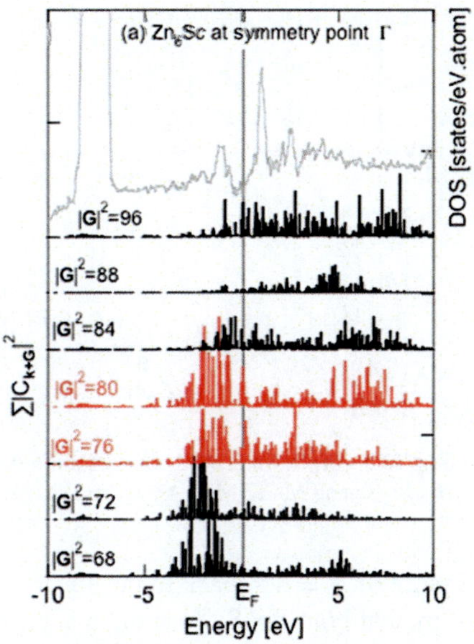

Fig. 15.3 FLAPW-Fourier spectrum at symmetry point G with $|\mathbf{G}|^2$s ranging over 68 to 96 for Zn_6Sc. The total DOS is superimposed

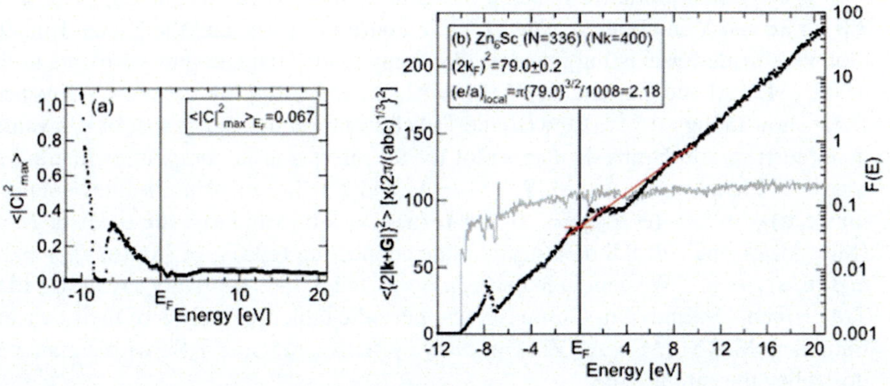

Fig. 15.4 (**a**) Energy dependence of $\langle |C|^2_{\max} \rangle$ and (**b**) Hume–Rothery plot for Zn_6Sc. The ordinate in (**b**) is expressed in units of that in the orthorhombic structure by ignoring a slight deviation of an angle g from 90° in the monoclinic structure [12]. $F(E)$ represents non-dimensional standard deviation [14]

well with the *critical* $|\mathbf{G}|^2\text{s} = 76$ and 80 mentioned above. This explains the origin of the pseudogap at the Fermi level in terms of Fs–Bz interactions involved.

The effective $\mathbf{e/a}$ value can be immediately calculated by inserting $(2k_F)^2$ obtained above into $\mathbf{e/a} = [\pi\{(2k_F)^2\}^{3/2}]/3N$, where N is the number of atoms per unit cell. It turns out to be 2.18, in a good agreement with a composition average of 2.14 under the condition that $(\mathbf{e/a})_{Zn} = 2.0$ and $(\mathbf{e/a})_{Sc} = 3.0$. We showed that the matching condition for Cd_6Ca approximant containing 168 atoms per cubic cell is satisfied with $(2k_F)^2 = |\mathbf{G}|^2 = 50$ [15]. An insertion of $\mathbf{e/a} = 2.18$ and $N = 168$ in

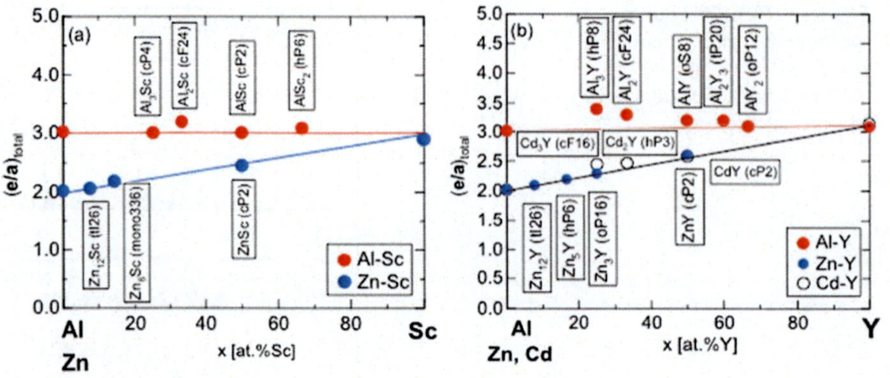

Fig. 15.5 (**a**) Sc and (**b**) Y concentration dependence of $(\mathbf{e/a})_{total}$ for intermetallic compounds in the M–Sc and M–Y (M = Al, Zn and Cd) alloy systems, respectively

place of $N = 336$ back to the equation above results in $(2k_F)^2 = 50$. Hence, we can say that both Zn_6Sc forming a superlattice with $N = 336$ and Cd_6Ca approximant are isoelectronic and obey the Hume–Rothery stabilization mechanism.

It is of great importance to study the universality for the assignment of $(\mathbf{e/a}) = 3.0$ for Sc and Y in Group 3. The FLAPW-Fourier analysis and the Hume–Rothery plot were made for existing intermetallic compounds [16] and pure elements in the M–Sc (M = Al and Zn) and M–Y (M = Al, Zn and Cd) alloy systems. We respectively show in Figs. 15.5(a) and (b) the TM concentration dependences of **e/a** values deduced from the Hume–Rothery plot for all intermetallic compounds mentioned above. The data points for M–Sc (M = Al and Zn) fall on straight lines connecting $(\mathbf{e/a})_{Al} = 3.0$, $(\mathbf{e/a})_{Zn} = 2.0$ and $(\mathbf{e/a})_{Sc} = 3.0$. Similarly, the data for M–Y (M = Al, Zn and Cd) fall on a straight line connecting $(\mathbf{e/a})_{Al} = 3.0$, $(\mathbf{e/a})_{Cd} = 2.0$ and $(\mathbf{e/a})_{Y} = 3.1$. We conclude that $\mathbf{e/a} = 3.0$ and 3.1, respectively, are assigned to TM elements Sc and Y in Group 3 in the periodic table, regardless of their concentrations in M–TM (M = Al, Zn, Cd) alloy systems and that Zn_6Sc with a giant cell does obey this simple rule.

Acknowledgements One of the authors (UM) is grateful for the financial support of the Grant-in-Aid for Scientific Research (Contract No. 23560793) from the Japan Society for the Promotion of Science.

References

1. Tsai AP, Inoue A, Yokoyama Y, Masumoto T (1990) Mater Trans Jpn Inst Met 31:98
2. Yokoyama Y, Tsai AP, Inoue A, Masumoto T, Chen HS (1991) Mater Trans Jpn Inst Met 32:421
3. Raynor GV (1949) Prog Met Phys 1:1
4. Tsai AP, Guo JQ, Abe E, Takakura H, Sato TJ (2000) Nature 408:538
5. Kaneko Y, Arichika M, Ishimasa T (2001) Philos Mag Lett 81:777

6. Andrusyak RI, Kotur BYa, Zavodnik VE (1989) Kristallografiâ 34:996
7. Ishimasa T, Kaneko Y, Kaneko H (2004) J Non-Cryst Solids 334– 335:1
8. Pauling L (1960) The nature of the chemical bond, 3rd edn. Cornell University Press, Ithaca
9. Mizutani U (2010) Hume–Rothery rules for structurally complex alloy phases. CRC Press/Taylor & Francis Group, Boca Raton
10. Inukai M, Zijlstra ES, Sato H, Mizutani U (2011) Philos Mag 91:4247
11. Mizutani U, Inukai M, Sato H, Zijlstra ES (2012) Philos Mag 92:1691
12. Ishimasa T, Kasano Y, Tachibana A, Kashimoto S, Osaka K (2007) Philos Mag 87:2887
13. Blaha P, Schwarz K, Madsen G, Kvasnicka D, Luitz J (2012) WIEN2k. http://www.wien2k.at/, last accessed on August 12, 2012
14. Mizutani U, Inukai M, Sato H, Zijlstra ES (2012) Chem Soc Rev 41:6799
15. Mizutani U, Inukai M, Sato H, Nozawa K, Zijlstra ES (2012) Presented at aperiodic 2012
16. Villars P (1997) Pearson's handbook, crystallographic data. ASM, Materials Park

Chapter 16
Analysis of Dislocations in Quasicrystals Composed of Self-assembled Nanoparticles

Liron Korkidi, Kobi Barkan, and Ron Lifshitz

Abstract We analyze transmission electron microscopy (TEM) images of self-assembled quasicrystals composed of binary systems of nanoparticles. We use an automated procedure that identifies the positions of dislocations and determines their topological character. To achieve this, we decompose the quasicrystal into its individual density modes, or Fourier components, and identify their topological winding numbers for every dislocation. This procedure associates a Burgers function with each dislocation, from which we extract the components of the Burgers vector after choosing a basis. The Burgers vectors that we see in the experimental images are all of lowest order, containing only 0s and 1s as their components. We argue that the density of the different types of Burgers vectors depends on their energetic cost.

16.1 Dislocations in Self-assembled Soft-Matter Quasicrystals

Self-assembled soft-matter quasicrystals have been observed in recent years in a wide variety of different systems, in all cases but one with dodecagonal (12-fold) point-group symmetry. First discovered by Zeng et al. [21] in liquid crystals made of micelle-forming dendrimers, self-assembled soft-matter quasicrystals have since appeared in other systems such as ABC-star polymers [9], in binary systems of nanoparticles [16], in block co-polymer micelles [6], and in mesoporous silica [20]. These newly-realized systems not only provide exciting platforms for the fundamental study of the physics of quasicrystals [2], they also hold the promise for new and exciting applications, especially in the field of photonics. An overview of soft matter quasicrystals, including many references relevant to these systems, is given by Lifshitz and Diamant [14] as well as by Ungar et al. [17, 18] and Dotera [4, 5].

Here we concentrate on the systems of nanoparticles studied by Talpin et al. [16], consisting typically of two types of particles, such as PbS, Au, Fe_2O_3, and Pd,

L. Korkidi · K. Barkan · R. Lifshitz (✉)
Raymond and Beverly Sackler School of Physics and Astronomy, Tel Aviv University, Tel Aviv 69978, Israel
e-mail: ronlif@tau.ac.il

S. Schmid et al. (eds.), *Aperiodic Crystals*, DOI 10.1007/978-94-007-6431-6_16,

with different diameters. These binary systems of particles, when placed in a solution, self-assemble into structures with long-range order, including 12-fold symmetric quasicrystals. The dimensions of the particles—typically a few nanometers in diameter—are such that they can be imaged directly using a transmission electron microscope (TEM). This allows one to study effects that are inaccessible with atomic-scale quasicrystals. Here we present a quantitative analysis of the dislocations that are naturally formed in these quasicrystals as they self-assemble.

In periodic crystals in d-dimensions, one can usually identify the position of a dislocation rather easily by the termination of a plane of atoms in three dimensions, or a line of atoms in two dimensions. One then chooses a basis for the periodic lattice; encircles each dislocation with a Burgers loop, or a Burgers circuit, of basis vectors; and counts the accumulated difference between the number of steps taken forward and backward in the direction of each of the d basis vectors. The d integers thus obtained define the Burgers vector which encodes the topological character of the dislocation. A similar real-space procedure can be used on a quasiperiodic crystal by overlaying it with a quasiperiodic tiling of rank $D > d$ (for a definition, see below), yielding a D-component Burgers vector [19]. A tiling-based analysis [10] of binary systems of nanoparticles was indeed recently carried out by Bodnarchuk et al. [3]. Here we propose an alternative approach for analyzing dislocations in Fourier space that we believe is useful when dealing with aperiodic crystals.

16.2 Density Modes, Winding Numbers, and the Burgers Function

Let us describe the density of nanoparticles in a self-assembled crystal by a function $\rho(\mathbf{r})$. The Fourier expansion of such a function is given by

$$\rho(\mathbf{r}) = \sum_{\mathbf{k}\in L} \rho(\mathbf{k}) e^{i\mathbf{k}\cdot\mathbf{r}}, \tag{16.1}$$

where the (reciprocal) lattice L is a finitely generated $\mathbb{Z}$-module, which means that it can be expressed as the set of all integral linear combinations of a finite number D of d-dimensional wave vectors, $\mathbf{b}^{(1)}, \ldots, \mathbf{b}^{(D)}$. In the special case where the smallest possible D, called the *rank* of the crystal, is equal to the physical dimension d, the crystal is periodic. More generally, for quasiperiodic crystals $D \geq d$, and we refer to all quasiperiodic crystals that are not periodic as *quasicrystals* [11, 12].

As explained elsewhere [13], the topological nature of a dislocation is related to the fact that it cannot be made to disappear by local structural changes. For this to be the case, as one follows a loop around the position of a dislocation and returns to the point of origin, one sees a crystal that is everywhere only-slightly distorted from the perfectly ordered state, except near the core of the dislocation. In particular, the complex amplitudes $\rho(\mathbf{k})$ of the density modes maintain their magnitudes along the loop, each accumulating at most a phase, which upon return to the point of origin must be an integer multiple of 2π. The collection of all such integers, or so-called *winding numbers*, for a given dislocation defines a linear function $\mathcal{N}_{\mathcal{B}}(\mathbf{k})$ from the lattice L to the set of integers $\mathbb{Z}$, which we call the *Burgers function*.

The Burgers function of a given dislocation associates a particular winding number $\mathcal{N}_{\mathcal{B}}(\mathbf{k})$ with every wave vector $\mathbf{k} \in L$. Because this function is linear, after choosing a basis $\{\mathbf{b}^{(i)}\}$ for the lattice, it is uniquely specified by a set of only D integers $n_i \equiv \mathcal{N}_{\mathcal{B}}(\mathbf{b}^{(i)})$, forming the *Burgers vector* $(n_1, \ldots, n_D)$. Thus,

$$\forall \mathbf{k} = \sum_{i=1}^{D} a_i \mathbf{b}^{(i)} \in L: \quad \mathcal{N}_{\mathcal{B}}(\mathbf{k}) = \sum_{i=1}^{D} a_i \mathcal{N}_{\mathcal{B}}(\mathbf{b}^{(i)}) = \sum_{i=1}^{D} a_i n_i, \tag{16.2}$$

where $a_i \in \mathbb{Z}$. This implies that in order to fully characterize a dislocation in an experimental image it suffices to isolate the D density modes associated with a chosen basis, and obtain their corresponding winding numbers. This is the basis of the approach presented below for analyzing dislocations (for more detail, see [1, 7, 8]).

16.3 Analysis of the Dislocations in a Quasicrystal of Nanoparticles

We begin with a high-resolution TEM image of one of the dodecagonal quasicrystals grown by Talapin [15], a section of which is shown in Fig. 16.1(a). This particular quasicrystal is self-assembled from 11.2 nm PbS and 5.2 nm Au nanoparticles, and contains a distribution of dislocations that are formed naturally during the self assembly. We Fourier transform the TEM image, to obtain the diffraction image shown in Fig. 16.1(b), and then choose four of the Bragg peaks in the 12-fold ring containing the strongest reflections as a basis $\mathbf{b}^{(1)}, \ldots, \mathbf{b}^{(4)}$ for the reciprocal lattice. These are labeled in the schematic representation of the lattice in Fig. 16.2.

For each of the four pairs of Bragg peaks, associated with the chosen basis vectors and their negatives, we then carry out the following procedure:

1. We filter out small regions around the two opposite Bragg peaks, as indicated by a pair of red circles in Fig. 16.1(b) for the case of the density mode associated with the wave vectors $\pm\mathbf{b}^{(1)}$.
2. We invert the Fourier transform in the filtered regions resulting in a real-space image of a single density mode. Dislocations appear as discontinuities in the stripes. We use a routine that identifies all the discontinuities and marks their positions, as shown in Fig. 16.1(c) for this density mode.
3. For each dislocation, a second routine then extracts the ith component $n_i = \mathcal{N}_{\mathcal{B}}(\mathbf{b}^{(i)})$ of the Burgers vector. This is done by enclosing a counter-clockwise loop around its position and calculating the accumulated phase in units of 2π. Practically what we do is count the number of stripes crossed moving in the direction of the wave vector $\mathbf{b}^{(i)}$ on one side of the dislocation, and subtract the number of stripes crossed moving against the direction of $\mathbf{b}^{(i)}$ while returning on the other side, as demonstrated in Fig. 16.3.

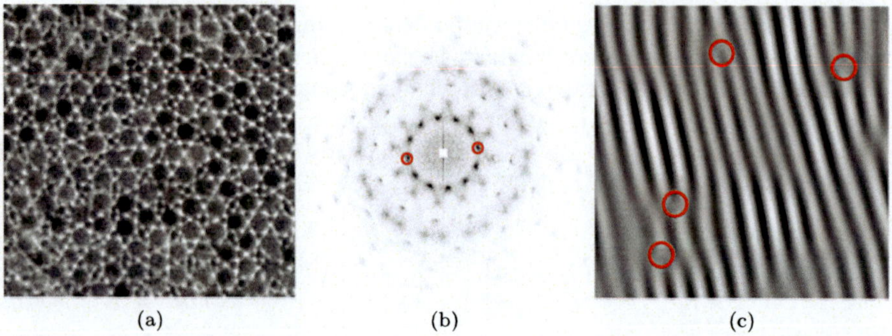

Fig. 16.1 (**a**) A section of a TEM image of a dodecagonal quasicrystal, self-assembled from 11.2 nm PbS and 5.2 nm Au showing several dislocations (courtesy of Dmitri Talapin 2012). (**b**) Fourier transform of the TEM image in (**a**), with the central peak blocked. A pair of Bragg peaks, associated with one of the basis vectors and its negative, is marked in *red*. (**c**) The corresponding section of the inverse Fourier transform of the Bragg peaks marked in (**b**) with *red circles* marking the positions of four dislocations

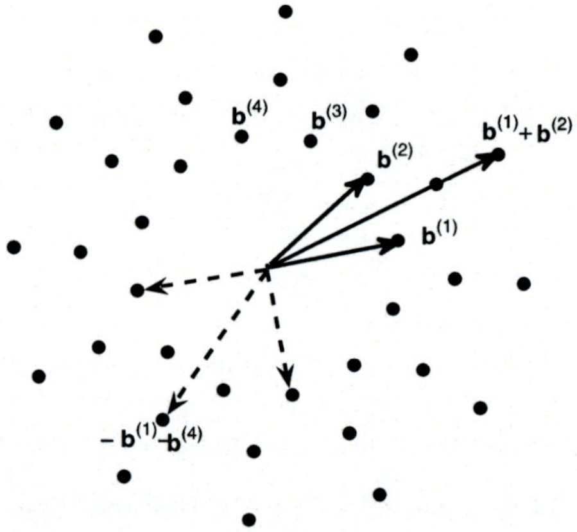

Fig. 16.2 Schematic representation of the three strongest rings in the Fourier transform of our dodecagonal quasicrystal. The inner ring is the strongest, containing the four basis vectors $\mathbf{b}^{(1)}, \ldots, \mathbf{b}^{(4)}$. The second strongest ring is the outer one, obtained from all the sums of two adjacent vectors in the inner ring, as indicated by *solid arrows*. The third strongest ring lies in between, obtained from all sums of vector pairs in the inner ring that are separated by 90 degrees, as indicated by *dashed arrows*

Finally, we verify the correctness of the calculation by extracting the values $\mathcal{N}_{\mathcal{B}}(\mathbf{k})$ for additional wave vectors $\mathbf{k}$ and checking that they satisfy the linearity requirement given by Eq. (16.2).

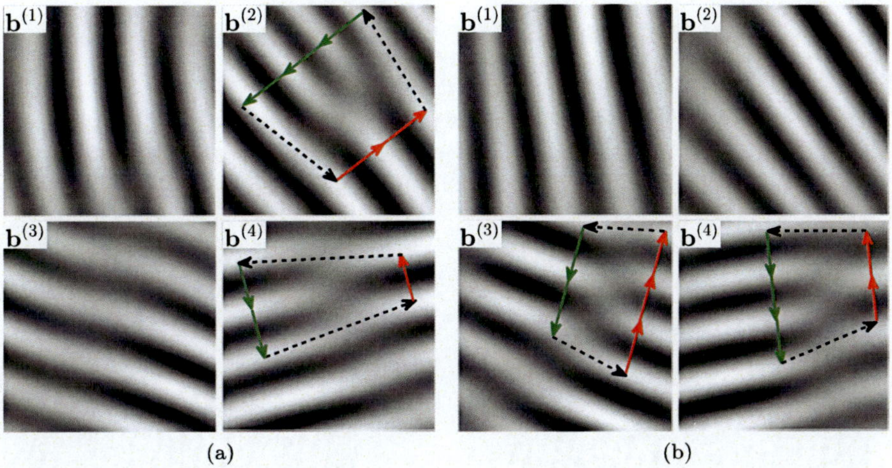

Fig. 16.3 Two typical examples for the dislocations found using our procedure. In each counter-clockwise loop, the winding number n_i is given by the number of *red arrows*, counting stripes crossed in the positive direction of $\mathbf{b}^{(i)}$, minus the number of *green arrows*, counting stripes in the negative direction. (**a**) A *single-subset* dislocation with Burgers vector $(0, -1, 0, -1)$ and (**b**) a *dual-subset* dislocation with Burgers vector $(0, 0, 1, -1)$ (see text for definitions)

16.4 Results and Discussion

We typically find a density of a few dozen dislocations per $\mu\mathrm{m}^2$ in the nanoparticle quasicrystals of [15]. All of these dislocations are of lowest order in the sense that $n_i = 0$, 1, or -1. To understand the topological nature of these dislocations, it is useful to classify them by dividing the four basis vectors into two hexagonal subsets, $\{\mathbf{b}^{(1)}, \mathbf{b}^{(3)}\}$ and $\{\mathbf{b}^{(2)}, \mathbf{b}^{(4)}\}$ (see Fig. 16.2). By doing so, we find that the density of dislocations with non-zero components in only one of the subsets, which we call *single-subset* dislocations, is five times larger than that of dislocations with non-zero components in both subsets, which we call *dual-subset* dislocations. Examples of the two types of dislocations are shown in Fig. 16.3.

To try and explain these observations, consider the free energy of the self-assembled crystal as an expansion in products of density mode amplitudes $\rho(\mathbf{k})$ (see [13]),

$$\mathcal{F}\{\rho\} = \sum_{n=2}^{\infty} \sum_{\mathbf{k}_1 \ldots \mathbf{k}_n} A(\mathbf{k}_1, \ldots, \mathbf{k}_n)\rho(\mathbf{k}_1) \cdots \rho(\mathbf{k}_n), \tag{16.3}$$

where one can show that $A(\mathbf{k}_1, \ldots, \mathbf{k}_n) = 0$ unless $\mathbf{k}_1 + \cdots + \mathbf{k}_n = 0$. We argue that products in the sum (16.3) that contain high-intensity modes with large winding numbers have a greater contribution to increasing the free energy away from its minimum value in the perfect crystal. Accordingly, high-intensity modes tend to exhibit smaller winding numbers. Indeed, we find that all the winding numbers associated with the two brightest rings (see Fig. 16.1(b) and Fig. 16.2) are either 0 or

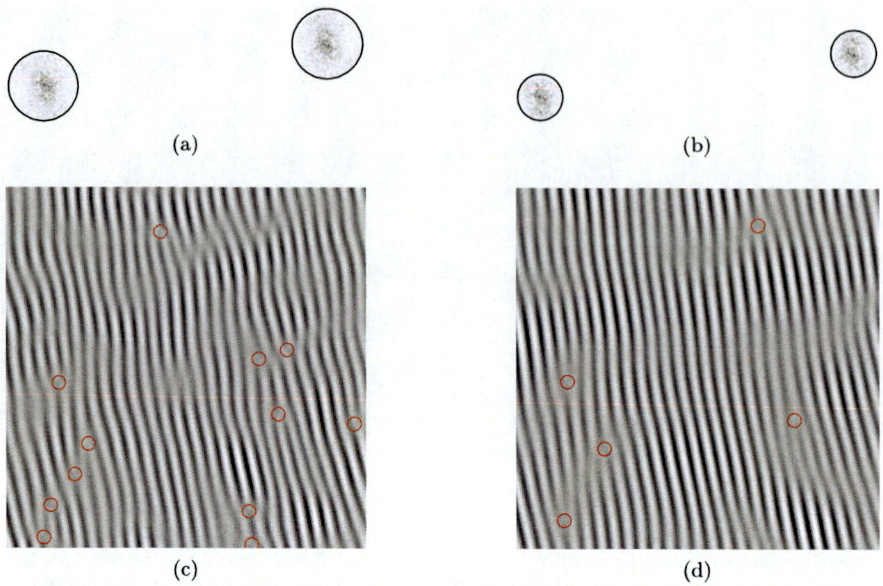

Fig. 16.4 Inverse Fourier transform of a pair of Bragg peaks using filters of different size. For a circular filter with a radius of 25 pixels, used in (**a**), we find 12 dislocations, marked with red circles in (**c**). As we decrease the filter radius to 15 pixels, in (**b**), we find only 5 dislocations in the inverse Fourier transform in (**d**)

± 1, whereas it is only on the 3rd ring that we begin to see winding numbers that are either 0, ± 1, or ± 2. Moreover, owing to the linearity of the Burgers function, the fact that the winding numbers on the second brightest ring are at most of magnitude 1 prevents two adjacent peaks from the different subsets in the first ring from having non-zero winding numbers of the same sign. Because the ring of Bragg peaks, obtained by adding pairs of wave vectors separated by 60 degrees, is extremely weak (see Fig. 16.1(b)), there is no such constraint on the winding numbers belonging to the same subset of basis vectors. The fact that this constraint applies only to dual-subset dislocations reduces their possible combinations and overall relative density.

A word of caution is in order regarding our approach for analyzing dislocations. Because the density of the dislocations is relatively high, the Bragg peaks are not point-like but are rather spread as can be seen in Fig. 16.1(b). This means that some of the information about the dislocations may lie between Bragg peaks and may be lost if the filters are too small. Therefore, our approach is sensitive to the shape and size of the filter that we use around each Bragg peak. As we increase the filter size we obtain more information and potentially find more dislocations, as demonstrated in Fig. 16.4. We thus try to optimize the filter by gradually enlarging its size until the number of dislocations stops increasing substantially.

Our approach for analyzing dislocations should be easily adapted to other systems even when the density of the dislocations is quite large, as one may expect for soft matter systems. Moreover, for dynamical systems that can be imaged in real time one can use our automated method to follow and quantitatively analyze the dynamics of the dislocations.

Acknowledgements We are very grateful to Dmitri Talapin for providing the TEM images. This research is supported by the Israel Science Foundation through grant No. 556/10.

References

1. Barak, G, Lifshitz, R (2006) Dislocation dynamics in a dodecagonal quasiperiodic structure. Philos Mag 86:1059. doi:10.1080/14786430500256383
2. Barkan, K, Diamant, H, Lifshitz, R (2011) Stability of quasicrystals composed of soft isotropic particles. Phys Rev B 83:172201. doi:10.1103/PhysRevB.83.172201
3. Bodnarchuk, MI, Shevchenko, EV, Talapin, DV (2011) Structural defects in periodic and quasicrystalline binary nanocrystal superlattices. J Am Chem Soc 133:20837. doi:10.1021/ja207154v
4. Dotera, T (2011) Quasicrystals in soft matter. Isr J Chem 51:1197–1205. doi:10.1002/ijch.201100146
5. Dotera, T (2012) Toward the discovery of new soft quasicrystals: from a numerical study viewpoint. J Polym Sci, Polym Phys Ed 50:155–167. doi:10.1002/polb.22395
6. Fischer, S, Exner, A, Zielske, K, Perlich, J, Deloudi, S, Steurer, W, Lindner, P, Förster, S (2011) Colloidal quasicrystals with 12-fold and 18-fold diffraction symmetry. Proc Natl Acad Sci USA 108:1810–1814. doi:10.1073/pnas.1008695108
7. Freedman, B, Bartal, G, Segev, M, Lifshitz, R, Christodoulides, DN, Fleischer, JW (2006) Wave and defect dynamics in nonlinear photonic quasicrystals. Nature 440:1166–1169. doi:10.1038/nature04722
8. Freedman, B, Lifshitz, R, Fleischer, JW, Segev, M (2007) Phason dynamics in nonlinear photonic quasicrystals. Nat Mater 6:776–781. doi:10.1038/nmat1981
9. Hayashida, K, Dotera, T, Takano, A, Matsushita, Y (2007) Polymeric quasicrystal: mesoscopic quasicrystalline tiling in ABC star polymers. Phys Rev Lett 98:195502. doi:10.1103/PhysRevLett.98.195502
10. Leung PW, Henley CL, Chester GV (1989) Dodecagonal order in a two-dimensional Lennard–Jones system. Phys Rev B 39:446–458. doi:10.1103/PhysRevB.39.446
11. Lifshitz, R (2003) Quasicrystals: a matter of definition. Found Phys 33:1703–1711
12. Lifshitz, R (2007) What is a crystal? Z Kristallogr 222:313–317. doi:10.1524/zkri.2007.222.6.313
13. Lifshitz, R (2011) Symmetry breaking and order in the age of quasicrystals. Isr J Chem 51:1156. doi:10.1002/ijch.201100156
14. Lifshitz, R, Diamant, H (2007) Soft quasicrystals—why are they stable? Philos Mag 87:3021. doi:10.1080/14786430701358673
15. Talapin, DV (2012) Private communication
16. Talapin, DV, Shevchenko, EV, Bodnarchuk, MI, Ye, X, Chen, J, Murray, CB (2009) Quasicrystalline order in self-assembled binary nanoparticle superlattices. Nature 461:964. doi:10.1038/nature08439
17. Ungar, G, Zeng, X (2005) Frank–Kasper, quasicrystalline and related phases in liquid crystals. Soft Matter 1:95–106. doi:10.1039/B502443A
18. Ungar, G, Percec, V, Zeng, X, Leowanawat, P (2011) Liquid quasicrystals. Isr J Chem 51:1206–1215. doi:10.1002/ijch.201100151

19. Wollgarten M, Franz V, Feuerbacher M, Urban K (2002) In: Suck J-B, Schreiber M, Hussler P (eds) Quasicrystals: an introduction to structure, physical properties, and applications. Springer, Berlin, Chap. 13
20. Xiao, C, Fujita, N, Miyasaka, K, Sakamoto, Y, Terasaki, O (2012) Dodecagonal tiling in mesoporous silica. Nature 487:349. doi:10.1038/nature11230
21. Zeng, X, Ungar, G, Liu, Y, Percec, V, Dulcey, AE, Hobbs, JK (2004) Supramolecular dendritic liquid quasicrystals. Nature 428:157–160. doi:10.1038/nature02368

Chapter 17
Average Unit Cell in Fourier Space and Its Application to Decagonal Quasicrystals

B. Kozakowski and J. Wolny

Abstract This paper describes a new technique for solving the structure of quasicrystals. The technique is based on transformations between an average unit cell (AUC) and an envelope of diffraction peaks. For centrosymmetric structures like the Penrose tiling, the envelope makes it possible to determine the sign of the phase straight from the diffraction pattern. A Fourier transform of an envelope leads to a distribution of atomic positions within an AUC. Apart from theoretical and modeling aspects of the technique, the paper also presents the results of applying it to the well-known decagonal quasicrystal Al–Ni–Co.

17.1 Introduction

The most commonly used techniques for recovering the phase of diffraction reflections are Low Density Elimination [12] and Charge Flipping [10]. Their comparison and efficiency in the examination of decagonal quasicrystalline structures were discussed in [4, 5]. The results of such analyses are entry points to a refinement process which uses a structure factor derived for a chosen structure model. The best-known structure model of 2D quasicrystals which proved to be an excellent starting point for the structure refinement of real decagonal quasicrystals is the Penrose tiling [1, 13–15]. Structure factors based on this tiling can operate in a higher-dimensional space—"cut-and-project" [2, 7, 9] method or solely in the physical space: average unit cell (AUC) approach [18]. In this paper, we use the AUC approach and, as the model structure, the Penrose tiling with rhombuses of the edge length equal to one. Based on those two, we developed a new technique for the phase recovery. It is achieved by using "envelope" curves. They allow us to determine the sign of the phase. The technique is still being developed but the initial results obtained on the widely studied alloy $Al_{72}Ni_{20}Co_8$ [8, 14–16] prove that the method is effective.

B. Kozakowski (✉) · J. Wolny
Faculty of Physics and Applied Computer Science, AGH University of Science and Technology, Krakow, Poland
e-mail: Kozakowski@interia.pl

S. Schmid et al. (eds.), *Aperiodic Crystals*, DOI 10.1007/978-94-007-6431-6_17,

The paper is structured as follows: first, we define some basic terms we use in the diffraction analysis based on the average unit cell approach—among them, the structure factor and envelope function. Then, we show how we use the envelope function in the phase retrieval process and the reverse Fourier transform. Finally, we discuss an application of this method to $Al_{72}Ni_{20}Co_8$ alloy.

17.2 Average Unit Cell Approach

The Reference Grid is a regular set of points **r** such that $r_i = N_i\lambda$, where N_i is an integer and $\boldsymbol{\lambda}$ denotes the vector of lengths between two neighboring points in the grid. For a 1D example, we simply have $x_i = N_i\lambda_x$. The vector $\boldsymbol{\lambda}$ defines also the *elementary cell* of a grid. The *reduced coordinate* $\boldsymbol{u}$ is the distance of an arbitrary position. For a 1D reference frame **X**, we have $X_j = u_{xj} + N_{xj} \cdot \lambda_x$. An *average unit cell* is a *distribution* of reduced coordinates $P(u)$. *The choice of* $\boldsymbol{\lambda}$ is determined by the diffraction pattern that is studied and is associated with the base vectors of the reciprocal space $\lambda = 2\pi/k_0$, where k_0 is the length of a chosen base vector.

It is shown [3, 11] that for the Fibonacci chain the average unit cell is non-zero only within a limited space where it assumes a constant value and the distribution becomes uniform. Similarly, in case of the Penrose tiling, the non-zero area of the AUC consists of four pentagonal areas within which the density is constant [17].

The Base Vectors of the Reciprocal Space define the coordinate system of the reciprocal space. For a 1D grid, we have $k = nk_0 + mq_0$, where k_0 is called the main vector and q_0 is the modulated vector, n and m are integers, $q_0 = k_0/\tau$ and the value of $\tau = 0.5(1 + \sqrt{5}) \cong 1.618$.

Structure Factor We can apply the definitions set above to a derivation of the structure factor:

$$F(k) = \sum_j \exp i(k \cdot r_j) = \sum_j \exp i(nk_0u_j + mq_0v_j)$$
$$= \iint P(u, v) \exp i(nk_0u + mq_0v)\, dv\, du.$$

The sum is over a very large set of points. In the first step of the derivation, we use reduced coordinates written in two reference grids u, v associated with the main and modulated vectors. In the second step, we exchange the sum for an integral over a density function $P(u, v)$ of a 2D AUC.

$\boldsymbol{v(u)}$ **Relationship** For both the Fibonacci chain and the Penrose tiling, this function assumes non-zero values only along a line segment defined by the equation $v = -\tau^2 \cdot (u - b) + b$. For the Fibonacci chain $b = 0$, and for the Penrose tiling $b = j \cdot \lambda/5$ [6, 18].

Envelope Function Let's examine the implication of employing the $v(u)$ relationship to the structure factor. For the Fibonacci chain, we have

$$F(w) = \int P(u) \cdot \exp i(w \cdot u)\,du = \exp i(-w \cdot u_0) \cdot \frac{\sin(w \cdot u_0)}{w \cdot u_0}$$

where $w = k_0(n - \tau \cdot m)$, $u_0 = 1/2\tau$, and the integral is over a uniform distribution limited by $(0, u_0)$. In general, the structure factor of the Fibonacci chain is a complex function. However, it is possible to cancel out the complex part of the function by introducing a shift in atomic positions Δu. This shift should result in moving the $v(u)$ relationship to one of the symmetry points of the rectangular reference frame (u, v). It can be easily proved that if we choose as the main vector $k_0 = 2\pi/(3 - \tau)$ and at the same time we shift the whole chain by u_0, we get

$$F(w) = \cos(m \cdot \pi) \cdot \frac{\sin(w \cdot u_0)}{w \cdot u_0}.$$

Even though this is a special case for the Fibonacci chain and generally, for a freely decorated chain, it is impossible to make this sequence symmetric, the equation obtained above will prove very useful for a demonstration of a technique that allows us to establish the phases of the diffraction peaks for more symmetrical structures, such as decagonal, freely decorated Penrose tilings.

We can relate w to k by combining the definition of τ, $v(u)$ relationship, and $q = k/\tau$. As a result we obtain $w = k - m\sqrt{5}k_0$,

$$F(k) = \cos(m \cdot \pi) \cdot \frac{\sin(k - m\sqrt{5}k_0 \cdot u_0)}{k - m\sqrt{5}k_0 \cdot u_0}.$$

The structure factor can be used not only for the positions of peaks ($k = nk_0 + mq_0$) but for any other continuous value of k. In such a case, it is called the *envelope function*. An envelope connects the tops of peaks that share the same value of m. For other peaks we observe a shift of an envelope by $m\sqrt{5}k_0$ (plus a reversion caused by the cosine factor for odd m's); see Fig. 17.1 (bottom, left).

Note that for $m = 0$ we obtain exactly the same equation as for $F(w)$. This allows us to construct an envelope based on the experimental data. Namely, if we plot the function $F(w)$ for the experimental peaks, the tops of those peaks will approximate the curve of the envelope function; see Fig. 17.1 (top, right).

Obviously, experimental data gives us little indication about the phase. For symmetrical structures, however, the relationship between theoretical envelops and envelops obtained from experimental data allows us to recover the phase. It is so because, firstly, those two types of envelops cross the k-axis at the same points and, secondly, because the phase is constant between two neighboring zeros of the function. After the phase is estimated and we have the model curve of the envelope function, we use the reverse Fourier transform to obtain the distribution of atoms within an AUC; see Fig. 17.1 (bottom, right)

$$P(u) = \int F(w) \cdot \exp i(-w \cdot u)\,dw.$$

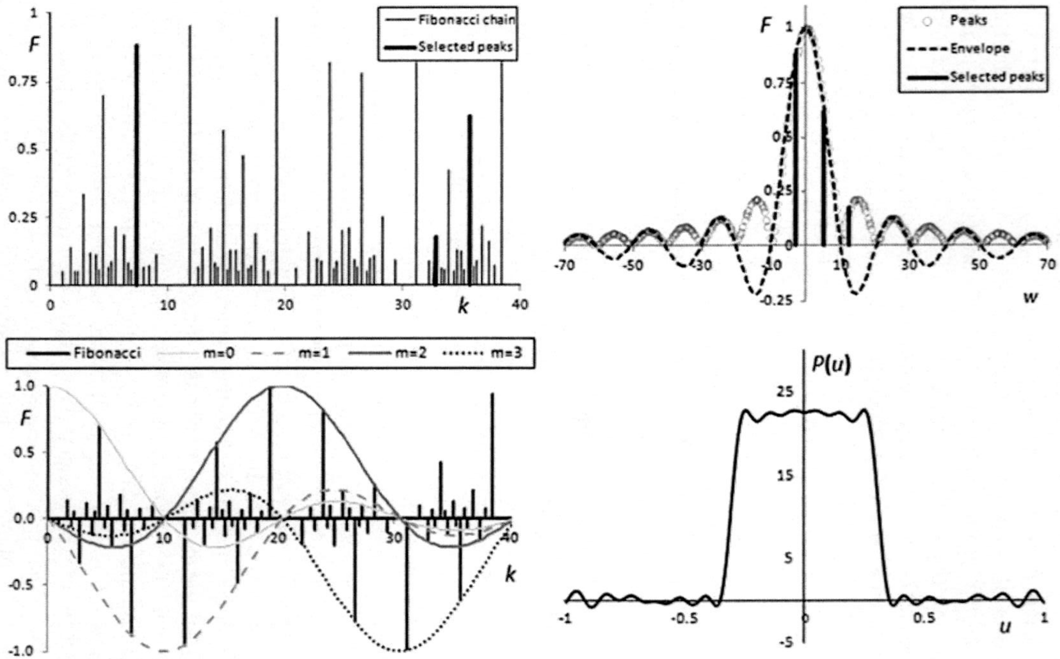

Fig. 17.1 Fibonacci chain analysis. *Top, left*: diffraction peaks; *top, right*: envelopes obtained from the peaks and the model; *bottom, left*: the diffraction peaks with phases retrieved from the envelops; *bottom, right*: the result of a Fourier reverse transform of the envelope curve

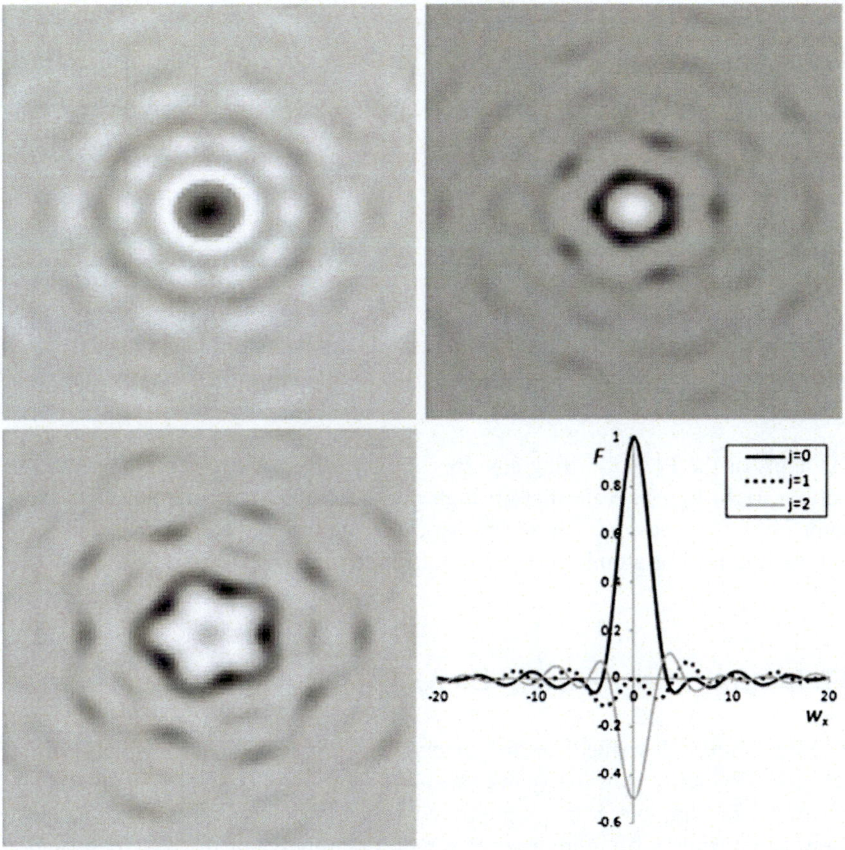

Fig. 17.2 Intensity maps of the three types of Penrose tiling envelopes ($j = 0$: *top, left*; $j = 1$: *top, right*; $j = 2$: *bottom left*) and their cross-section along the X-axis (*bottom, right*)

A very similar reasoning can lead us to the envelope function for the Penrose tiling:

$$F(w_x, w_y) = \sum_{j=1,2,3,4} \operatorname{expi}\bigl(J(n_x, n_y, m_x, m_y) \cdot j/5\bigr) \times \int P_j(u_x, u_y) \cdot \operatorname{expi}(w_x \cdot u_x + w_y \cdot u_y)\, du_x\, du_y.$$

The integral over the pentagonal density functions has a simple closed-form solution. It is provided in [6]. The sum goes over all 4 distributions. The factor in front of the integral is a function of the diffraction peaks indices. An important property of function J is that it returns only integers. Consequently, the whole complex factor containing this function assumes only 5 different values. If, additionally, we take the symmetry into consideration this number reduces to only 3 values. As a result, for a given pair (w_x, w_y) we can find three different sets of indices that approximate this pair, and as a result, three different types of envelope functions. Figure 17.2 shows

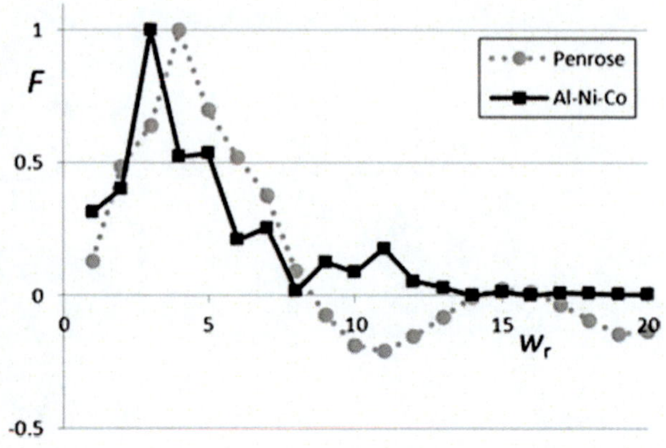

Fig. 17.3 2D envelopes ($j = 0$) curves aggregated radially. A comparison of the experimental data and the Penrose tiling

2D maps of the Penrose tiling envelopes. Their cross-sections are presented in the bottom, right figure. Note that all high peaks belonging to one envelope have the same phase. The reverse Fourier transform over those envelopes gives the pentagonal distributions of atomic positions reduced to an AUC.

17.3 Application—Al–Ni–Co Alloy

The technique described in the paper was applied to the widely discussed $Al_{72}Ni_{20}Co_8$ alloy [8, 14–16]. We used 264 unique reflections. Due to a low number of peaks available, it is not possible to show envelopes as in Fig. 17.2. Instead, we took advantage of an approximate radial symmetry of the envelopes and combined peaks into a radial function $F(w_r)$, where $w_r^2 = w_x^2 + w_y^2$. The results for the envelope indexed as $j = 0$ (see Fig. 17.2 for a reference) are presented in Fig. 17.3 where the envelope obtained from the experimental data is compared to the combined envelope of the Penrose tiling. The curve proves that the envelope is present and that its zeros are very close to the zeros of the Penrose tiling envelope. It means we can use Penrose tiling envelopes as a source of the phase sign. After applying the signs to the experimental data, we calculated the initial distribution, and afterwards, we used the LDE algorithm to validate and correct the phase signs. It turned out that out of 264 peaks only 22 required further phase modification. Those were only very low peaks. The final results are presented in Fig. 17.4. The density and the shape of the distributions retrieved are in accordance to other analyses [4, 5, 15].

It is important to point out that those are initial results. There are some challenges to overcome. Firstly, the method assumes centrosymmetric structures. It is able to predict only the sign of the phase. Secondly, well-shaped envelopes appear only for structures that closely resemble the Penrose tiling. Additionally, the amplitudes of very low peaks do not form an envelope of clearly exposed zeros. And only by examining zeroes we can deduce the sign of a group of peaks. Currently, whenever we were uncertain, we used the phase obtained from the Penrose model. Finally,

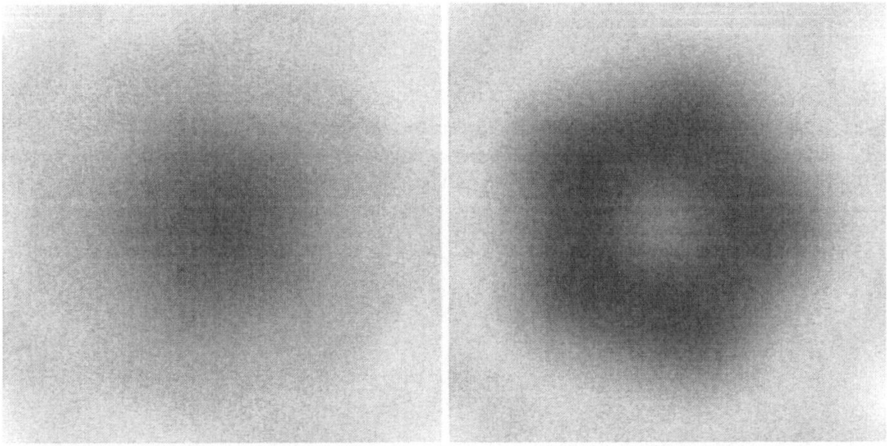

Fig. 17.4 Distributions of atomic positions of the Al–Ni–Co alloy obtained as a result of a reverse Fourier transform and the technique described in the paper. *Left*: small pentagon; *right*: large pentagon

with only a few hundred peaks we were unable to reconstruct the whole envelope. Therefore, here we used only peaks and not continuous curves.

17.4 Conclusions

The paper presents a fast technique for the retrieval of the phase sign straight from the experimental diffraction data. The technique is based on the analysis of the envelope functions. The reverse Fourier transform performed on envelopes results in distributions of atomic positions written in boundaries of an average unit cell. We applied the technique to the experimental data for the $Al_{72}Ni_{20}Co_8$ alloy and obtained results that are in accordance with the generally accepted view of the structure of this alloy.

Acknowledgements The authors thank H. Takakura for providing us with experimental data. This work is supported by the Polish Ministry of Science and Higher Education and its grant for Scientific Research (N N202 326440).

References

1. Baake M, Schlottmann M, Jarvis PD (1991) J Phys A 24:4637
2. de Bruijn NG (1981) Proc K Akad Wet, Ser A, Indag Math 43:38–66
3. Buczek P, Sadun L, Wolny J (2005) Acta Phys Pol A 36:919
4. Fleischer F, Weber T, Steurer W (2010) J Phys Conf Ser 226:012002
5. Fleischer F Weber T et al. (2010) J Appl Crystallogr 43:89–100
6. Kozakowski B, Wolny J (2006) Philos Mag 86:549–555

7. Kramer P, Neri R (1984) Acta Crystallogr, Ser A 40:580
8. Kuczera P, Wolny J, Fleischer F, Steurer W (2011) Philos Mag 91:2500–2509
9. Levine D, Steinhardt PJ (1984) Phys Rev Lett 53:2477
10. Oszlanyi G, Suto A (2004) Acta Crystallogr, Ser A 60:134–141
11. Senechal M (1995) Quasicrystals and geometry. Cambridge University Press, Cambridge
12. Shiono M, Woolfson MM (1992) Acta Crystallogr, Ser A 48:451–456
13. Steurer W, Haibach T (1999) Acta Crystallogr, Ser A 55:48
14. Steurer W, Cervellino A (2001) Acta Crystallogr, Ser A 57:333
15. Takakura H, Yamamoto A, Tsai AP (2001) Acta Crystallogr, Ser A 57:576–585
16. Tsai AP, Inoue A, Matsumoto T (1989) Mater Trans, JIM 30:463–473
17. Wolny J, Kozakowski B (2003) Acta Crystallogr, Ser A 59:54
18. Wolny J (1998) Philos Mag A 77:395–412

Chapter 18
A Study of Phase Equilibria in the Al–Pd–Co System at 700 °C

I. Černičková, R. Čička, P. Švec, D. Janičkovič, P. Priputen, and J. Janovec

Abstract $Al_{68}Pd_{14.6}Co_{17.4}$, $Al_{69.8}Pd_{13.8}Co_{16.4}$, $Al_{72}Pd_{12.8}Co_{15.2}$, $Al_{73.8}Pd_{11.9}Co_{14.3}$, and $Al_{76}Pd_{11}Co_{13}$ alloys annealed at 700 °C for 2000 h were studied. In the investigation, scanning electron microscopy including energy dispersive X-ray spectroscopy and electron backscatter diffraction, X-ray diffraction, and transmission electron microscopy were used. Altogether five near-equilibrium phases (β, U, Al_5Co_2, ε, Al_9Co_2) were identified. Transitions between β, U, and ε phases were also determined dependent on the alloy bulk metal composition. The experimental results were used to propose the partial isothermal section of the Al–Pd–Co phase diagram at 700 °C. The maximum solubilities at 700 °C of Pd in Al_9Co_2 and Al_5Co_2 were determined as 1.7 and 2.69 at.%, respectively.

18.1 Introduction

Structurally complex phases consisting of large cluster-base unit cells are attributed to complex metallic alloys (CMA) inclusive of ternary Al-base CMAs [1–5]. In the Al–Pd–Co alloys, more structurally complex phases were observed [6–11], e.g., the ternary monoclinic U-phase or orthorhombic phases of the ε-family. The latter phases can be classified as either binary phases alloyed with the third element (ε_6 and ε_{28}) or ternary phases (ε_{22} and ε_{34}) [12]. Yurechko et al. [6, 7, 9, 11] studied phase equilibria in the Al–Pd–Co system and published isothermal sections of the related phase diagram at 1050, 1000, 940, and 790 °C.

In the present work, five ternary alloys ($Al_{68}Pd_{14.6}Co_{17.4}$, $Al_{69.8}Pd_{13.8}Co_{16.4}$, $Al_{72}Pd_{12.8}Co_{15.2}$, $Al_{73.8}Pd_{11.9}Co_{14.3}$, and $Al_{76}Pd_{11}Co_{13}$) were long-term annealed at 700 °C and the near-equilibrium phases formed were characterized. The exper-

I. Černičková (✉) · R. Čička · P. Švec · P. Priputen · J. Janovec
Faculty of Materials Science and Technology in Trnava, Slovak University of Technology in Bratislava, Paulínska 16, 917 24 Trnava, Slovak Republic
e-mail: ivona.cernickova@stuba.sk

P. Švec · D. Janičkovič
Institute of Physics, Slovak Academy of Sciences, Dúbravská 9, 845 11 Bratislava, Slovak Republic

S. Schmid et al. (eds.), *Aperiodic Crystals*, DOI 10.1007/978-94-007-6431-6_18,

iment was done with the intention to propose a partial isothermal section of the Al–Pd–Co phase diagram at 700 °C still missing in the literature.

18.2 Experimental Procedures

The investigated alloys were prepared by arc melting of pure components under argon atmosphere. After casting, the samples were annealed at 700 °C for 2000 h and rapidly cooled in water to fix their high-temperature microstructure.

In the investigation, scanning electron microscopy (SEM) including energy dispersive X-ray spectroscopy (EDX) and electron backscatter diffraction (EBSD), X-ray diffraction (XRD), and transmission electron microscopy (TEM) were used. For XRD a Philips PW 1830 diffractometer with Bragg–Brentano geometry was selected using iron filtered $Co_{K\alpha 1}$ radiation, scattering angle 2θ ranged between 5 and 70°, step size was 0.02°, and exposure time was 10 s per step. For SEM a JEOL JSM-7600F microscope was used equipped with an EDX spectrometer X-max working with INCA software and an EBSD Nordlys detector working with FLAMENCO software. At least 10 measurements per microstructure constituent were performed to determine their metal compositions. To calculate volume fractions of microstructure constituents an ImageJ software was used. The identification of phase by selected-area electron diffraction (SAED) was performed in a JEOL 2000FX microscope operating at 200 kV.

18.3 Results

The results of the characterization of near-equilibrium phases present in particular alloys are summarized in Table 18.1. Each of the observed microstructure constituents was found to consist of a single phase. For instance, the single-phase constituents observed in the $Al_{76}Pd_{11}Co_{13}$ alloy (Fig. 18.1) were identified as ε and Al_9Co_2 (Fig. 18.2). Altogether five phases were found in the investigated alloys. Monoclinic U, cubic β, orthorhombic ε, and monoclinic Al_9Co_2 were identified by XRD because their volume fractions were detectable for this technique (compare Table 18.1 and Fig. 18.2). These phases were also identified by SAED/TEM as documented in Fig. 18.3 for β. For the identification of hexagonal Al_5Co_2 appearing in extremely small amounts in $Al_{68}Pd_{14.6}Co_{17.4}$ and $Al_{69.8}Pd_{13.8}Co_{16.4}$ alloys, the EBSD/SEM technique was used (Fig. 18.4). Metal compositions and volume fractions of the identified near-equilibrium phases are also given in Table 18.1.

18.4 Discussion

The experiments were done with the aim to find correlations between the microstructure constituents (SEM), the identified near-equilibrium phases (XRD,

Table 18.1 Overview of experimental results. Observed microstructure constituents, identified phases, and metal compositions and volume fractions of the phases are given in respective columns for all the investigated alloys. Experimental techniques used are related to the obtained results (see two bottom rows)

Microstructure constituent	Phase	Atomic content in %			Volume fraction in %
		Al	Co	Pd	
$Al_{68}Pd_{14.6}Co_{17.4}$					
grey	U	68.85 ± 0.12	16.14 ± 0.12	15.01 ± 0.09	92.8
white	β	57.61 ± 0.26	8.13 ± 0.13	34.26 ± 0.18	6.6
dark	Al_5Co_2	72.52 ± 0.1	25.42 ± 0.11	2.06 ± 0.06	0.6
$Al_{69.8}Pd_{13.8}Co_{16.4}$					
grey	U	69.09 ± 0.16	15.13 ± 0.14	15.78 ± 0.11	99.4
dark	Al_5Co_2	72.47 ± 0.19	24.84 ± 0.19	2.69 ± 0.05	0.6
$Al_{72}Pd_{12.8}Co_{15.2}$					
darker grey	ε	72.78 ± 0.81	14.60 ± 0.93	12.62 ± 0.49	87.5
lighter grey	U	69.92 ± 0.21	15.47 ± 0.64	14.62 ± 0.52	12.5
$Al_{73.8}Pd_{11.9}Co_{14.3}$					
grey	ε	73.0 ± 0.14	14.1 ± 0.12	12.9 ± 0.13	100
$Al_{76}Pd_{11}Co_{13}$					
grey	ε	74.2 ± 0.21	12.7 ± 0.19	14.1 ± 0.08	71.5
dark	Al_9Co_2	81.8 ± 0.05	16.5 ± 0.05	1.7 ± 0.03	28.5
Experimental technique used					
SEM	XRD, SAED/TEM EBSD/SEM	EDX/SEM			SEM

SAED/TEM, EBSD/SEM), as well as metal compositions (EDX/SEM) and volume fraction (SEM) of the identified phases. Bulk metal compositions of the investigated alloys were selected with the intention to hold the Pd/Co-ratio constant. Thus, an increase in the Al bulk content was accompanied by a decrease in both Pd- and Co-bulk contents. As follows from Fig. 18.5, there is a correlation between the bulk Al content and the occurrence of dominant phases in the investigated alloys. The increase in the Al bulk content from 68 to 76 at.% evoked $\beta + U \rightarrow U \rightarrow U + \varepsilon \rightarrow \varepsilon$ phase transitions. This shows that Al stabilizes mainly ε and Pd + Co stabilize mainly β in the investigated alloys. Moreover, the Al contents in both U and ε were found to increase slightly with increasing the bulk Al content in agreement with [6–8].

The partial isothermal section of the Al–Pd–Co phase diagram (in the next text shortly "diagram") at 700 °C (Fig. 18.6) was proposed with respect to both own experimental results and the diagram published by Yurechko for 790 °C [7]. Positions and sizes of single-phase ε and U areas were modified only slightly. The double-phase $\varepsilon + U$ area became wider. The positions of $\varepsilon + Al_9Co_2$ and $U + \beta + Al_5Co_2$

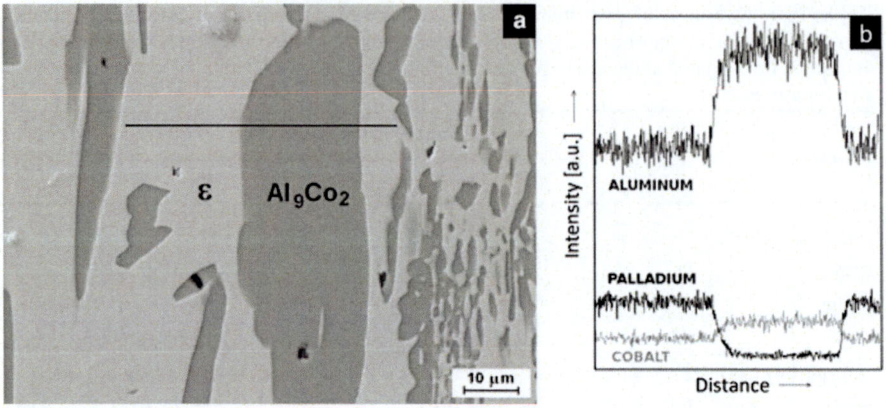

Fig. 18.1 SEM micrograph showing single-phase microstructure constituents (the phases were determined by XRD) in $Al_{76}Pd_{11}Co_{13}$ alloy (**a**) and compositional changes along the indicated line determined by EDX (**b**)

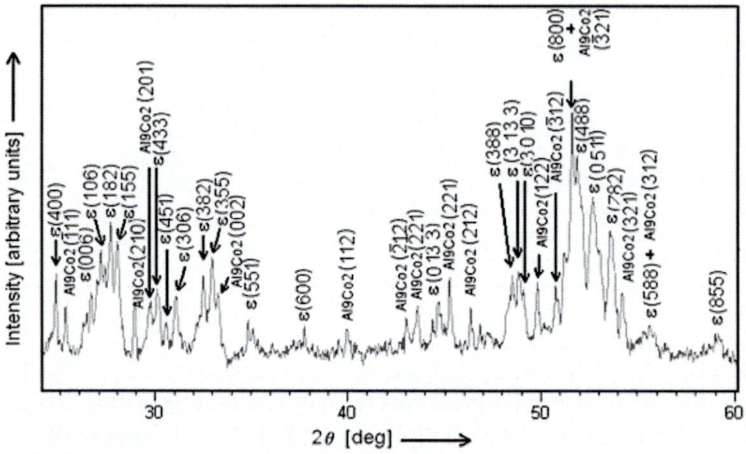

Fig. 18.2 Powder X-ray diffraction pattern corresponding to $Al_{76}Pd_{11}Co_{13}$ alloy

areas are extended towards the higher Pd contents. Dashed lines were used when experimental results were not available. The F-phase was not found experimentally after long-term annealing at 700 °C. It confirms the trend reported in [7, 8] that F disappears gradually with decreasing temperature. This happens probably between 790 and 700 °C. Several binary Al–Co and Al–Pd phases exhibit extensions into the ternary compositional area. At 700 °C, the maximum solubilities of Pd in Al_9Co_2 and Al_5Co_2 were determined as 1.7 and 2.69 at.%, respectively. Congruent AlCo and AlPd phases of the CsCl-type were found to form continuous set of β-Al (Pd, Co) solid solutions [7–9].

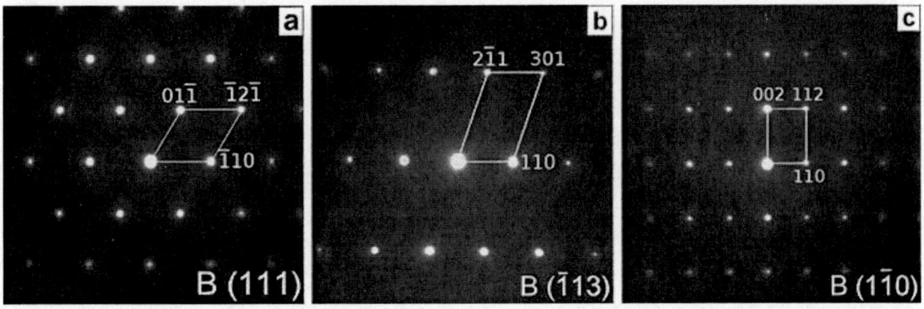

Fig. 18.3 SAED/TEM patterns of β in $Al_{68}Pd_{14.6}Co_{17.4}$ alloy with three different zone axes

Fig. 18.4 EBSD/SEM pattern of Al_5Co_2 in $Al_{68}Pd_{14.6}Co_{17.4}$ alloy

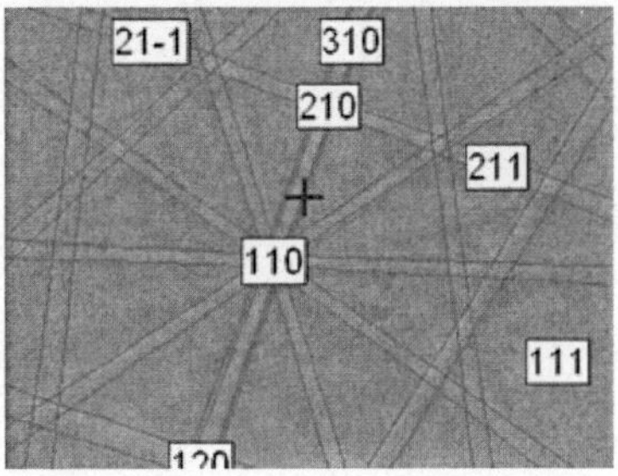

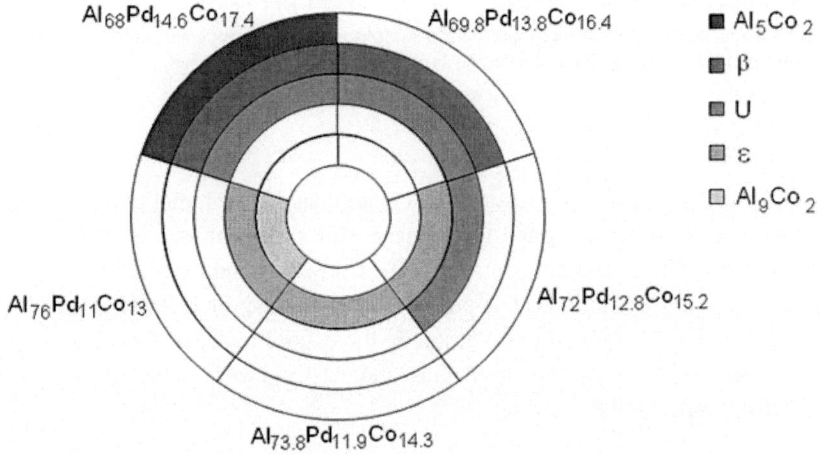

Fig. 18.5 Schematic diagram showing transitions in phase occurrence in dependence on alloy bulk composition

18.5 Conclusions

In the investigated $Al_{68}Pd_{14.6}Co_{17.4}$, $Al_{69.8}Pd_{13.8}Co_{16.4}$, $Al_{72}Pd_{12.8}Co_{15.2}$, $Al_{73.8}Pd_{11.9}Co_{14.3}$, and $Al_{76}Pd_{11}Co_{13}$ complex metallic alloys, altogether five near-equilibrium phases (β, U, Al_5Co_2, ε, Al_9Co_2) were identified after annealing at

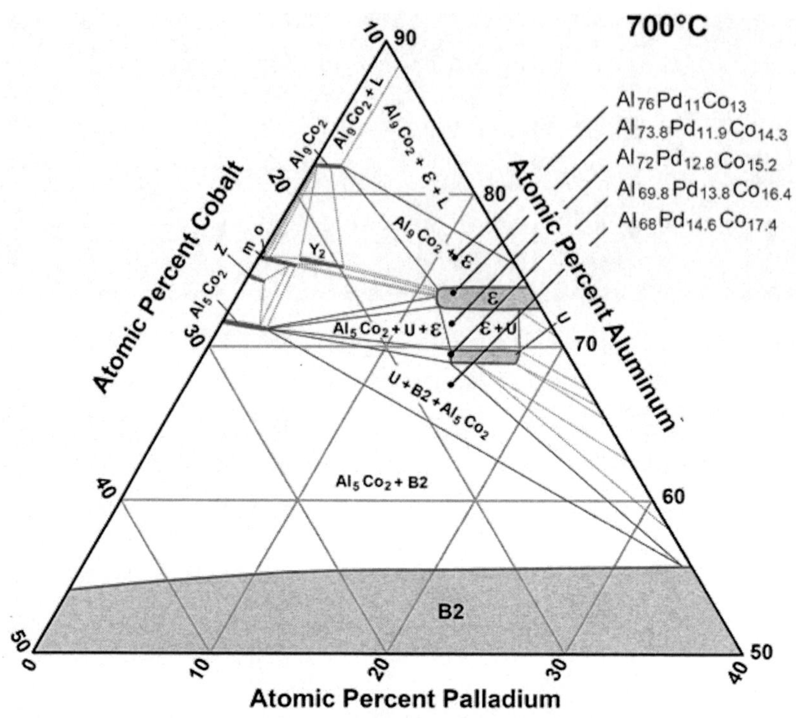

Fig. 18.6 Partial isothermal section of Al–Pd–Co phase diagram proposed for 700 °C with respect to experimental results. In the diagram the symbol B2 is used for β-phase

700 °C for 2000 h. The microstructure constituents, their metal compositions, and volume fraction were assigned to the near-equilibrium phases identified. The increase in the bulk Al content from 68 to 76 at.% was found to evoke $\beta + \mathrm{U} \rightarrow \mathrm{U} \rightarrow \mathrm{U} + \varepsilon \rightarrow \varepsilon$ phase transitions. The partial isothermal section of the Al–Pd–Co phase diagram at 700 °C was proposed based on the experimental results. The maximum solubilities at 700 °C of Pd in Al_9Co_2 and Al_5Co_2 were determined as 1.7 and 2.69 at.%, respectively.

Acknowledgements The authors wish to thank to the European Regional Development Fund (ERDF) for financial support of the project ITMS:26220120014 "Center for development and application of advanced diagnostic methods in processing of metallic and non-metallic materials" funded within the Research & Development Operational Programme, to the Slovak Academy of Sciences for the support in the frame of the "Center of Excellence for functional multiphase materials" (FUN-MAT), to the Slovak Research and Development Agency (APVV) for the financial support under the contract APVV-0076-11, as well as to the Grant Agency of the Ministry of Education, Science, Research, and Sport of the Slovak Republic and the Slovak Academy of Sciences (VEGA) for the financial support under the contracts 1/0143/12, 2/0111/11, and 1/0339/11.

References

1. Wang N, Fung K, Kuo KH (1988) Symmetry study of the Mn–Si–Al octagonal quasicrystal by convergent beam electron diffraction. Appl Phys Lett 52:2120–2121
2. Beeli C, Nissen HU, Robadey J (1991) Stable Al–Mn–Pd quasi-crystals. Philos Mag Lett 63:87–95
3. Dubois JM (2001) Quasicrystals J Phys Condens Matter 13:7753–7762
4. Okabe T, Furihata JI, Morishita K, Fujimori H (1992) Decagonal phase and pseudo-decagonal phase in the Al–Cu–Cr system. Philos Mag Lett 66:259–264
5. Mihalkovič M, Zhu WJ, Henley CL, Phillips R (1996) Icosahedral quasicrystal decoration models, II: optimization under realistic Al–Mn potential. Phys Rev B 53:9022–9044
6. Yurechko M, Grushko B (2000) A study of the Al–Pd–Co alloy system. Mater Sci Eng A 294:139–142
7. Yurechko M, Grushko B, Velikanova T, Urban K (2002) Isothermal sections of the Al–Pd–Co alloy system for 50–100 at% Al. J Alloys Compd 337:172–181
8. Raghavan V (2008) Al–Pd–Co. J Phase Equilib 29:54–56
9. Yurechko M, Fattaha A, Velikanova T, Grushko B (2001) A contribution to the Al–Pd phase diagram. J Alloys Compd 329:173–181
10. Matsuo Y, Hiraga K (1994) The structure of Al3Pd—close relationship to decagonal quasi-crystals. Philos Mag Lett 70:155–161
11. Yurechko M, Grushko B, Velikanova T, Urban K (2004) A comparative study of the Al–Co–Pd and Al–Co–Ni alloy system. J Alloys Compd 367:20–24
12. Frigan B, Santana A, Engel M, Schopf D, Trebin H-R, Mihalkovič M (2011) Low-temperature structure of ξ-Al–Pd–Mn optimized by ab initio methods. Phys Rev B 84:184203

Chapter 19
Evolution of Phases in Selected Al–Co–Cu Complex Metallic Alloys Under Near-Equilibrium Conditions at 800–1150 °C

P. Priputen, T.Y. Liu, I. Černičková, D. Janičkovič, P. Švec, E. Illeková, M. Drienovský, R. Čička, and J. Janovec

Abstract This work is focused on the experimental investigation of intermetallic phases in $Al_{60}Co_{29}Cu_{11}$, $Al_{63}Co_{24}Cu_{13}$, and $Al_{67}Co_{20}Cu_{13}$ complex metallic alloys at near-equilibrium conditions. The alloys were long-term annealed at temperatures between 800 and 1150 °C and subsequently rapidly cooled to fix their high-temperature microstructures. Annealing temperatures were chosen with respect to the results of differential thermal analysis. Particular samples were studied by X-ray diffraction, scanning electron microscopy including energy dispersive X-ray spectroscopy and electron backscatter diffraction, and transmission electron microscopy. In the microstructures of particular samples, various combinations of D, B2, m, Al_5Co_2, and Θ-Al_2Cu phases were identified depending on both bulk metal composition and thermal history.

19.1 Introduction

Binary and ternary systems based on aluminium as a dominant component and transition metals as complementary components belong to the most investigated metallic systems in the last decades [1]. The reason resides in looking for new complex intermetallic phases including quasicrystals, all characterized by giant unit cells containing up to thousands of atoms ordered in clusters and exhibiting interesting physical properties [2, 3].

Al–Cu–Co is one of the systems where stable decagonal quasicrystalline phase (D-phase) was observed and studied in several alloys [4–8]. Beside the D-phase, ternary tetragonal (T) and hexagonal (H) phases and several binary phases extending into the ternary region (e.g., B2, $Al_{13}Co_4$) were found in this system [7]. The binary

P. Priputen (✉) · T.Y. Liu · I. Černičková · P. Švec · M. Drienovský · R. Čička · J. Janovec
Faculty of Materials Science and Technology in Trnava, Slovak University of Technology in Bratislava, Paulínska 16, 917 24 Trnava, Slovak Republic
e-mail: pavol.priputen@stuba.sk

D. Janičkovič · P. Švec · E. Illeková
Institute of Physics, Slovak Academy of Science, Dúbravská 9, 845 11 Bratislava, Slovak Republic

S. Schmid et al. (eds.), *Aperiodic Crystals*, DOI 10.1007/978-94-007-6431-6_19,

B2-phase in the Al–Cu–Co diagram separates Al-rich and Al-poor areas. In this phase, a relevant amount of Co can be replaced with Cu [1]. The monoclinic m-phase of the $Al_{13}Co_4$ family can also dissolve a significant amount of Cu. Contrary to B2, Cu atoms in m replace mostly the positions of Al atoms and contribute in this way to the reduction of Al content. Besides m the $Al_{13}Co_4$ family contains at least three other phases denoted as Z, Y, and o [9]. The latter phases occur within the compositional range Al_5Co_2–Al_9Co_2 of the Al–Co phase diagram and are considered as quasicrystalline approximants of the D-phase. Even if the binary phases from the $Al_{13}Co_4$ family show structural similarities with m, they do not dissolve any significant amounts of Cu. Al_5Co_2 and Al_9Co_2 exhibit the same feature [1].

In the Al–Cu–Co system, depending on the bulk alloy composition, a direct formation of D from liquid (L) is possible between 700 and 1100 °C [4–7]. On the other hand, Zhang and Gille [8] proposed the partial reaction scheme where D is expected to be formed by peritectic reaction involving L, m, and B2 phases. The present paper is focused therefore on the study of phase evolution under near-equilibrium conditions between 800–1150 °C in three complex metallic alloys ($Al_{60}Co_{29}Cu_{11}$, $Al_{63}Co_{24}Cu_{13}$, and $Al_{67}Co_{20}Cu_{13}$) showing metal compositions suitable for the D formation by peritectic reaction. The attention was paid to the precise characterization of particular phases in the samples after differential thermal analysis (DTA) as well as after annealing at 800, 1050, and 1150 °C. The annealing temperatures were chosen with respect to the results of DTA analysis.

19.2 Experimental Procedures

The investigated $Al_{60}Co_{29}Cu_{11}$, $Al_{63}Co_{24}Cu_{13}$, and $Al_{67}Co_{20}Cu_{13}$ alloys were prepared from pure elements (>99.99 wt.%) by arc melting in argon atmosphere and subsequently cast to form button ingots. Each of the ingots was then sectioned into several pieces (samples). One of the samples was used for DTA performed with rate of 10 °C/min under argon atmosphere from ambient temperature up to 1550 °C and vice versa. The mass of the DTA samples was about 120 mg. Based on the results of DTA measurements, 800, 1050, and 1150 °C were chosen as annealing temperatures. The other three samples were placed into evacuated silica capsule (the capsules used for this purpose were several times purged with argon and subsequently evacuated before being sealed up) and annealed at 800 °C for 1200 h, 1050 °C for 200 h, and 1150 °C for 50 h. To prevent contact with silica and a possible contamination with silicon, the samples annealed at 1050 °C and 1150 °C were inserted into silica capsules placed in small alumina crucibles. Every annealing was followed with immediate water quenching to fix the high-temperature microstructures of the samples.

An X-ray diffractometer Philips PW 1830, a scanning electron microscope JEOL JSM-7600F (SEM) equipped with both an energy dispersive X-ray spectrometer X-max (EDX) using INCA software and an electron backscattered diffraction detector Nordlys (EBSD) using FLAMENCO software, and a differential thermal analyser

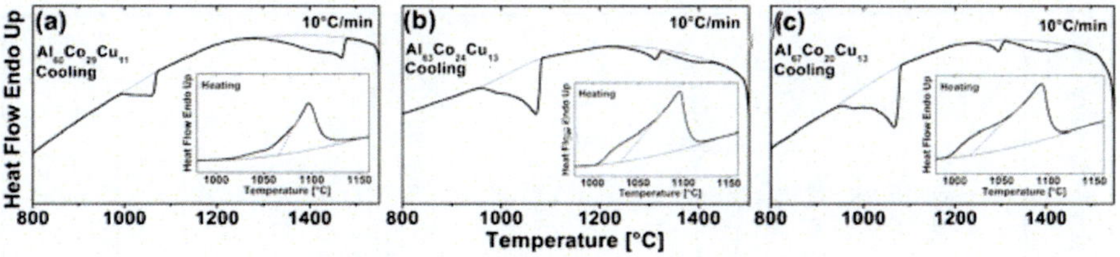

Fig. 19.1 DTA records for $Al_{60}Co_{29}Cu_{11}$ (**a**), $Al_{63}Co_{24}Cu_{13}$ (**b**), $Al_{67}Co_{20}Cu_{13}$ (**c**) alloys. Larger and *smaller plots* correspond to cooling and heating regimes, respectively

STA 409 CD Netzsch calibrated for the heating regime were used to characterize the samples. The X-ray diffraction (XRD) was performed on powder samples with the iron filtered Co-Kα1 radiation, between 20 and 70° of 2θ, and with the rate of 0.02°/step. For the identification of minor phases with volume fractions lower than 5 %, the highly sensitive EBSD technique was used in addition to the XRD technique.

19.3 Results and Discussion

In Figs. 19.1(a)–(c), DTA records are illustrated for the originally as-cast samples of the respective $Al_{60}Co_{29}Cu_{11}$, $Al_{63}Co_{24}Cu_{13}$, and $Al_{67}Co_{20}Cu_{13}$ alloys. To see high temperature transformations, the cooling regime was preferred in this work. Two pronounced temperature ranges were observable for all the alloys. In the first temperature range between 1500 and 1200 °C, smooth and sharp peaks were observable. The former peaks appearing at approximately 1450 °C are not any solidification peaks. They can originate from long rising time of temperature after switching between heating and cooling regimes of relatively heavy samples. The latter sharp peaks are shifted to higher temperatures with decreasing the bulk Al content. These peaks seem to be solidification peaks. In the second temperature range between 1100 and 950 °C, two overlapping peaks were observable for all the alloys. The higher peaks are located close to 1100 °C and the lower peaks appear at slightly lower temperatures. To see the correct temperatures of phase transformations related to both the peaks, the records corresponding to heating regime were also inserted into Figs. 19.1(a)–(c). To find a correct interpretation of the observed peaks, annealing experiments at 800, 1050, and 1150 °C were done.

The microstructure (SEM/BSE mode), the corresponding XRD pattern, and EBSD patterns identifying all the present phases are documented in Fig. 19.2 for the $Al_{60}Co_{29}Cu_{11}$ alloy annealed at 800 °C. This is the characteristic illustrative presentation applicable to all the investigated alloys.

In the samples investigated, altogether five phases (B2, D, m, Al_5Co_2, and Θ-Al_2Cu) were identified. The presence of phases in particular conditions of all the alloys is summarized in Table 19.1.

The annealing temperature 1150 °C is positioned between the high- (1500–1200 °C) and low-temperature (950–1100 °C) ranges of the peak occurrence in the

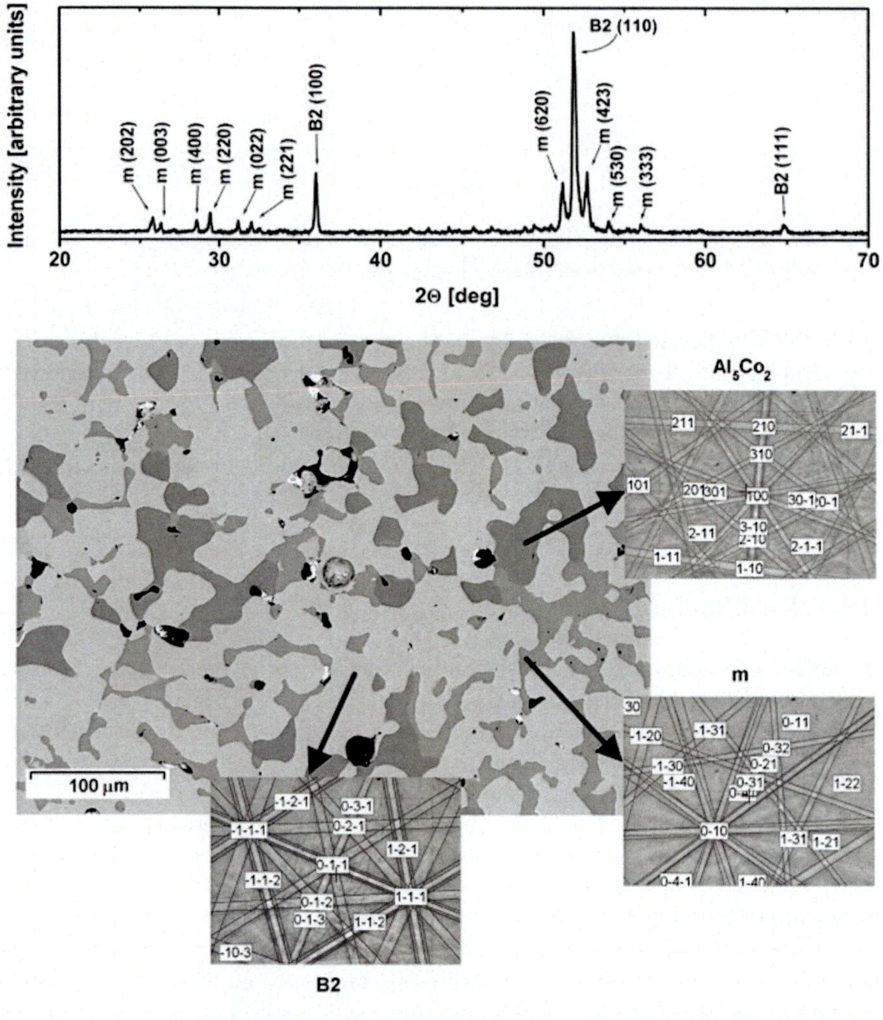

Fig. 19.2 Microstructure constituents and diffraction patterns (XRD, EBSD) of corresponding phases presented for $Al_{60}Co_{29}Cu_{11}$ alloy annealed at 800 °C

DTA records corresponding to the cooling regime (Fig. 19.1). As follows from Table 19.1, three phases (B2, D, and Θ-Al_2Cu) were identified in the samples annealed at 1150 °C and subsequently quenched. However, in the high-temperature range, one solidification peak only was found related to the formation of a solid phase from the liquid. The explanation of this apparent discrepancy follows from Fig. 19.3 showing the characteristic microstructure and the corresponding XRD pattern of the condition annealed at 1150 °C. Based on both, XRD pattern and metal composition, the white-colour microstructure constituent forming discrete areas with sharp boundaries was identified as B2. Two other phases (Θ-Al_2Cu and D) forming fine-grained

Table 19.1 Phases identified in $Al_{60}Co_{29}Cu_{11}$, $Al_{63}Co_{24}Cu_{13}$, and $Al_{67}Co_{20}Cu_{13}$ alloys

Sample	Condition			
	DTA 10°/min	Annealed at 800 °C, quenched	Annealed at 1050 °C, quenched	Annealed at 1150 °C, quenched
$Al_{60}Co_{29}Cu_{11}$	m, B2	m, Al_5Co_2, B2	m, B2	D, B2, Θ-Al_2Cu,
$Al_{63}Co_{24}Cu_{13}$	m, D, B2	m, B2	D, B2, Θ-Al_2Cu	D, B2, Θ-Al_2Cu
$Al_{67}Co_{20}Cu_{13}$	m, D, B2	m, D	D, B2, Θ-Al_2Cu	D, B2, Θ-Al_2Cu

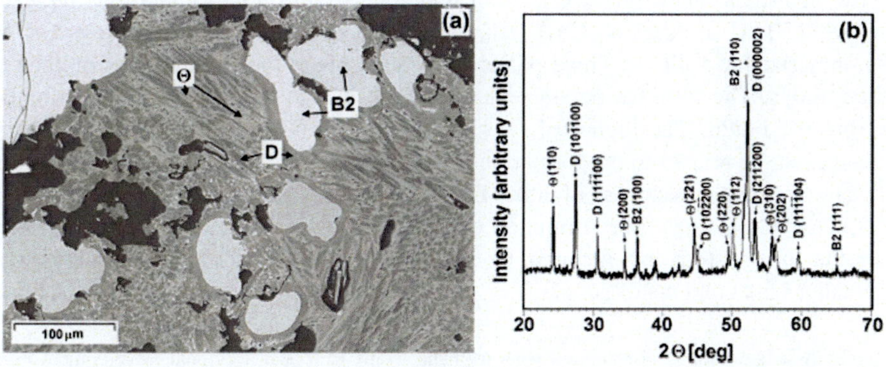

Fig. 19.3 Microstructure of $Al_{67}Co_{20}Cu_{13}$ alloy annealed at 1150 °C and subsequently quenched (**a**) and corresponding XRD pattern (**b**). D-phase is indexed following decagonal notation by Mukhopadhyay

and well-mixed grey areas of different shades of grey correspond to the liquid phase co-existing at 1150 °C with B2. They are products of the final step of sample solidification during quenching. Θ-Al_2Cu is the typical low-temperature phase forming at around 592 °C [10].

In the samples annealed at 1050 °C and subsequently quenched, the same phases were identify in the $Al_{63}Co_{24}Cu_{13}$, and $Al_{67}Co_{20}Cu_{13}$ alloys as it is documented for the samples annealed at 1150 °C (B2, D, and Θ-Al_2Cu). However, beside B2 also D was found to form discrete areas with sharp boundaries. Between D-grains a mixture of D and Θ-Al_2Cu phases representing the original liquid was found in a small amount. Contrary of the $Al_{63}Co_{24}Cu_{13}$, and $Al_{67}Co_{20}Cu_{13}$ alloys, m and B2 were observed in the $Al_{60}Co_{29}Cu_{11}$ alloy after annealing at 1050 °C and quenching. The reason resides in positions of the overlapped peaks from the low-temperature range of the DTA records (see curves corresponding to the heating regime in Fig. 19.1). For the $Al_{63}Co_{24}Cu_{13}$ and $Al_{67}Co_{20}Cu_{13}$ alloys, the onset temperature of the higher peak is slightly below 1050 °C (Figs. 19.1(b)–(c)), however, that corresponding to the $Al_{60}Co_{29}Cu_{11}$ alloy occurs at a bit higher temperature than 1050 °C (Fig. 19.1(a)). Taking into account the fact that m was identified in both the condition after DTA and the condition annealed at 800 °C for all the alloys (Table 19.1), the higher peak has to correspond to the direct formation of D from the liquid and the lower peak represents the formation of m at the expense of either D ($Al_{60}Co_{29}Cu_{11}$ and $Al_{63}Co_{24}Cu_{13}$ alloys) or B2 ($Al_{67}Co_{20}Cu_{13}$). In the $Al_{60}Co_{29}Cu_{11}$ alloy, a small amount of Al_5Co_2 was also formed besides m at temperatures between 800 and 1050 °C.

19.4 Conclusions

The investigation of phase evolution under near-equilibrium conditions between 800–1150 °C in $Al_{60}Co_{29}Cu_{11}$, $Al_{63}Co_{24}Cu_{13}$, and $Al_{67}Co_{20}Cu_{13}$ alloys can be summarized as follows: Three peaks are observable in the DTA records of each of the alloys. The high-temperature peak corresponds to the direct formation of B2 from the liquid. The higher of two overlapping peaks from the low-temperature range can be related to the formation of D from the liquid and the lower to the formation of m at the expense of either D ($Al_{60}Co_{29}Cu_{11}$ and $Al_{63}Co_{24}Cu_{13}$ alloys) or B2 ($Al_{67}Co_{20}Cu_{13}$). The formation of D by peritectic reaction was not confirmed in the current study, but it cannot be excluded in the case of Al–Co–Cu alloys with smaller bulk contents of copper.

Acknowledgements The authors wish to thank to the European Regional Development Fund (ERDF) for financial support of the project ITMS:26220120014 "Center for development and application of advanced diagnostic methods in processing of metallic and non-metallic materials" funded within the Research & Development Operational Programme, to the Slovak Academy of Sciences for the support in the frame of the "Center of Excellence for functional multiphase materials", as well as to the Grant Agency VEGA for the financial support under the contracts 1/0143/12, 2/0111/11, and 1/0339/11, and to the Slovak Research and Development Agency for the financial support under the contract APVV-0076-11.

References

1. Grushko B, Velikanova T (2007) Formation of quasiperiodic and related periodic intermetallics in alloy systems of aluminum with transition metals. Calphad 31:217–232
2. Dubois JM (2008) An introduction to complex metallic alloys and to the CMA network of excellence. In: Belin-Ferré E (ed) Basics of thermodynamics and phase transitions in complex intermetallics. World Scientific, Singapore, pp 1–29
3. Urban K, Feuerbacher M (2004) Structurally complex alloy phases. J Non-Cryst Solids 334–335:143–150
4. Grushko B, Urban K (1991) Solidification of $Al_{65}Cu_{20}Co_{15}$ and $Al_{65}Cu_{15}Co_{20}$ alloys. J Mater Res 6:2629–2636
5. Dong C, Dubois JM, De Boissieu M, Janot C (1991) Phase transformations and structure characteristics of the $Al_{65}Cu_{17.5}Co_{17.5}$ decagonal phase. J Phys Condens Matter 3:1665–1673
6. Grushko B (1993) Phase equilibrium and transformation of stable quasicrystals—decagonal Al–Cu–Co phase. Mater Trans, JIM 34:116–121
7. Grushko B (1993) A study of the Al–Cu–Co phase diagram and the solidification of alloys containing decagonal phase. Phase Transit 44:99–110

8. Zhang LM, Gille P (2004) Solidification study of Al–Co–Cu alloys using the Bridgman method. J Alloys Compd 370:198–205
9. Grushko B, Wittenberg R, Bickmann K, Freiburg C (1996) The constitution of aluminum–cobalt alloys between Al_5Co_2 and Al_9Co_2. J Alloys Compd 233:279–287
10. Goedecke T, Sommer F (1996) Solidification behavior of the Al_2Cu phase. Z Metkd 87:581–586

Chapter 20
Superspace Description of the System $Bi_{2(n+2)}Mo_nO_{6(n+1)}$ ($n = 3$, 4, 5 and 6)

P.J. Bereciartua, F.J. Zuñiga, J.M. Perez-Mato, V. Petříček, E. Vila, A. Castro, J. Rodríguez-Carvajal, and S. Doyle

Abstract The system $Bi_{2(n+2)}Mo_nO_{6(n+1)}$ is described with the superspace formalism. Considering the cationic distribution of the member with $n = 3$, a superspace model is constructed beginning with a model previously proposed for the compound Bi_2MoO_6. The description of even members requires additional modifications. As a result, two superspace models are proposed for the different members of this system, depending on the parity of the parameter n. Both models have been checked through the Rietveld method combining synchrotron and neutron powder diffraction data.

20.1 Introduction

Four new compounds have been synthesized within the binary system Bi_2O_3–MoO_3 ($Bi_{10}Mo_3O_{24}$, $Bi_6Mo_2O_{15}$, $Bi_{14}Mo_5O_{36}$ and $Bi_8Mo_3O_{21}$) [7]. They can be considered as members of the system $Bi_{2(n+2)}Mo_nO_{6(n+1)}$ (members $n = 3$, 4, 5 and 6). As shown by electron diffraction patterns obtained for each phase, the unit cells of these compounds are related to the fluorite-type unit cell ($\{\mathbf{a}_F, \mathbf{b}_F, \mathbf{c}_F\}$) of the com-

P.J. Bereciartua (✉) · V. Petříček
Institute of Physics of the Academy of Sciences of the Czech Republic, Praha, Czech Republic
e-mail: bereciart@fzu.cz

F.J. Zuñiga · J.M. Perez-Mato
Department of Condensed Matter Physics, University of the Basque Country, Bilbao, Spain

E. Vila · A. Castro
Instituto de Ciencia de Materiales de Madrid (ICMM, CSIC), Madrid, Spain

J. Rodríguez-Carvajal
Institut Laue-Langevin, Grenoble, France

S. Doyle
Institute for Synchrotron Radiation (ISS), ANKA, Eggenstein-Leopoldshafen, Germany

S. Schmid et al. (eds.), *Aperiodic Crystals*, DOI 10.1007/978-94-007-6431-6_20,

pound δ-Bi_2O_3 (member with $n = 0$) through the equation

$$\begin{pmatrix} \mathbf{a} \\ \mathbf{b} \\ \mathbf{c} \end{pmatrix} = \begin{pmatrix} n+1 & 0 & 1 \\ 0 & 1 & 0 \\ -\frac{1}{2} & 0 & \frac{3}{2} \end{pmatrix} \begin{pmatrix} \mathbf{a}_\mathrm{F} \\ \mathbf{b}_\mathrm{F} \\ \mathbf{c}_\mathrm{F} \end{pmatrix}. \tag{20.1}$$

The atomic structure of the compound $Bi_{10}Mo_3O_{24}$ ($n = 3$) presents space group $C2$. It is described as a stacking of puckered layers along **b** direction combined with groups of three isolated $\{MoO_4\}$ tetrahedra [5]. Considering Eq. (20.1), this model can be extended to the whole family. The member $n = 5$ keeps the same $C2$ space group with 5 tetrahedra and members $n = 4$ and 6 exhibit 4 and 6 tetrahedra, respectively, with space group $P2_1/a$. The cationic skeletons of these phases were confirmed by a high-resolution transmission electron microscopy (HRTEM) study [6].

In the superspace formalism, the obtained diffraction patterns can be described as consisting of a common set of main reflections and satellite reflections, associated with a modulation vector which depends on the composition. The modulation vector $\mathbf{q}_V = \frac{1}{3n+4}(6\mathbf{a}_F^* + 2\mathbf{c}_F^*)$ is proposed in [7].

In the present work, we develop the superspace description for this family of compounds. After choosing the proper average unit cell and modulation vector, a superspace model has been constructed. However, the application of this model to the even members requires some modifications. A more comprehensive description of this work is published in [1].

20.2 Experimental Section

Samples of the four studied compounds were obtained through the wet-chemistry method described in [7]. The XPD data were collected on the PDIFF beamline at the ANKA Synchrotron in Bragg–Brentano geometry. The NPD data were measured on the D2B diffractometer at the ILL (Grenoble) in Debye–Scherrer geometry.

20.3 Odd Members

The 3D structure proposed for the member $n = 3$ is considered as a starting point for the superspace construction.

20.3.1 Choice of the Average Unit Cell and the Modulation Vector

The cationic distribution of this 3D structure can be described with the above mentioned fluorite-like unit cell, which associates all cations with the same average position. The embedding process gives rise to certain order along internal space. Using

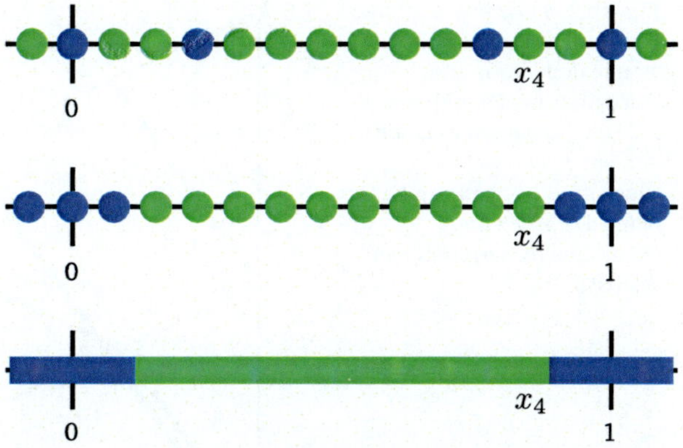

Fig. 20.1 Occupational modulation of Bi (*green*) and Mo (*blue*) atoms. Two different orderings are obtained with modulation vectors $\mathbf{q}_V$ (*top*) and $\mathbf{q}_o$ (*middle*). The second one allows a simple description with two complementary crenel functions (*bottom*). (Reprinted by permission of the IUCr)

the proposed modulation vector $\mathbf{q}_V$ leads to the ordering indicated in Fig. 20.1. A more simple ordering can be obtained with the modulation vector

$$\mathbf{q}_o = \frac{1}{3n+4}\left(-2\mathbf{a}_F^* + 2(n+1)\mathbf{c}_F^*\right). \tag{20.2}$$

In this case, the occupational modulation can be represented by only two complementary crenel functions (see Fig. 20.1).

The embedding also establishes the symmetry group $F2(\alpha 0\gamma)$ for this superspace model. The t-section which allows recovering the correct cationic distribution in 3D structure is defined by the value $t = 0$ (or its equivalent).

20.3.2 Construction of the Superspace Model

The compound Bi_2MoO_6 ($n = \infty$) belongs to the space group $Pca2_1$ [4] with a structure close to the ideal Aurivillius. Within superspace approach, Aurivillius phases have been characterized as cation-deficient perovskites with formula $AB_{1-\gamma}O_3$ within the range $0 < \gamma < \frac{1}{2}$ [2, 3]. Thus, the compound Bi_2MoO_6 can be described as a modulated structure with $\gamma = \frac{1}{2}$. The superspace group of this model is $X2cm(00\gamma)000$ with centring vectors $(\frac{1}{2}\frac{1}{2}\frac{1}{2}0)$, $(\frac{1}{2}00\frac{1}{2})$ and $(0\frac{1}{2}\frac{1}{2}\frac{1}{2})$. This model contains four atoms: Bi, Mo, O1 and O2. Bi and Mo atomic domains present occupational modulation given by the corresponding crenel functions. The fcc cationic distribution of the 3D structure can be obtained by including appropriate sawtooth functions for these atoms (see Fig. 20.2). The resulting Bi and Mo atomic domains are aligned along direction [0 0 1 2] with an ordering equivalent to the one obtained in Fig. 20.1. Thus, there is a suitable change of the unit cell which allows to obtain a model with the desired cationic occupational modulation. This transformation leads to the average unit cell used in the previous embedding. This change also implies a new modulation vector $\mathbf{q}_F = 2\mathbf{c}_A^* - 2\mathbf{q}_A = \mathbf{c}_A^* = \frac{2}{3}\mathbf{c}_F^*$, corresponding to the limit for $n = \infty$ of the propose modulation vector $\mathbf{q}_o$ (see Eq. (20.2)).

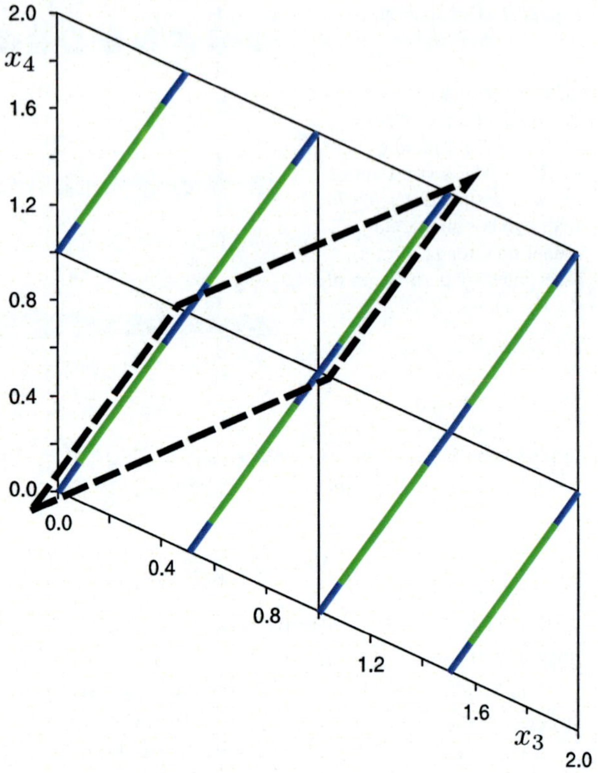

Fig. 20.2 Distribution of the Bi (*green*) and Mo (*blue*) atomic domains in the Aurivillius model with the indicated sawtooth functions. A more convenient superspace unit cell can be defined (*dashed line*) (Reprinted by permission of the IUCr)

The oxygen atomic domains resulting from the unit cell transformation can also be described in a more suitable way by considering neighbouring cations. They are divided, giving rise to several discontinuous atomic domains described by the corresponding crenel functions. Finally, the symmetry of this model can be reduced to the group $F2(\alpha 0\gamma)$. Note that this constructed model corresponds to the member with $n = \infty$ (Bi_2MoO_6), but can be extrapolated to any value of n parameter by just including the oblique modulation vector $\mathbf{q}_o$ (Eq. (20.2)) and modifying accordingly the widths of the atomic domains.

The positional modulations of the different atoms have been determined through the embedding of the 3D structure. This superspace model has been checked through the Rietveld method combining the XPD and NPD data. The displacive modulations obtained in this model are large and rather irregular, especially for the O atoms associated with the environments of Mo atoms (O1 and O2), as shown in Fig. 20.3.

20.3.3 Member with $n = 5$

The initial superspace model for member $n = 5$ can be derived from the general model. This structure presents the expected cation distribution and a rather accurate

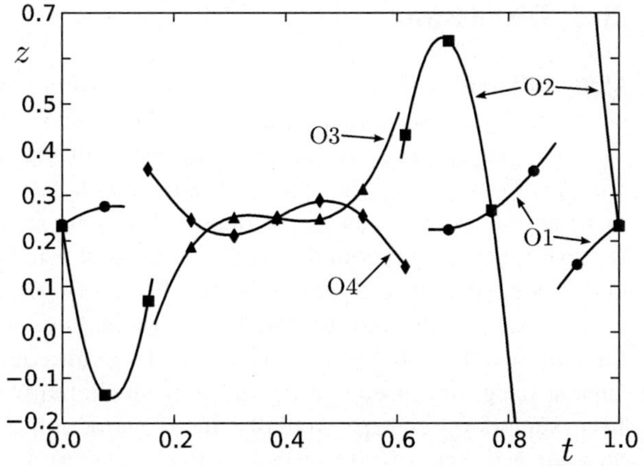

Fig. 20.3 Positional modulation along z coordinate for O1, O2, O3 and O4 atomic domains (*continuous lines*). Only some points of the atomic domains are relevant (*discrete points*) (Reprinted by permission of the IUCr)

description of the O atoms related to Bi atoms. However, with this model, there appear corner-sharing $[MoO_6]$ octahedra instead of the expected $\{MoO_4\}$ tetrahedra. Actually, to obtain the correct positional modulations it is necessary to perform the embedding of the 3D structure. This model has been checked using the Rietveld method combining XPD and NPD data. As in previous case, the displacive modulations of the O1 and O2 atoms are very irregular.

20.4 Even Members

For the superspace description of the even members, we first consider the compound with $n = 4$. The main modification implies using a different modulation vector since neither $\mathbf{q}_o$ nor $\mathbf{q}_V$ are suitable choices to index the diffraction pattern. It is necessary to introduce another modulation vector given by

$$\mathbf{q}_e = \frac{1}{3n+4}\left(\mathbf{a}_F^* + (2n+3)\mathbf{c}_F^*\right). \tag{20.3}$$

With this modulation vector the reflection condition $h0lm: m = 2n$ should be assumed, pointing out to space group $F2/m(\alpha 0\gamma)0s$ or $Fm(\alpha 0\gamma)s$, and the centrosymmetric group is chosen. Maintaining the same average unit cell, this new modulation vector $\mathbf{q}_e$ also implies changes in the superspace unit cell, the atomic domains distribution and the t value which determines the 3D commensurate superstructure.

The 3D structure obtained from this superspace model again leads to Mo coordinations very different from the expected ones. As previously, it is necessary to carry out the embedding of the 3D structure to determine the positional modulations. The superspace models for the compounds $n = 4$ and 6 were obtained and checked through the Rietveld method using XPD and NPD data. The atomic domains of the final model also present very strong and irregular positional modulations, especially O1 and O2 atomic domains.

20.5 Discussion

Many different systems with flexible composition have been described in the superspace formalism, where each family is described by a unique model. However, two superspace models are introduced in the present work for the system $Bi_{2(n+2)}Mo_nO_{6(n+1)}$. The symmetry, the modulation vector and the atomic domain distribution of each model are different. The main reason is that the diffraction patterns of the even compounds cannot be indexed using the modulation vector of the odd model ($\mathbf{q}_o$). And the modulation vector of even model ($\mathbf{q}_e$) does not lead to the correct 3D space group for the corresponding superstructure when used with the odd model. Probably, a general model would require combining one of the two superspace groups used in this work with an alternative modulation vector, different from vectors $\mathbf{q}_o$ and $\mathbf{q}_e$. Actually, the construction of such a model for both odd an even members of this system is an open question which would require further investigations.

Notwithstanding, the application of the superspace formalism to this system shows both the possibilities and the limitations of this approach. The cationic distribution is represented in a rather concise manner, using complementary crenel functions whose widths depends on the n parameter. On the other hand, the O atoms forming $\{MoO_4\}$ tetrahedra present large and irregular positional modulations, so it is not straightforward to describe their distribution through an average position and the corresponding distortion. In any case, both superspace models make visible the structure-composition relationship of the family.

Acknowledgements The authors are grateful to J. Galy, A.R. Landa-Cánovas and J. Hernández-Velasco for their helpful discussions. This work was supported by the Basque Government (project IT-282-07). This work was also supported by the European Community—Research Infrastructure Action under the FP6: "Structuring the European Research Area" ("Integrating Activity on Synchrotron and Free Electron Laser Science" (IA-SFS) RII3-CT-2004-506008). PJB also thanks Basque Government for financial support. Adapted from [1] with permission of the IUCr.

References

1. Bereciartua PJ, Zuñiga FJ, Perez-Mato JM, Petříček V, Vila E, Castro A, Rodríguez-Carvajal J, Doyle S (2012) Structure refinement and superspace description of the system $Bi_{2(n+2)}Mo_nO_{6(n+1)}$ ($n = 3, 4, 5$ and 6). Acta Crystallogr, Sect B 68(4):323–340
2. Boullay P, Trolliard G, Mercurio D, Perez-Mato JM, Elcoro L (2002) Toward a unified approach to the crystal chemistry of Aurivillius-type compounds, I: the structural model. J Solid State Chem 164:252–260
3. Boullay P, Trolliard G, Mercurio D, Perez-Mato JM, Elcoro L (2002) Toward a unified approach to the crystal chemistry of Aurivillius-type compounds, I: $Bi_7Ti_4NbO_{21}$, a case study. J Solid State Chem 164:261–271
4. van den Elzen AF, Rieck GD (1973) The crystal structure of $Bi_2(MoO_4)_3$. Acta Crystallogr, Sect B 29:2433–2436
5. Galy J, Hernández-Velasco J, Landa-Cánovas AR, Vila E, Castro A (2009) Ab initio structure determination and Rietveld refinement of $Bi_{10}Mo_3O_{24}$ the member $n = 3$ of the $Bi_{(2n+4)}Mo_nO_{6(n+1)}$ series. J Solid State Chem 182:1177–1187

6. Landa-Cánovas AR, Vila E, Hernández-Velasco J, Galy J, Castro A (2009) Structural elucidation of the $Bi_{2(n+2)}Mo_nO_{6(n+1)}$ ($n = 3$, 4, 5 and 6) family of fluorite superstructures by transmission electron microscopy. Acta Crystallogr, Sect B 65:458–466
7. Vila E, Landa-Canovas AR, Galy J, Iglesias JE, Castro A (2007) $Bi_{2n+4}Mo_nO_{6(n+1)}$ with $n =$ 3, 4, 5, 6: a new series of low-temperature stable phases in the $m Bi_2O_3$–MoO_3 system ($1.0 < m < 1.7$): structural relationships and conductor properties. J Solid State Chem 180:661–669

Chapter 21
Pseudo-Commensurate $GdBaCo_2O_{5+\delta}$ and Its Phase Transition at Elevated Temperatures

N. Ishizawa, T. Asaka, T. Kudo, K. Fukuda, N. Abe, and T. Arima

Abstract An in-situ single-crystal X-ray diffraction study on tetragonal $GdBaCo_2 O_{5+\delta}$ with $\delta \sim 0.38$ revealed that the crystal is pseudo-commensurate at room temperature with the magnitudes of the modulation vectors $\boldsymbol{q}_1$ and $\boldsymbol{q}_2$ parallel to the basal axes increasing gradually from the nearly commensurate value close to 1/3 upon heating. The basic structure of the compound is a double-layered perovskite type, having an alternating layer sequence $[GdO_\delta]$–$[CoO_2]$–[BaO]–$[CoO_2]$ along the c axis. The oxygen deficiency of the crystal occurs only in the $[GdO_\delta]$ layer, though it causes many positional modulations of constituent atoms in association with the valence fluctuation of Co cations between +2 and +3. Because of its pseudo-commensurate nature, the room temperature structure was also investigated by the commensurately-modulated approach as well as the conventional three-dimensional ones assuming a $3 \times 3 \times 2$ supercell of the $P4/mmm$ symmetry. These approaches successfully reproduced a prime structure of the compound, consisting of intersecting CoO_5 pyramidal arrays parallel to $\boldsymbol{a}$ and $\boldsymbol{b}$ axes. The incommensurate approach, on the other hand, also suggested a presence of a local disorder having a structural similarity with the high-temperature modification.

21.1 Introduction

Double perovskite-type cobaltates $LnBaCo_2O_{5+\delta}$ (Ln = lanthanide or yttrium) have attracted considerable attention because they exhibit strong correlation among their crystallographic, magnetic, and electric properties [1, 2]. They are also proposed as potential candidates for intermediate-temperature solid state oxides fuel cells because of their excellent oxygen transport properties [3–5]. The structure consists of layers with a stacking sequence $[CoO_2]$–[BaO]–$[CoO_2]$–$[LnO_\delta]$ along the c axis. The basic structure is tetragonal with unit cell dimensions $1a_p \times 1a_p \times 2a_p$ (abbre-

N. Ishizawa (✉) · T. Asaka · T. Kudo · K. Fukuda
Nagoya Institute of Technology, Nagoya, Japan
e-mail: ishizawa@nitech.ac.jp

N. Abe · T. Arima
Department of Advanced Materials Science, The University of Tokyo, Tokyo, Japan

S. Schmid et al. (eds.), *Aperiodic Crystals*, DOI 10.1007/978-94-007-6431-6_21,

viated as 112 hereafter), where a_p corresponds to the edge length of the perovskite-type cube [6]. In addition to 112, another three structural modifications for the compound are known to date, $3a_p \times 3a_p \times 2a_p$(332), $1a_p \times 2a_p \times 2a_p$(122), and $2a_p \times 2a_p \times 2a_p$(222) [2, 7–9].

Recently, the transmission electron microscope analysis revealed that the 332 phase of $GdBaCo_2O_{5+\delta}$ ($\delta \sim 0.38$) is incommensurate under the magnetic field with $\boldsymbol{q}_1 = (\sim 1/3, 0, 0)$ and $\boldsymbol{q}_2 = (0, \sim 1/3, 0)$ with respect to the fundamental 112 tetragonal cell, and that the compound undergoes a phase transition to commensurate 122 at elevated temperatures through different routes depending on the magnetic field applied [10]. In this paper, we investigated the temperature dependence of the $\boldsymbol{q}$ vectors of the 332 phase of $GdBaCo_2O_{5+\delta}$ ($\delta \sim 0.38$) by the single-crystal X-ray diffraction under zero magnetic fields, determined the incommensurately-modulated structure at room temperature, and compared the results with those obtained by the commensurate and supercell approaches [11].

21.2 Experimental

Crystals of $GdBaCo_2O_{5+\delta}$ grown by the floating-zone method were investigated by the in-situ single-crystal X-ray diffraction at temperatures 296–393 K using Mo Kα. The crystal has a tetragonal symmetry at room temperature and its superspace group was determined to be $P4/mmm(\alpha 00)0000(0\alpha 0)0000$ with $\alpha = 0.3368(1)$ according to the notation for five-dimensional superspace groups [12]. The temperature dependence of the $\boldsymbol{q}$ vectors and cell dimensions of the 332 phase were determined at elevated temperatures through the phase transition.

The average structure of 332 belongs to the space group $P4/mmm$, containing 6 crystallographically independent atoms Gd1, Ba1, Co1, O1, O2, and O3 (Fig. 21.1). Only the O3 site is oxygen-deficient. The occupational modulation waves for the O3 site and the positional modulation waves for all atom sites were then examined step by step in the 5-dimensional superspace using Jana2006 [13]. The incommensurately-modulated structure model converged with $R = 0.0285$ using 67 parameters for 2216 reflections with $I > 3\sigma(I)$. The composition was determined as $GdBaCo_2O_{5+\delta}$, $\delta = 0.38(1)$ in the final stage of refinement.

Since the magnitude of the modulation vectors are close to 1/3, the structure was also examined by the supercell models and the commensurately-modulated model. One supercell model assuming split-atom sites for part of Gd atoms converged with the best R factor of 0.0240 using 77 parameters. Another supercell model assuming the 3rd order anharmonic atomic displacement parameters (ADPs) for part of Gd atoms also converged with $R = 0.0269$ using 78 parameters. The commensurately-modulated model assuming the 3rd order ADP for Gd converged with $R = 0.0279$ using the minimum number of 63 parameters.

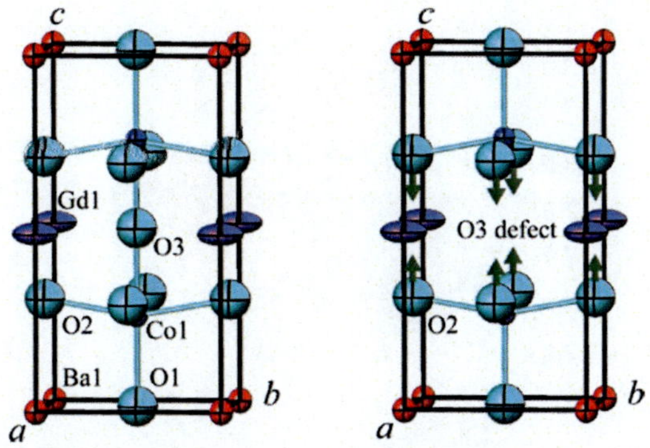

Fig. 21.1 The average structure (*left*) of the 332 phase in the 112 fundamental cell and the local distortion around the O3 defect (*right*) with the displacement direction of O2 indicated by arrows

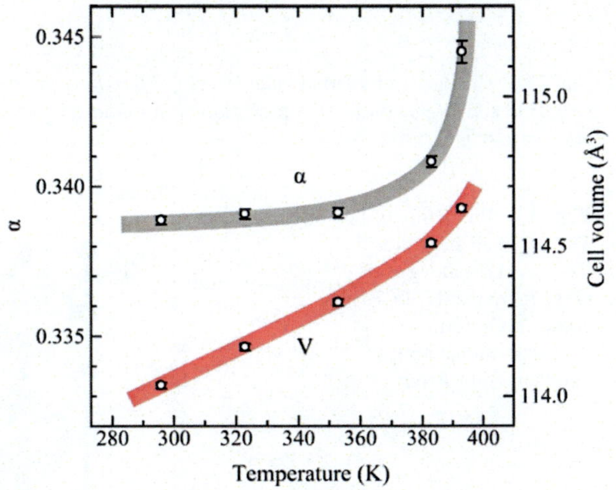

Fig. 21.2 Temperature changes in the magnitude α of the modulation vector $\boldsymbol{q}$ and the fundamental unit cell volume V of the 332 phase

21.3 Results and Discussion

As shown in Fig. 21.2, the magnitude α of the modulation vector $\boldsymbol{q}$ increased rapidly when the temperature approached ~393 K around which the commensurate 122 phase commenced to appear in addition to 332. Since α is so close to 1/3 at room temperature, the incommensurate nature of the 332 phase tends to be overlooked. The present study first confirmed the incommensurability of the 332 phase quantitatively through the measurement of temperature dependence of the $\boldsymbol{q}$ vector by X-ray diffraction under zero magnetic fields.

Crystals contain the oxygen deficiency in the $[GdO_\delta]$ layer, which causes positional modulations of constituent atoms in association with the valence modulation of nearby Co cations between +2 and +3. The O3 defect in the GdO layer invokes the displacement of not only the neighboring Gd on the same layer but also the Co and O2 atoms in the adjacent [CoO] layer. Resultant z shifts of O2, as shown by arrows in Fig. 21.1 (right), get larger a pair of electrical dipole moments of the CoO_5 pyramids opposing each other along c in the fundamental cell. The dipoles in tail to tail arrangement are mitigated by reduction in the oxidation state of Co in the

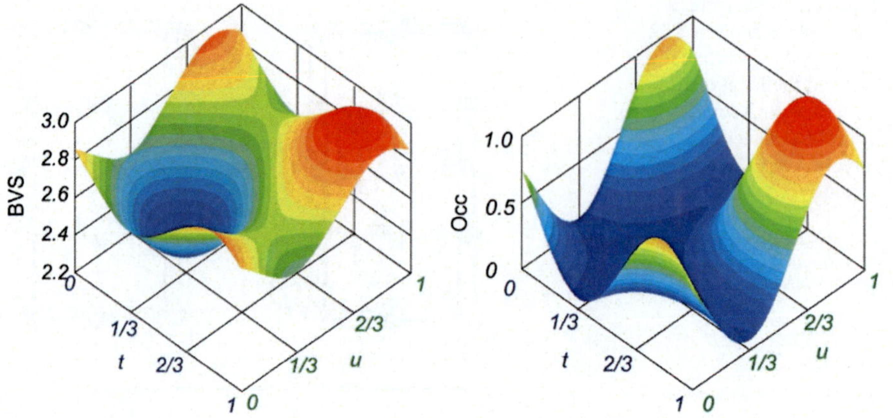

Fig. 21.3 Changes in the bond valence sum of Co (*left*) and the occupation parameter of O atom at the O3 site (*right*) on the $t-u$ section with minimum and maximum extrema at $t=u=1/3$, and $t=u=5/6$, respectively

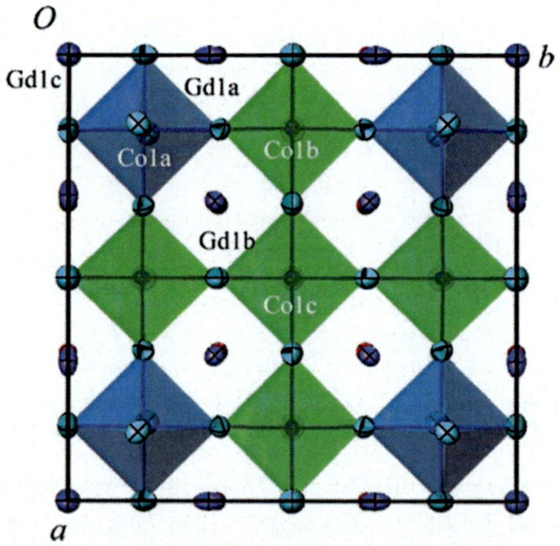

Fig. 21.4 Projection of the 332 supercell structure ($0<z<1/2$) along c. Oxygen pyramids of $Co1bO_5$ and $Co1cO_5$ form one-dimensional arrays parallel to a or b axis

pyramidal coordination. As such, the changes in Co valency, occupational, and positional modulations occur in a concerted manner. A close relationship between the first two, for example, is exemplified in the topological similarity of the $t-u$ plots [14] of the bond valence sum [15] of Co and the occupation parameter of O atoms at the O3 site, having extrema at the same positions $t=u=1/3$ and $5/6$ (Fig. 21.3).

The 332 supercell and the commensurately-modulated approaches succeeded in obtaining a clear-cut image of the structure of prime importance (Fig. 21.4), which contains one-dimensional CoO_5 pyramidal arrays running along the a and b axes. At the intersection of the pyramidal arrays, the occupation factor of the apical O3 is minimized and the cobalt cation (Co1c in Fig. 21.4) takes the lowest oxidation state close to +2.

The incommensurately-modulated approach not only revealed the typical structure (Fig. 21.4) as obtained by the other approaches, but also succeeded in describing another approximant which has a structural similarity with the high-temperature orthorhombic 122 modification in a point that the CoO_5 pyramidal arrays run parallel only along one direction, either ***a*** or ***b*** [11]. This sort of intrinsic local disorder would append important information to the phase transition mechanism of the compound at elevated temperatures. In this respect, the incommensurately-modulated approach seems to have an advantage over the other ones.

Acknowledgements The authors are grateful to Dr Vaclav Petricek, Institute of Physics, Academy of Science, Czech Republic, for his valuable comments and calculations, and Dr. Terutoshi Sakakura, Institute of Multidisciplinary Research for Advanced Materials, Tohoku University, Japan, for supplying us a Python script for data conversion. This work was supported by the Grant-in-Aids for Scientific Research No. 22360272 from the Japan Society for the Promotion of Science.

References

1. Moritomo Y, Takeo M, Liu XJ, Akimoto T, Nakamura A (1998) Metal–insulator transition due to charge ordering in $R_{1/2}Ba_{1/2}CoO_3$. Phys Rev B, Condens Matter Mater Phys 58(20):R13334–R13337
2. Maignan A, Martin C, Pelloquin D, Nguyen N, Raveau B (1999) Structural and magnetic studies of ordered oxygen-deficient perovskites $LnBaCo_2O_{5+d}$, closely related to the "112" structure. J Solid State Chem 142(2):247–260
3. Taskin AA, Lavrov AN, Ando Y (2005) Achieving fast oxygen diffusion in perovskites by cation ordering. Appl Phys Lett 86(9):091910
4. Tarancón A, Morata A, Dezanneau G, Skinner SJ, Kilner JA, Estradé S, Hernández-Ramírez F, Peiró F, Morante JR (2007) $GdBaCo_2O_{5+x}$ layered perovskite as an intermediate temperature solid oxide fuel cell cathode. J Power Sources 174(1):255–263. doi:10.1016/j.jpowsour.2007.08.077
5. Peña-Martínez J, Tarancón A, Marrero-López D, Ruiz-Morales JC, Núñez P (2008) Evaluation of $GdBaCo_2O_{5+\delta}$ as cathode material for doped lanthanum gallate electrolyte IT-SOFCs. Fuel Cells 8(5):351–359. doi:10.1002/fuce.200800026
6. Er-Rakho L, Michel C, Lacorre P, Raveau B (1988) $YBaCuFeO_{5+\delta}$: a novel oxygen-deficient perovskite with a layer structure. J Solid State Chem 73(2):531–535. doi:10.1016/0022-4596(88)90141-7
7. Zhou W, Lin CT, Liang WY (1993) Synthesis and structural studies of the perovskite-related compound $YBaCo_2O_{5+x}$. Adv Mater 5(10):735–738. doi:10.1002/adma.19930051010
8. Akahoshi D, Ueda Y (2001) Oxygen nonstoichiometry, structures, and physical properties of $YBaCo_2O_{5+x}$ ($0.00 \le x \le 0.52$). J Solid State Chem 156(2):355–363. doi:10.1006/jssc.2000.9006
9. Frontera C, García-Muñoz JL, Llobet A, Aranda MAG (2002) Selective spin-state switch and metal–insulator transition in $GdBaCo_2O_{5.5}$. Phys Rev B, Condens Matter Mater Phys 65(18):180405
10. Asaka T, Abe N, Kudo T, Fukuda K, Kimoto K, Matsui Y, Ishizawa N, Arima T (2013) Structural phase transition and magnetic-field effect on the modulated structure in $GdBaCo_2O_{5+\delta}$ ($\delta < 0.5$). Phys Rev Lett, in press
11. Ishizawa N, Asaka T, Kudo T, Fukuda K, Abe N, Arima T (2013) Pseudo-commensurate $GdBaCo_2O_{5+\delta}$. J Solid State Chem 198:532–541. doi:10.1016/j.jssc.2012.11.004

12. Stokes HT, Campbell BJ, van Smaalen S (2011) Generation of $(3 + d)$-dimensional superspace groups for describing the symmetry of modulated crystalline structures. Acta Crystallogr, Sect A, Found Crystallogr 67(1):45–55. doi:10.1107/S0108767310042297
13. Petricek V, Dusek M, Palatinus L (2006) Jana2006, structure determination software programs. Institute of Physics, Praha
14. Smaalen Sv (2007) Incommensurate crystallography. Oxford University, Oxford
15. Brown ID (2002) The chemical bond in inorganic chemistry: the bond valence model. Oxford University Press, Oxford

Chapter 22
Al_4(Cr,Fe): A Structure Survey

B. Bauer, B. Pedersen, and F. Frey

Abstract A single crystal of Al_4(Cr,Fe) grown by the Czochralski method has been investigated using X-ray and also neutron diffraction. The average structure of this crystal with a composition of $Al_{79.1}Cr_{17.8}Fe_{3.1}$ was found to be body-centered orthorhombic with the space group *Immm*. Using neutrons it was possible to distinguish the Fe and Cr positions within the structure. However, the diffraction patterns of both, X-ray as well as neutrons, showed additional reflections beyond the Bragg reflections violating the body-centering and reducing the space group symmetry to *Pmm*2. A renewed structure analysis taking also these additional reflections into account exhibits significant changes of about 30 % of the atomic positions. Features related to extra diffraction phenomena beyond the Bragg reflections are discussed in some detail.

22.1 Introduction

Al_4(Cr,Fe) is considered as an approximant of the decagonal quasicrystal in the appropriate system. Several different structure models related to this composition have been published during the last decades. In the following, a short summary of the different models will be given. In an alloy of nominal composition $Al_{12}Fe_2Cr$, Sui et al. [1] identified an orthorhombic phase with lattice constants $a = 1.234$ nm, $b = 1.241$ nm and $c = 3.071$ nm. Using high-resolution electron microscopy, a structure model was derived and the space group *Imm*2 was determined (Pearson symbol: *oI*316-10.85). Additionally, on ordering a *C*-centered monoclinic superstructure occurred ($a_M = c/\sin\beta = 3.31$ nm, $b_M = a = 1.23$ nm, $c_M = 2b = 2.48$ nm and $\beta = 112°$), leading to superlattice reflections in the elec-

B. Bauer (✉) · F. Frey
Department of Earth and Environmental Sciences, Crystallography Section,
Ludwig-Maximilians-Universität München, Theresienstr. 41, 80333 Munich, Germany
e-mail: birgitta.bauer@lrz.uni-muenchen.de

B. Pedersen
Forschungs-Neutronenquelle Heinz Maier-Leibnitz (FRM II), TU München, Lichtenbergstr. 1,
85747 Garching, Germany

S. Schmid et al. (eds.), *Aperiodic Crystals*, DOI 10.1007/978-94-007-6431-6_22,

tron diffraction patterns [2]. Deng et al. [3] investigated a crystal with composition $Al_{80.6}Cr_{10.7}Fe_{8.7}$ using single crystal X-ray diffraction. The space group *Immm* rather than *Imm*2 was obtained with the lattice constants $a = 1.2500(6)$ nm, $b = 1.2617(2)$ nm and $c = 3.0651(8)$ nm (Pearson symbol: *oI*366-59.56). In this structure model, there are 28 Al and 11 TM (TM = Cr/Fe) sites in a unit cell, where two Al sites are statistically occupied by Al and TM. Positions of Cr and Fe atoms are not distinguished [3]. No superlattice reflections are mentioned. On the basis of powder X-ray diffraction data and electron diffraction patterns, a recent publication by Pavlyuchkov et al. [4] suggested that Al_4(Cr,Fe) is actually the ternary extension of the binary phase η-$Al_{11}Cr_2$. This phase was reported to have a body-centered orthorhombic structure with $a = 1.24$ nm, $b = 1.26$ nm and $c = 3.05$ nm based on single crystal X-ray diffraction [5]. Though further investigations using transmission electron microscopy revealed additional information, and finally, a monoclinic C-centered structure with $a = 1.76$ nm, $b = 3.05$ nm, $c = 1.76$ nm and $\beta \approx 90°$ was deduced [5]. From a study of anisotropic transport measurements of $Al_{80}Cr_{15}Fe_5$, a relaxed version of the structural model by Deng et al. [3] is concluded where atomic sites of mixed and partial occupancies are discarded [6]. A preliminary study using X-ray and neutron diffraction on a single crystal of Al_4(Cr,Fe) grown via the Czochralski method supports the body-centered orthorhombic model [7, 8]. But the diffraction patterns of X-ray as well as neutron data showed additional reflections which were mentioned in a former publication [7] but were not included in the analysis up to now. Two of the previously mentioned publications also report the occurrence of additional reflections, which are in both cases referred to monoclinic superstructures [2, 5]. The authors of [5] relate their findings to a layered two-domain structure with a twin relation between the domains. The aims of this work are to give a renewed structure analysis using extended X-ray data sets including all reflections and also neutron data towards:

(a) The space group of the average structure and the distinction of Cr and Fe sites in comparison with the structure model published by Deng et al. [3],
(b) The true structure (superstructure), i.e., including all reflections, and
(c) A qualitative discussion of the Bragg and diffuse scattering phenomena.

22.2 Experimental

The investigated single crystal was grown from an off-stoichiometric melt using the Czochralski method. Further details have been given elsewhere [9]. According to previous studies investigating the 1000 °C isothermal section of the Al–Cr–Fe phase diagram, it could be shown that the existence region of Al_4(Cr,Fe) decomposes into four regions of structurally different phases depending on the Cr/Fe ratio [9]. The composition of the analyzed crystal was measured to be $Al_{79.1}Cr_{17.8}Fe_{3.1}$ using electron probe microanalysis. For the experimental details of the neutron diffraction work such as the used wavelength $\lambda = 0.15391$ nm (without $\lambda/2$-contamination), we refer to [7]. The main X-ray work is well documented in [8]. In order to inspect the diffraction patterns in more detail up to high Q-values, supplementary X-ray diffraction was carried out. The measurement was performed at room temperature using Mo-K_α radiation ($\lambda = 0.071073$ nm, rotating anode, graphite monochromator) and a MAR345 image plate detector. All the neutron data, Bragg reflections as well as additional scattering phenomena, were recorded at the instrument RESI of the FRM II reactor facility; cf. [7]. For a rough estimate of the relative contributions of the different atomic species to the diffraction intensities, the ratio of the respective X-ray scattering power (Z^2 with $Z =$ number of electrons at $\theta = 0°$), i.e., neglecting the form factor influence, is compared to the neutron scattering cross-sections (assuming a natural isotope abundance of the species) given in barns: XRD: Al: $0.79 \cdot 13^2 \approx 133$; Cr: $0.18 \cdot 24^2 \approx 103.7$; Fe: $0.03 \cdot 26^2 \approx 20$; i.e., **Al:Cr:Fe** $\approx$ 6:5:1; ND: Al: $0.79 \cdot 1.5 \approx 1.2$; Cr: $0.18 \cdot 1.66 \approx 0.3$; Fe: $0.03 \cdot 11.3 \approx 0.3$; i.e., **Al:Cr:Fe** $\approx$ 4:1:1. Generally, the overall relative influence of the Cr and Fe species in the diffraction patterns should therefore be better contrasted in X-ray diffraction (XRD) than in neutron diffraction (ND). On the other hand, the heavy metals have a similar weight in ND. These statements relate both to Bragg-scattering and "extra"-diffraction phenomena.

22.3 Observations

Apart from the Bragg reflections, extra reflections and diffuse phenomena are clearly observable. Figures 22.1(a)–(b) and 22.2(b) show different X-ray diffraction patterns, i.e., the $(0\ k\ l)$, $(h\ 1.5\ l)$ and the $(h\ 3.5\ l)$ planes, respectively. Note that the planes where $k = n/2$ ($n =$ odd) are away from any regular Bragg positions. Figures 22.2(a)–(b) show sections of neutron diffraction patterns again of the $(0\ k\ l)$ and of the $(h\ 3.5\ l)$ planes, respectively. The main observations can be summarized as follows:

(a) Additional well-defined strong and weak but somewhat diffuse "extra"-reflections are observable at Bragg-off positions $(h\ k'\ l)$ half-way between $k =$ integer positions which are denoted by k'. These k'-reflections $(h\ n/2\ l)$ ($n =$ integer, odd) are arranged along rows parallel to c^* (parallel to l). They are somewhat broader along the l-direction as compared to the main Bragg reflections $(h\ k\ l)$. No additional k'-reflections exist in the planes $(h\ k\ 0)$, $h =$ even or odd. Apart from different Bragg intensities, there is no general difference between X-ray and neutron patterns. The presence of these extra reflections clearly violates the extinction rule for a body-centered lattice.

(b) The strong reflections $(h\ k'\ l)$ are observed at the same positions in the X-ray and the neutron patterns. Moreover, $(h\ k'\ l)$ reflections are strong if neighboring regular $(h\ k\ l)$ Bragg reflections are also strong.

(c) The relative intensities of the k'-reflections increase with increasing Q-vectors (distance from the origin) which could be easily recognized comparing the spots in the interlayer sections: $(h\ n/2\ l)$, $n = +/-3, 5, 7, \ldots$ (XRD or ND). There are no or only very weak k'-reflections close to the reciprocal origin, i.e., where the Q-values are small.

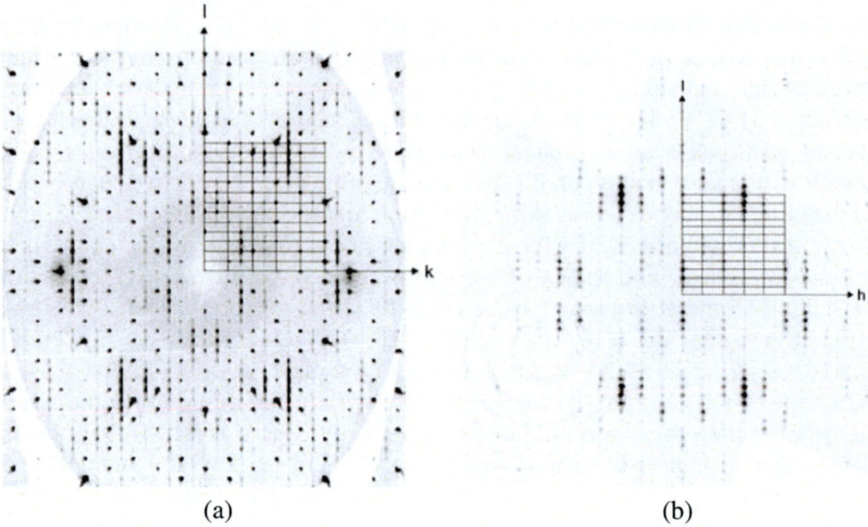

Fig. 22.1 (**a**) $0kl$-layer from X-ray diffraction data. (**b**) $h1.5l$-layer from X-ray diffraction data

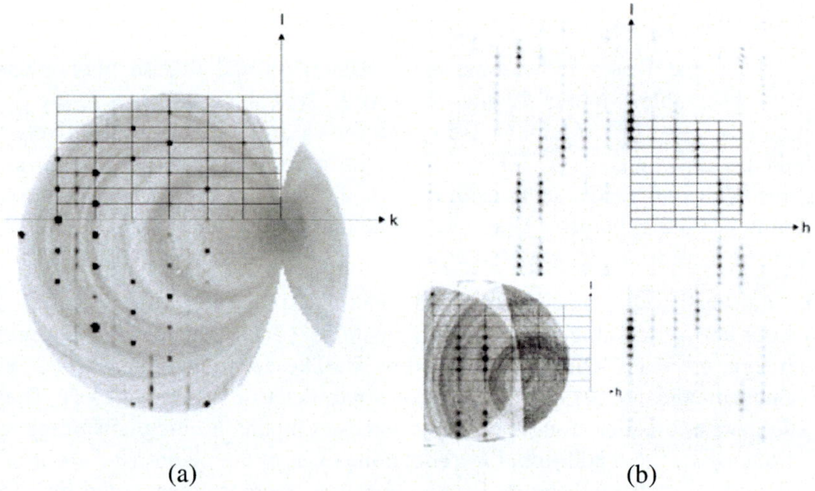

Fig. 22.2 (**a**) $0kl$-layer from neutron diffraction data. (**b**) $h3.5l$-layer from X-ray diffraction data; in the left hand corner, an extract of the same layer from neutron data is shown

(d) Beneath the rows of k'-reflections there are diffuse rods ("streaks") along c^* which are not due to a superposition of diffuse intensity parts of "smeared out" k'-reflections (cf., for example, Figs. 22.1(b) or 22.2(b)). This indicates a different origin of these "continuous" l-rods. The widths measured across these rods do not increase with increasing l-index. The intensity ratio between these diffuse rods and the superimposed rows of k'-reflections becomes much smaller with higher Q, enhancing the tentative interpretation of a different structural phenomenon.

From these observations it becomes clear that the k'-type reflections indicate a superstructure—at least in the particular Al–Cr–Fe sample with this chemical composition—with a doubled lattice constant along the b-direction (cf. Sect. 22.5 for a more detailed discussion). It should be pointed out that this superstructure may be changed or even absent dependent on only small changes of stoichiometry.

22.4 Analysis of Bragg Data

22.4.1 Average Structure from Neutron Data (k'-Reflections Omitted)

The refinement of the average structure using neutron diffraction data was done using the program JANA 2006 [10]. The orthorhombic body-centered structure model published by Deng et al. [3] was taken as a starting model. Out of 8303 measured reflections 2592 symmetrically independent reflections were averaged. The final refinement led to $R_1 = 0.07$ for 2578 reflections with $I > 3\sigma(I)$ in the space group *Immm* with $a = 1.2498(3)$ nm, $b = 1.2550(3)$ nm and $c = 3.0512(2)$ nm. According to Deng et al. [3], there exist 11 TM (TM = Cr/Fe) sites in the structure. The refinement of the neutron data showed that the sites TM 1, 4, 6, 9 and 11 are only occupied by Cr while the other sites are mixed sites occupied by both Cr and Fe atoms. The mixed position Al/TM5 is occupied by Al as well as Fe while a mixed occupation for Al15 could not be confirmed. The occupation of the Al sites Al19–28 is less than 100 %.

22.4.2 Superstructure from X-ray Data (k' Reflections Treated as Bragg Peaks)

Refining the X-ray data set that was measured up to high Q-values and including also the superstructure reflections, the structure model could be improved even more. However, in this refinement no distinction between Cr and Fe atoms is made. Including also the superstructure reflections in the refinement leads to a doubling of the b-axis and also to a reduction of the high symmetry space group *Immm* to *Pmm*2. Out of 79763 measured reflections, 30188 symmetrically independent reflections were averaged. The final refinement led to $R_1 = 0.15$ for 16252 reflections with $I > 3\sigma(I)$ with $a = 1.2480(1)$ nm, $b = 2.5029(1)$ nm and $c = 3.0581(1)$ nm. The significant larger R-value is due to the worse quality of the integrated intensities of the weaker reflections which are superimposed on the diffuse background due to the diffuse rods. Figure 22.3 shows a comparison of projections of the structure along c for the neutron (average structure,

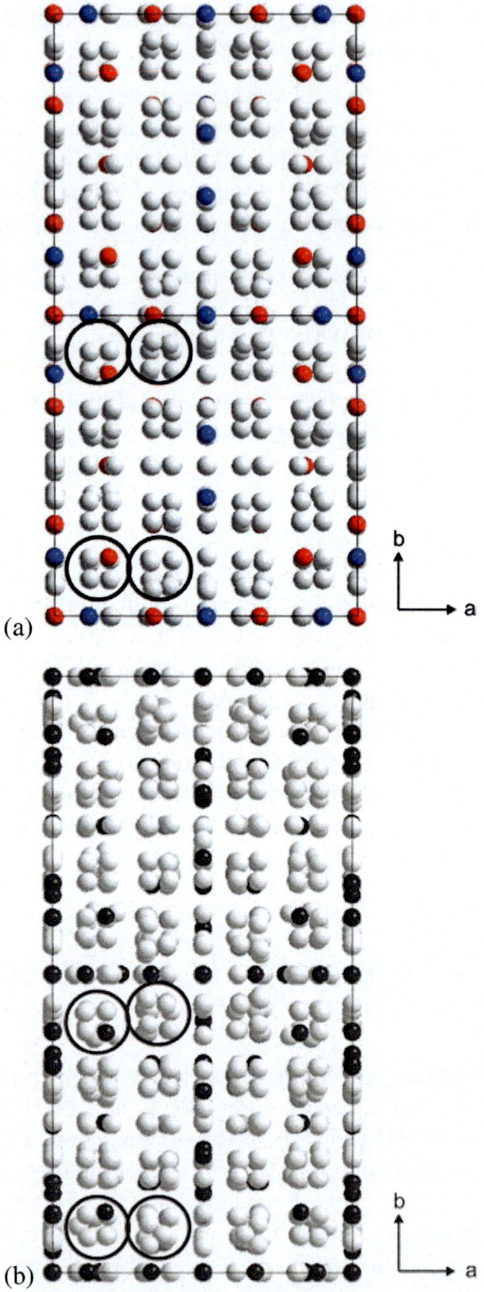

Fig. 22.3 (**a**) Projection of the average structure along *c* (two unit cells in *b*-direction) (*white*: Al, *red*: Fe/Cr , *blue*: Cr). (**b**) Projection of the real structure (i.e., including the 2*b*-superordering) along *c* (*white*: Al, *black*: TM (TM = Cr/Fe))

Fig. 22.3(a)) and the X-ray (superstructure, Fig. 22.3(b)) data. For example, four atomic groups which violate the body-centering lead to the doubling of the *b*-

axis, and finally, to the occurrence of the superstructure reflections which are encircled.

22.5 Discussion and Conclusions

A superstructure exhibiting the characteristics indicated in Sect. 22.3 may be described by a modulation wave along the b-direction which doubles the b-period. The relative intensity increase of k'-reflections with higher Q-values (and the absence of k'-reflections close to O^*) indicates a dominant displacive character of atoms from mean positions. Moreover, the general absence of this type of superstructure reflections in the $(h\ k\ 0)$ plane and their intensity increase with increasing Q are indicative of a transversal displacive modulation and are not due to a chemical modulation. The displacements ("polarization vectors" of the modulation) are basically perpendicular to the c-direction (in real space), and are restricted to the a–b-plane. From the (weak) broadening of the k'-reflections along c^*, one may conclude some limited correlation length, which may be, however, not adequately described by "short-range" correlations. From the arguments given above, we may separate the origin of the "continuous" diffuse rods from the discussion of the superstructure itself: Lamellar-type domain-slabs extended perpendicular to c, i.e., parallel to the a–b-plane, are formed within which the "$(a, 2b, c)$"-superstructure exists. These domains are relatively large in the a–b-plane as can be deduced from an only marginal increase of the widths of k'-reflections compared to those of the basic Bragg reflections. Domain walls might be due to a (irregular) mixing of these domains with regular "(a, b, c)"-domains. The average structure remains unaffected by these walls, long-range correlations still exist, and the widths of the "basic" Bragg reflections remain sharp. If the lateral distribution of walls between these two types of (a, b, c) and $(a, 2b, c)$ domains is irregular, there are no correlations (with respect to positions of the walls) between slabs separated in the c-direction. Continuous diffuse rods correspond to such a lamellar-domain structure. Therefore, a conventional twin-domain structure seems to be not realized. Moreover, the superstructure reflections could not be related to a formation of twins. It would be, however, worthwhile to check this model by TEM investigations.

Acknowledgement The authors would like to thank Dr. Sandra Altmannshofer for providing the X-ray data set.

References

1. Sui HX, Liao XZ, Kuo KH, Zou X, Hovmöller S (1997) Structural model of the orthorhombic non-Fibonacci approximant in the $Al_{12}Fe_2Cr$ alloy. Acta Crystallogr, Sect B 53:587–595
2. Sui HX, Liao XZ, Kuo KH (1995) A non-Fibonacci type of orthorhombic decagonal approximant. Philos Mag Lett 71:139–145

3. Deng DW, Mo ZM, Kuo KH (2004) Crystal structure of the orthorhombic Al_4(Cr, Fe) approximant of the Al–Cr–Fe decagonal quasicrystal. J Phys Condens Matter 16:2283–2296
4. Pavlyuchkov D, Bauer B, Kowalski W, Surowiec M, Grushko B (2010) On the constitution of Al_4(Cr, Fe). Intermetallics 18:22–26
5. Bendersky LA, Roth RS, Ramon JT, Shechtman D (1991) Crystallographic characterization of some intermetallic compounds in the Al–Cr system. Metall Trans A 22:5–10
6. Dolinšek J, Jeglič P, Komelj M, Vrtnik S, Smontara A, Smiljanić I, Bilušić A, Ivkov J, Stanić D, Zijlstra ES, Bauer B, Gille P (2007) Origin of anisotropic nonmetallic transport in the $Al_{80}Cr_{15}Fe_5$ decagonal approximant. Phys Rev B 76:174207
7. Bauer B, Pedersen B, Gille P (2009) Al_4(Cr, Fe): single crystal growth by the Czochralski method and structural investigation with neutrons at FRM II. Z Kristallogr 224:109–111
8. Bauer B (2010) Einkristallzüchtung und Charakterisierung Al-reicher komplexer metallischer Phasen in Al–Cr–Fe und benachbarten Systemen. Dissertation, LMU Munich
9. Bauer B, Gille P (2011) Crystal growth of Al-rich complex metallic phases in the system Al–Cr–Fe using the Czochralski method. Z Anorg Allg Chem 637:2052–2058
10. Petříček V, Dušek M, Palatinus L (2006) Jana2006. The crystallographic computing system. Institute of Physics, Praha

Chapter 23
Phase Transitions in Aperiodic Composite Crystals

P. Rabiller, B. Toudic, C. Mariette, L. Guérin, C. Ecolivet, and M.D. Hollingsworth

Abstract Aperiodic alkane/urea inclusion compounds (UIC) are prototype composites which exhibit complex sequences of phases that can clearly be described in the $(3 + d)$ dimension crystallographic superspace. By simply changing the length of the guest alkane molecules (C_nH_{2n+2}) which pile up in the channels of the host urea honeycomb-like framework, it is, for instance, possible to have phase-ordering phase transition from 3 to $(3 + 1)$ dimension in the case of n-heptane/urea ($n = 7$), or as in the case of n-hexadecane/urea ($n = 16$) or n-nonadecane/urea ($n = 19$), a generalization to higher dimensions of the phase transitions found in modulated structures. Such results are successfully obtained with the help of high resolution diffraction methods.

23.1 Introduction

Certain small molecules, such as urea, thiourea, and perhydrotriphenylene, can be co-crystallized with long-chain hydrocarbon molecules to form inclusion compounds. The guest chains are confined into narrow, approximately cylindrical channels created by the host small-molecules lattice. The stoichiometry and the conformation of the chains included inside the channels are function of internal interactions such as intra-chain interaction, but also of overall cooperative properties of the resulting three-dimensionally ordered single crystal [1]. These intergrowth compounds may form incommensurate composite crystals where the two subsystems present at least one incommensurate ratio of their respective sub-lattice parameters. Their diffraction pattern are characterized by a $(3 + d)$-dimensional module [2, 3] where the so-called "main reflections" are found in the reciprocal space at locations corresponding to the two separate sub-lattices (one of them being called

P. Rabiller (✉) · B. Toudic · C. Mariette · L. Guérin · C. Ecolivet
Institut de Physique de Rennes, UMR UR1-CNRS 6251, Université de Rennes 1, 35042 Rennes, France
e-mail: philippe.rabiller@univ-rennes1.fr

M.D. Hollingsworth
Department of Chemistry, Kansas State University, Manhattan, KS 66506-0401, USA

S. Schmid et al. (eds.), *Aperiodic Crystals*, DOI 10.1007/978-94-007-6431-6_23,

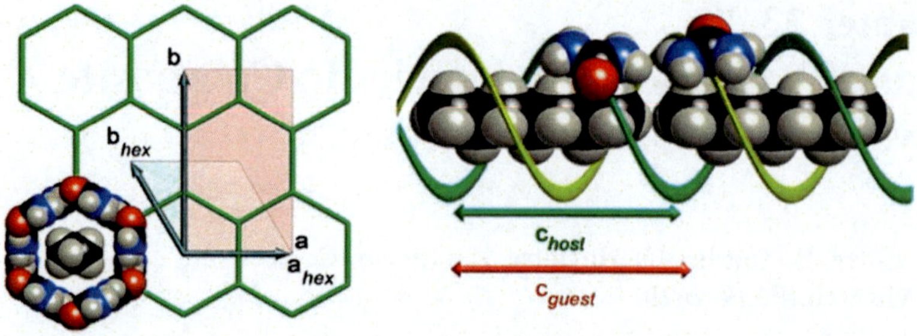

Fig. 23.1 *n*-alkane/urea inclusion compound (*n*-alkane UIC). Urea molecules are connected through hydrogen bonds network forming a honeycomb like hexagonal structure with *n*-alkane (C_nH_{2n+2}) chains filling the channels. In most alkane length cases (*n* ranging from 7 to more than 35), the ratio of the "host" and "guest" periodicities along the channel is incommensurate

"host" and the other one "guest") whereas additional satellite reflections arise due to the subsystem interactions. For the uniaxial composites with a single incommensurate direction (**c**-axis), the diffraction spots indexation reads:

$$\mathbf{Q}_{hklm} = h\mathbf{a}^* + k\mathbf{b}^* + l\mathbf{c}_h^* + m\mathbf{c}_g^*.$$

Bragg peaks may be separated into four classes. Peaks from the commensurate ($\mathbf{a}^*, \mathbf{b}^*$) plane are indexed ($h\ k\ 0\ 0$) and called common Bragg peaks; host peaks are indexed ($h\ k\ l\ 0$); guest peaks are indexed ($h\ k\ 0\ m$); and finally, satellite peaks, which characterize the intermodulation, are indexed as ($h\ k\ l\ m$) with l and $m \neq 0$.

A prototype example of such uniaxial intergrowth crystals is the family of *n*-alkane/urea inclusion compounds (UICs). In these supra-molecular systems, urea molecules ($CO(NH_2)_2$) are connected by H-bonds to form helical ribbons, which repeat every six urea molecules to form a series of linear, hexagonal tunnels that can accommodate linear alkanes (C_nH_{2n+2}) (Fig. 23.1, left).

Most of these compounds exhibit incommensurate composite feature since the guest alkane periodicity c_g along the channel does not fit with the helical parameter c_h of the host urea matrix which turns to be independent of the guest molecule (Fig. 23.1, right). An evidence of this intermodulation is given on the high resolution neutron scattering scan shown in Fig. 23.2 [4].

Because the channels (~0.53 nm) are larger than the hydrocarbon chains, guests are held loosely and typically undergo substantial motions. At low temperatures, these materials undergo continuous or weakly first order structural phase transitions. The literature first reported a single phase transition from a hexagonal phase to a sheared orthorhombic one combined with a preferred orientation of the guest molecules and a distortion of the host sub-lattice [5, 6]. As part of an effort to understand the extra degrees of freedom that aperiodic materials utilize when undergoing phase transitions, we have undertaken a systematic study of *n*-alkane UICs with alkane carbon atoms' number ranging from 7 to 28. We have found quite different results with at least nine different phases and eleven types of phase transitions, some of which referring only to degrees of freedom associated to the internal part of the superspace. Almost all high symmetry and low symmetry phases are aperiodic

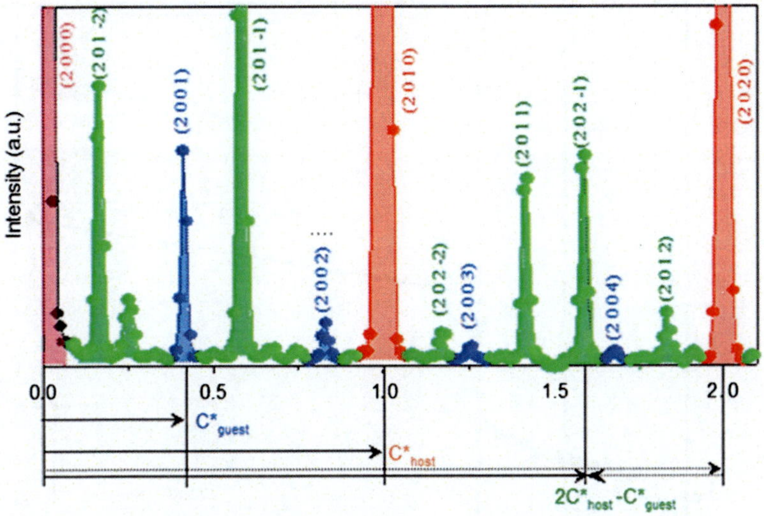

Fig. 23.2 Illustration of the modulated character of *n*-nonadecane UIC along a $(h\ 0\ l\ m)$ scan measured on a 3-axes cold neutron spectrometer [4]. $(h\ k\ 0\ 0)$ reflections (*pink*) are common two the host and guest sub-lattices, $(h\ k\ l\ 0)$ reflections in *red* and $(h\ k\ 0\ m)$ in *blue* respectively correspond to the host and guest lattices. $(h\ k\ l\ m)$ reflections in *green* are inter-modulation satellites

Table 23.1 Sequences of phases at ambient pressure for *n*-heptane, *n*-hexadecane, and *n*-nonadecane UICs with associated superspace groups. The dimension of the superspace is given in parentheses in the upper left corner of each cell

	n-nonadecane	*n*-nonadecane 0.5 GPa	*n*-hexadecane	*n*-heptane
c_h/c_g	0.418	0.428	0.486	0.981
Phase I	Hexagonal (3 + 1) $P6_122(00\gamma)$	Hexagonal (3 + 1) $P6_122(00\gamma)$	Hexagonal (3 + 1) $P6_122(00\gamma)$	Hexagonal (3 + 0) $P6_122$
$T_{c1} \sim 150$ K				
Phase II	Orthorhombic (3 + 2) $C222_1(00\gamma)(10\delta)$	Orthorhombic (3 + 2) $C222_1(00\gamma)(10\delta)$	Orthorhombic (3 + 1) $P2_12_12_1(00\gamma)$	Monoclinic (3 + 0) $P2_111$
$T_{c2} \sim 130$ K				
Phase III	Orthorhombic (3 + 2) $P2_12_12_1(00\gamma)(00\delta)$	– –	Orthorhombic (3 + 1) $P2_12_12_1(00\gamma)(00\delta)$	Monoclinic (3 + 1) $P2_111(0\beta\gamma)$

and these complex sequences of phases can be discussed in term of group/subgroup relationship within crystallographic superspace formalism.

We report here on the phase transitions exhibited by three *n*-alkane UICs, namely *n*-heptane [7], *n*-hexadecane [8, 9], and *n*-nonadecane [10, 11] UICs. Although the

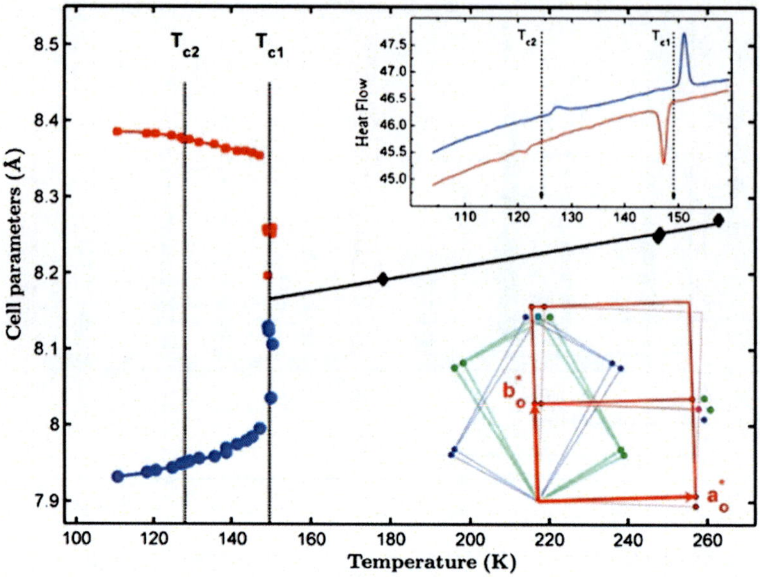

Fig. 23.3 Temperature evolution of the cell parameters in the scattering diffraction plane (a, b), as obtained by cold neutron diffraction analysis of the splitting of the Bragg peak (2 0 0 0) in one of the six induced domains. In the orthorhombic phase, the value of the b_o parameter (*blue*) is divided by $\sqrt{3}$ for extrapolation to the high symmetry phase degenerated value. Upper right inset gives DSC scans performed at increasing (in *blue*) and decreasing (in *red*) temperature, revealing two structural instabilities

first two have a $\gamma = c_h/c_g$ ratio very close to simple low index commensurate ratios ($0.981 \sim 1:1$ and $0.486 \sim 1:2$) no *lock-in* is observed at ambient pressure. Their sequences of phases are given in Table 23.1 with explicit superspace groups and all present two different phase transitions the first one around 150 K and the second one at about 130 K.

n-hexadecane UIC is one of those presenting a phase transition from high temperature hexagonal symmetry to low temperature orthorhombic symmetry. A significant shearing occurs as revealed by the changes of a and b parameters and the existence of domains (Fig. 23.3) below $T_{c_1} = 149$ K. It nevertheless differs from what was reported since a second phase transition occurs at $T_{c_2} = 128$ K with small anomaly of specific heat and cell parameters. The intermediate state is characterized by an increase of the superspace dimension from $(3 + 1)$ to $(3 + 2)$. A *lock-in plateau* towards a 3-dimensional structure with $\gamma = 1/2$ is observed when applying pressure of a few tenths of GPa and which evolution from 60 K to room temperature gives an estimation of the intermodulation energy of about 1 kcal mol^{-1} [9]. A lowering of symmetry is expected in this pressure induced 3-dimensional structure to accommodate the double helix construction of urea framework together with ordering of alkane molecules

The PT phase diagram is given in Fig. 23.4 for n-nonadecane/urea which exhibits an original phase transition. There, in phases II and IV, the symmetry breaking only

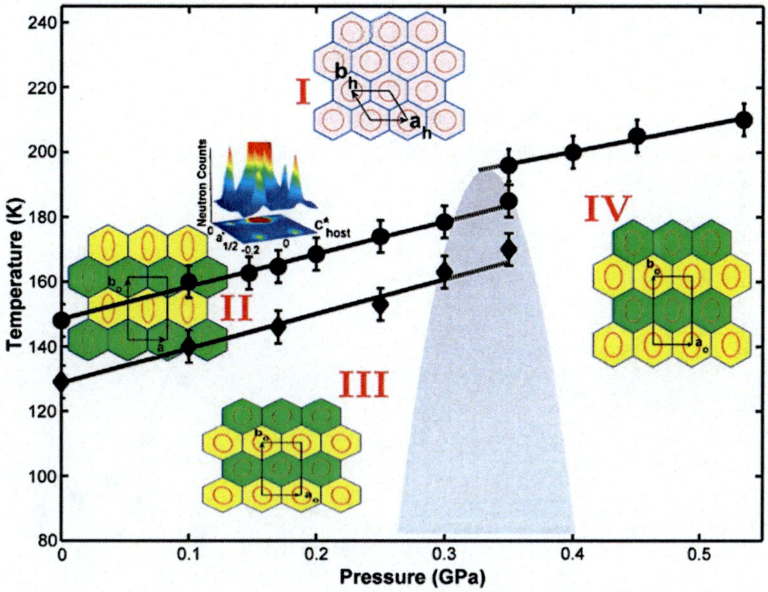

Fig. 23.4 Phase diagram (P, T) of the fully deuterated n-nonadecane/urea, as determined by neutron diffraction (4F1, LLB Saclay). All phases (I, II, III, IV) require a description within a crystallographic superspace: hexagonal for the phase I, orthorhombic for the phases II, III and IV. The *dashed region* indicates a metastable region, between phases II and III on one hand and phase IV on the other hand. The inset 3D plot of reciprocal section in phase II along the superstructure line $h = 1$, $k = 2$ (D10, ILL Grenoble) shows the absence of 3D antiferro-distorsive cell doubling signature (the latter case only occurring in phase III)

affects the internal variable of the superspace, as evidenced by the absence of Bragg spots at position $l = m = 0$ (Fig. 23.4, 3D inset) on superstructure lines ($h = 2n + 1$ in the low symmetry orthorhombic notation).

The phase transition from the phases I to II and II to III are found to be continuous or weakly discontinuous therefore allowing superspace group/subgroup symmetry breaking assertion from the hexagonal group $P6_122(00\gamma)$ to the C-centered orthorhombic maximum subgroup $C222_1(00\gamma, 10\delta)$ and then to the primitive orthorhombic maximum subgroup $P2_12_12_1(00\gamma, 00\delta)$. At pressure above 0.3 GPa, no distortion from the hexagonal metrics can be depicted through the single phase transition which thus leads to a C-centered orthorhombic group $C222_1(00\gamma, 10\delta)$ in phase IV. The value of the parameter δ characterizing the fifth dimension ranges from 0.09 to 0.12 when going from atmospheric pressure to 0.4 GPa which can be interpreted in terms of alkane sub-lattice elastic potential [11].

Last example is that of n-heptane UIC with a $\gamma = 0.981$ very close to one. The high temperature phase is not very stable and characterized by the absence of guest Bragg spots and intermodulation satellites. The alkane subsystem diffraction pattern consists only in sharp diffuse planes which width increases monotonically with their index. This is the signature of a one dimensional liquid-like system, confined, in the present case, inside a 3D host structure. The first transition is observed at

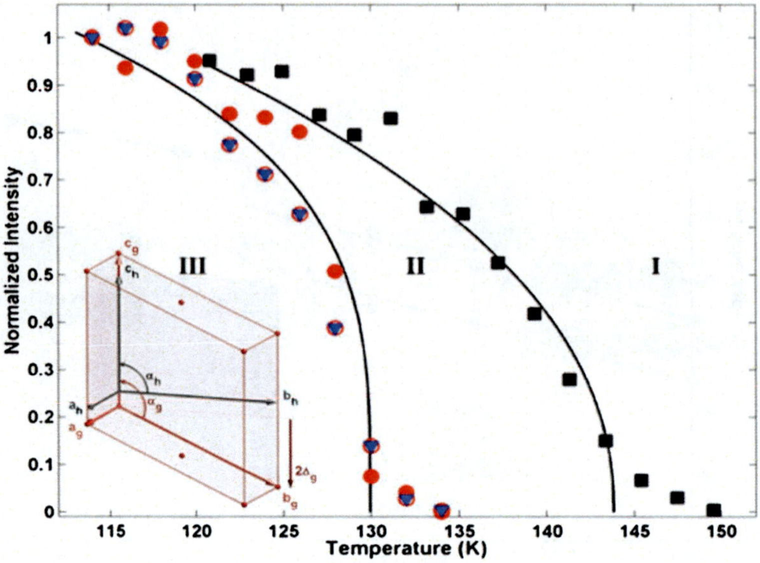

Fig. 23.5 Temperature evolution in n-heptane/urea of: (■) host Bragg peak $(1, 2, 2)$ appearing in phase II defining the first transition at $T_{c1} = 145$ K, (●) guest Bragg peaks $l = 0$ and $m = 1$, (○) guest Bragg peaks $l = 0$ and $m = 2$ and (▼) intermodulation satellites $l = 1$ and $m = 1$. The *inset* in phase III gives a schematic representation of the mean host and guest monoclinic cells

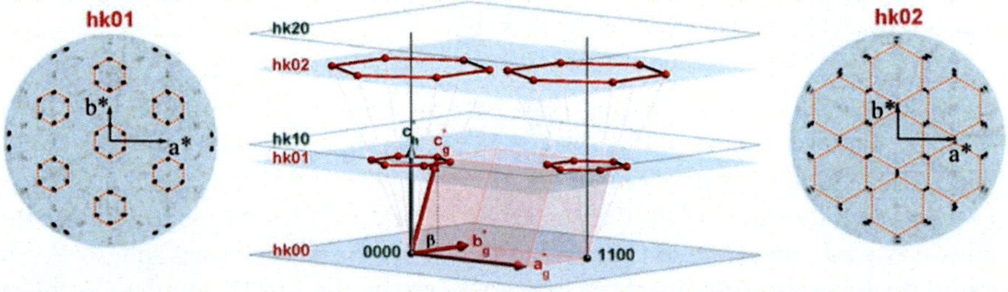

Fig. 23.6 Reconstruction of the $(h,\ k,\ 0,\ 1)$ and $(h,\ k,\ 0,\ 2)$ layers at 90 K on the *left* and *right*, respectively. *Hexagons* are guides for the eyes and refer to six equivalent domains ($2\pi/6$ rotation about $\mathbf{c}^*$). The corresponding 3D reciprocal pattern of these six domains is schematized in the center of the figure. Non-co-linearity of $c * h$ and $c * g$ is a signature of the incommensurate offset of the alkane chains from one channel to the adjacent one as illustrated in Fig. 23.5

$T_{c_1} = 145$ K (Fig. 23.5), far above the one that could be extrapolated from that of usual long alkane chains UICs. It corresponds to a distortion of the hexagonal cell towards monoclinic symmetry. Ordering of the alkane chains only takes place at the second phase transition at $T_{c_2} = 130$ K. This transition leads to a modulated monoclinic composite with a modulation vector having a nonzero component $\beta \lesssim 0.5$ along $\mathbf{b}^*$. This corresponds to a temperature dependant incommensurate offset $\Delta_g \sim 0.5c_g$ of the alkane chains from one channel to the adjacent one along **b** direction as illustrated in the inset of Fig. 23.6 [7].

The study of phase transitions in aperiodic composites is a complex and challenging field. The host-guest n-alkane/urea inclusion compounds constitute a prototype family where the single change of the number of carbon atoms in the alkane guest molecules totally modifies the sequences of phases in 3, 4 and 5 dimensional crystallographic superspaces.

References

1. Hollingsworth MD, Harris KDM (1996) In: MacNicol DD, Toda F, Bishop R (eds) Comprehensive supramolecular chemistry. Elsevier Science, Oxford, p 177
2. Janssen T, Chapuis G, de Boissieu M (2007) Aperiodic crystals. From modulated phases to quasicrystals. Oxford University Press, Oxford
3. van Smaalen S (2010) Incommensurate crystallography. Oxford University Press, Oxford
4. Lefort R, Etrillard J, Toudic B, Guillaume F, Breczewski T, Bourges P (1996) Incommensurate intermodulation of an organic intergrowth compound, observed by neutron scattering. Phys Rev Lett 77:4027–4030
5. Fukao K (1990) J Chem Phys 92:6867
6. Lynden-Bell RM (1993) Mol Phys 79:313
7. Mariette C, Huard M, Rabiller P, Nichols SM, Ecolivet C, Janssen T, Alquist KE, Hollingsworth MD, Toudic B (2012) A molecular "phase ordering" phase transition leading to a modulated aperiodic composite in n-heptane/urea. J Chem Phys 136:104507
8. Huard M, Toudic B, Rabiller P, Ecolivet C, Guérin L, Bourges P, Breczewski T, Hollingsworth MD (2011) Confined linear molecules inside an aperiodic supramolecular crystal: the sequence of superspace phases in n-hexadecane/urea. J Chem Phys 135:204505
9. Toudic B, Aubert F, Ecolivet C, Bourges P, Breczewski T (2006) Pressure-induced lock-in in an aperiodic nanoporous crystal. Phys Rev Lett 96:145503
10. Toudic B, Garcia P, Odin C, Rabiller P, Ecolivet C, Collet E, Bourges P, McIntyre GJ, Hollingsworth MD, Breczewski T (2008) Hidden degrees of freedom in aperiodic materials Science 319:69–71
11. Toudic B, Rabiller P, Bourgeois L, Huard M, Ecolivet C, McIntyre GJ, Bourges P, Breczewski T, Janssen T (2011) Temperature-pressure phase diagram of an aperiodic host guest compound. Europhys Lett 93:16003

Chapter 24
Pseudo-Symmetry in Tungsten Bronze Type $Sr_3TiNb_4O_{15}$

Thomas A. Whittle, William R. Brant, and Siegbert Schmid

Abstract The structure of $Sr_3TiNb_4O_{15}$ has been re-investigated using synchrotron X-ray powder diffraction data. Rietveld refinements of a structural model against these data were performed and confirmed a new unit cell and space group symmetry. $Sr_3TiNb_4O_{15}$ was found to possess $Pna2_1$ symmetry with a unit cell $a =$ 12.36081(2) Å, $b = 12.40288(2)$ Å, $c = 7.751270(10)$ Å.

24.1 Introduction

Compounds with tungsten bronze type structures have been shown to display technologically important properties. The properties common to these materials make them useful for applications in non-volatile memory and data storage, oscillators and transducers, thermistors and infrared radiation detectors as well as in non-linear optics and phase conjugated mirrors [1–8]. The determination of symmetry is essential to explain and predict properties for these materials, e.g. ferroelectric materials must lack an inversion centre and possess a unique, reversible, polar axis for the properties to arise. The point group symmetry of these compounds establishes which properties may be present [9, 10].

$Sr_3TiNb_4O_{15}$ is a compound which forms with a tungsten bronze type structure. $Sr_3TiNb_4O_{15}$ has been reported in several publications, yet questions remain concerning its unit cell dimensions and details of the symmetry it adopts (Table 24.1).

Most of the proposed structural models are very similar and result in a reasonable fit to the data. However, the published models were all tested by refinements against conventional X-ray powder diffraction data, which do not have the high resolution or intensity necessary to detect subtle symmetry changes or weak superstructure reflections. Nonetheless, Rao et al. [11] published a model with a doubled c-axis. However, the only reflection listed in support of this doubling was the (521) reflection, with an intensity that was 73 % of the strongest reflection. This is unusual for the only superstructure reflection observed and, in addition, this reflection is extinct

T.A. Whittle · W.R. Brant · S. Schmid (✉)
School of Chemistry, The University of Sydney, Sydney NSW 2006, Australia
e-mail: siegbert.schmid@sydney.edu.au

S. Schmid et al. (eds.), *Aperiodic Crystals*, DOI 10.1007/978-94-007-6431-6_24,

Table 24.1 Unit cell parameters and space groups reported for $Sr_3TiNb_4O_{15}$

Year	Space group	Cell parameters		
		a	*b*	*c*
1970 [12]	*P4bm*	12.48		3.99
1970 [13]	*P4bm*	12.36		3.89
1992 [14]	*Cmm2*	17.311	17.321	3.895
1997 [11]	*Cmm2*	17.461	17.811	7.776
2004 [15]	*Pba2*	12.3647	12.4039	3.8782
2005 [16]	*P4bm*	12.38		3.88
2011 [17]	*Pba2*	12.3556	12.3911	3.8757

in the *Cmm*2 symmetry they propose for $Sr_3TiNb_4O_{15}$. Chi et al. [15] published the most comprehensive paper on $Sr_3TiNb_4O_{15}$ in 2004, but the *R*-values for their refinements were somewhat high. The aim of the present paper is to report on our re-investigation of the structure of $Sr_3TiNb_4O_{15}$ using synchrotron X-ray powder diffraction data.

24.2 Experimental

Polycrystalline powder samples of $Sr_3TiNb_4O_{15}$ were prepared by conventional solid state synthesis techniques using the appropriate carbonates and oxides. The reagents were calcined at 950 °C for 36 h to remove CO_2 and were then annealed at 1300 °C in steps of 24 to 96 h.

Synchrotron X-ray powder diffraction patterns were collected across a range of temperatures at the powder diffraction beamline of the Australian Synchrotron with a wavelength of $\lambda = 0.82521(1)$ Å at room temperature and 0.82521(1) Å above. High temperature data were collected using a hot air blower (300–1248 K).

Structural refinements employing the Rietveld method were performed using the refinement program JANA2006 [18].

24.3 Results and Discussion

Three space group symmetries have been proposed in the literature for $Sr_3TiNb_4O_{15}$. The models based on these symmetries include a tetragonal, *P4bm*, model and two orthorhombic, *Cmm*2 and *Pba*2, models [12, 14, 15].

The first priority for this work was determining in which crystal system, orthorhombic or tetragonal, $Sr_3TiNb_4O_{15}$ forms. Peak splitting observed in the synchrotron X-ray powder diffraction pattern for $Sr_3TiNb_4O_{15}$ confirms that it crystallises in the orthorhombic crystal system. The inset in Fig. 24.1 clearly shows the splitting for the (140) and (410) reflections.

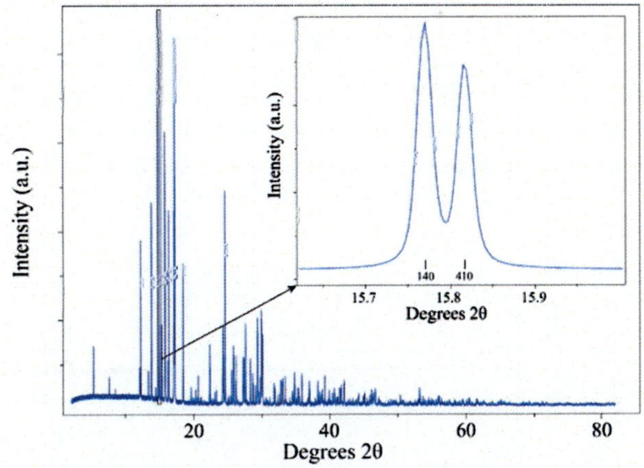

Fig. 24.1 Synchrotron X-ray diffraction pattern for $Sr_3TiNb_4O_{15}$. Peak splitting is clearly observed for the (140) and (410) reflections

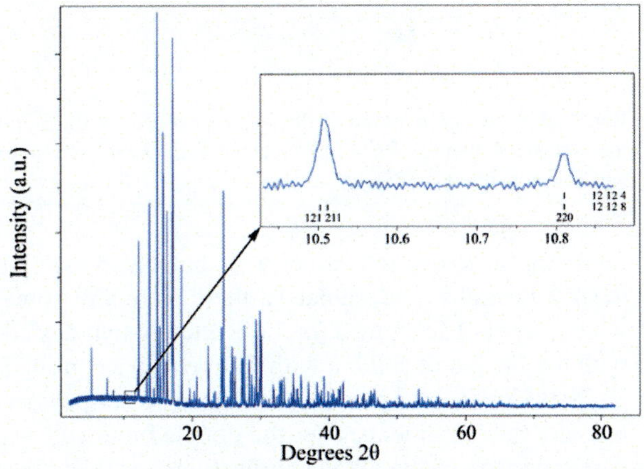

Fig. 24.2 Synchrotron X-ray diffraction pattern of $Sr_3TiNb_4O_{15}$. *Inset* shows an example of a weak reflection present which is indexed by a doubled unit cell but not by the cell proposed in the literature

After it had been confirmed that $Sr_3TiNb_4O_{15}$ crystallised in an orthorhombic crystal system, the two proposed models were tested against the data collected. The *Cmm*2 model, which has a doubled unit cell compared to the tetragonal *P*4*bm* cell, was relaxed and refined. This refined *Cmm*2 model gave a relatively poor fit to the data, as was found previously [15]. The structural model proposed by Chi et al. [15] with *Pba*2 symmetry and a 12 × 12 × 4 Å unit cell was also tested. This model gave a much better fit to the data and indexed all the strong reflections quite well. However, there remained a significant number of weak reflections at low angle which were not indexed by this model. A doubling of the unit cell in the *c* direction led to many of these weak reflections being indexed. An example of such reflections is the (121) reflection of the 12 × 12 × 8 Å unit cell, shown in Fig. 24.2.

A model of $Sr_3TiNb_4O_{15}$ with *Pba*2 symmetry and a doubled unit cell to 12 × 12 × 8 Å gave a reasonable fit to the data. This model significantly improved upon the *R*-factors published in the literature from Chi et al. [15] with $R_{wp} = 16.8$ to a value of 6.83 as well as a R_{Bragg} of 4.69. This unit cell and symmetry not only

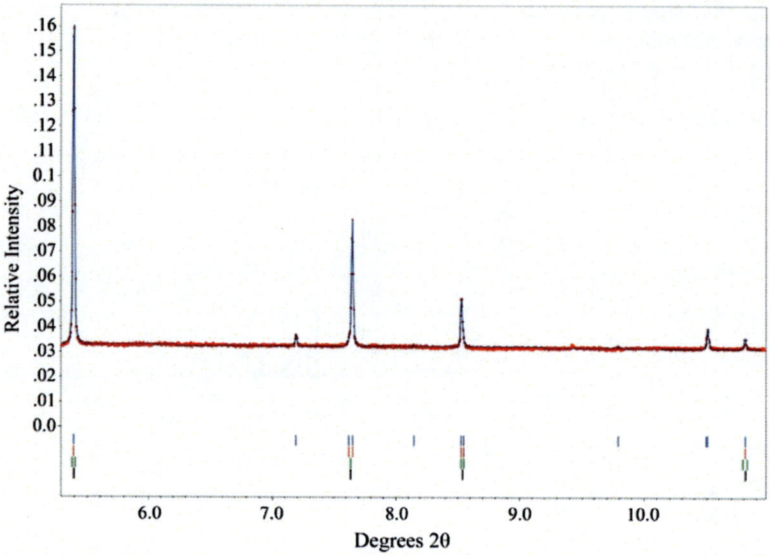

Fig. 24.3 Low angle region of the synchrotron X-ray diffraction pattern for $Sr_3TiNb_4O_{15}$. Peak markers are shown for $\boldsymbol{Pna2_1}$ symmetry (*blue*), $\boldsymbol{Pba2}$ symmetry (*red*), $\boldsymbol{Cmm2}$ symmetry (*green*) and $P4bm$ symmetry (*black*)

modelled the strong reflections well, but it also took into account the majority of the weak reflections. Unfortunately, there were still some weak, low angle reflections which were not accounted for. In an attempt to index all observed reflections, models with the same unit cell but with different space group symmetries were tested.

The refinements against synchrotron X-ray powder diffraction data showed that the best stable refinement for the doubled unit cell was in the $\boldsymbol{Pna2_1}$ space group. A comparison of the allowed reflections for this model and those suggested in the literature is shown in Fig. 24.3 for part of the powder diffraction pattern.

Lower symmetry space groups such as $P2_12_12$ and $P222$ were also considered. Refinements in these space groups were found to be less stable and problematic. In addition, no extra reflections were indexed and there was no appreciable improvement of the refinement statistics. Information from the refinement with $Pna2_1$ symmetry is provided in Table 24.2. The structural model obtained from this refinement is shown in Fig. 24.4.

Despite the improvement in the quality of refinement when it was performed with $\boldsymbol{Pna2_1}$ symmetry, there remained some very weak unindexed reflections. These reflections are most likely due to the presence of a small impurity phase in the powder.

Having confirmed that the structure adopted orthorhombic symmetry at room temperature, the high temperature phase behaviour of $Sr_3TiNb_4O_{15}$ was examined. Variable temperature synchrotron X-ray powder diffraction data were collected and confirmed that $Sr_3TiNb_4O_{15}$ underwent an orthorhombic to tetragonal phase transition. Figure 24.5 depicts the coalescence of the (140) and (410) reflections as $Sr_3TiNb_4O_{15}$ adopts the higher symmetry structure on heating. These data indicate

Table 24.2 Selected parameters from the refinement of $Sr_3TiNb_4O_{15}$ against synchrotron X-ray diffraction data

a (Å)	12.36081(2)
b (Å)	12.40288(2)
c (Å)	7.751270(10)
V (Å^3)	1188.344(2)
λ	0.82521 Å
Space group	$Pna2_1$
R_{Bragg} (%)	3.41
R_{wp} (%)	4.80

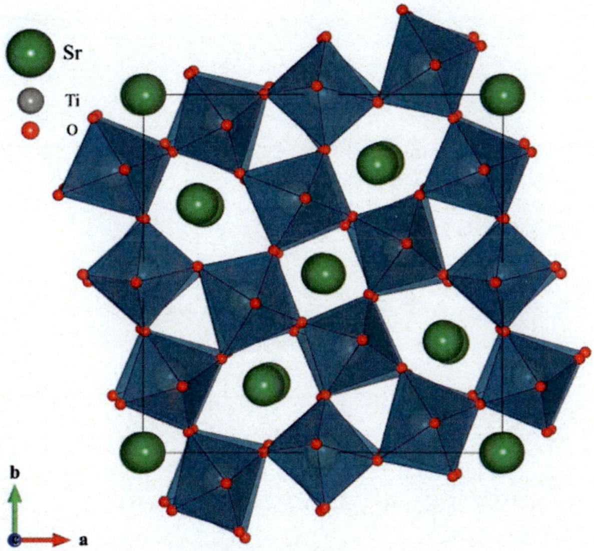

Fig. 24.4 Structure of $Sr_3TiNb_4O_{15}$ obtained from Rietveld refinement against synchrotron X-ray powder diffraction data

that the phase transition has been completed by about 648 K with possibly a small amount of orthorhombic character remaining.

Rietveld refinements of the high temperature structure were performed against the synchrotron data collected at 773 K. This is well above the phase transition temperature and therefore none of the low temperature orthorhombic phase should be present. Selected details for the refinement are provided in Table 24.3. The refinement against these data indicated that at this high temperature the structure adopted $P4bm$ symmetry with the 12 × 12 × 4 Å unit cell, similar to what was originally proposed for the room temperature structure [12]. Above the phase transition temperature, the superlattice reflections, including the (211) reflection, are no longer present. The R-factors for this refinement are somewhat higher, in part due to the lower signal-to-noise ratio of the data caused by the lower flux in the high temperature experimental set-up. The refinement profile shows that there are no major discrepancies across the profile. A close inspection of the low angle region of the profile, which was problematic in the room temperature refinement, shows that there

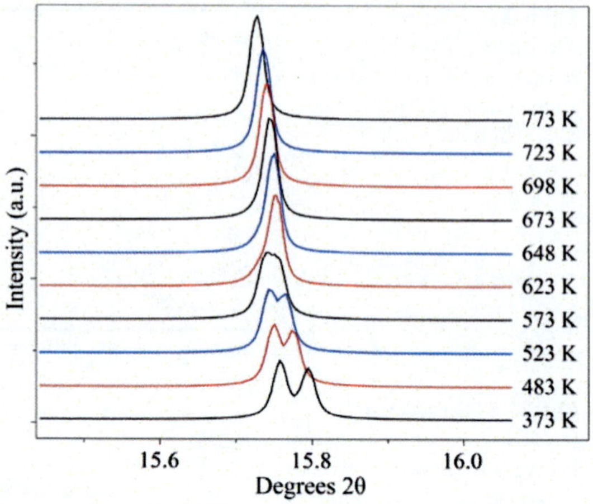

Fig. 24.5 Variable temperature diffraction patterns for $Sr_3TiNb_4O_{15}$. Depicted is the coalescence of the (140) and (410) reflections as the structure adopts tetragonal symmetry

Table 24.3 Information for refinement of $Sr_3TiNb_4O_{15}$ against synchrotron X-ray diffraction data at 773 K

a (Å)	12.45145(1)
c (Å)	3.913271(7)
V (Å^3)	606.7084(10)
λ	0.82664 Å
Space Group	*P*4*bm*
R_{Bragg} (%)	12.14
R_{wp} (%)	7.18

are no additional reflections observed and that all are successfully indexed by the 12 × 12 × 4 Å unit cell in *P*4*bm* symmetry.

24.4 Conclusions

The room temperature structure of $Sr_3TiNb_4O_{15}$ is structurally complicated due to the presence of nearly observed symmetry elements. The observation of weak symmetry breaking reflections makes the refinement of a model for $Sr_3TiNb_4O_{15}$ difficult. The major difficulty arises because low symmetry space groups are required to correctly model the structure of $Sr_3TiNb_4O_{15}$, but there are very few reflections on which to base a lower symmetry model. The best model of the room temperature structure determined from synchrotron X-ray powder diffraction data suggests the presence of an *n*-glide plane as well as a screw axis making the overall symmetry $Pna2_1$. A possible solution to the problems arising due to pseudo-symmetry would be a symmetry mode analysis of the structure as it is distorted from the high temperature *P*4*bm* symmetry to the room temperature structure. This type of analysis would allow for modelling in lower symmetry but would provide the appropriate constraints, which would stop the refinement from becoming unstable.

Acknowledgements A part of this work was carried out on the powder diffraction beamline at the Australian Synchrotron, Victoria, Australia, and the authors would like to thank Dr. Qinfen Gu for his assistance. The authors would also like to thank AINSE Ltd for providing financial assistance through the postgraduate award scheme.

References

1. Stephenson N (1965) Acta Crystallogr 18:496
2. Dickens PG, Whittingham MS (1968) Q Rev, Chem Soc 22:30
3. Jamieson PB, Abrahams SC, Bernstein JL (1968) J Chem Phys 48:5048
4. Oliver JR, Neurgaonkar RR, Cross LE (1989) J Am Ceram Soc 72:202
5. Neurgaonkar RR, Cory WK, Oliver JR, Sharp EJ, Wood GL, Salamo GJ (1993) Ferroelectrics 142:167
6. Neurgaonkar RR, Oliver JR, Cory WK, Cross LE, Viehland D (1994) Ferroelectrics 160:265
7. Fang L, Zhang H, Yang JF, Hong XK, Meng FC (2004) J Mater Sci, Mater Electron 15:355
8. Massarotti V, Capsoni D, Bini M, Azzoni CB, Mozzati MC, Galinetto P, Chiodelli G (2006) J Phys Chem B 110:17798
9. Halasyamani PS, Poeppelmeier KR (1998) Chem Mater 10:2753
10. Halasyamani PS (2004) Chem Mater 16:3586
11. Rao KS, Subrahmanyam ASV, Rao SMM (1997) Ferroelectrics 196:109
12. Ainger FW, Brickley WP, Smith GV (1970) Proc Br Ceram Soc 18:221
13. Ikeda T, Haraguchi T (1970) Jpn J Appl Phys 9:2
14. Neurgaonkar RR, Nelson JG, Oliver JR (1992) Mater Res Bull 27:677
15. Chi EO, Gandini A, Ok KM, Zhang L, Halasyamani PS (2004) Chem Mater 16:3616
16. Yuan Y, Chen XM, Wu YJ (2005) J Appl Phys 98:084110
17. Nair VG, Subramanian V, Santhosh PN (2011) AIP Conf Proc 1349:133
18. Petříček V, Dušek M, Palatinus L (2006) Jana2006. The crystallographic computing system. Institute of Physics, Praha

Chapter 25
Structural Investigation of the Incommensurate Modulated $Ta_2O_5{\cdot}Al_2O_3$ System

Denissa T. Murphy, Veronica Fung, and Siegbert Schmid

Abstract The $(1-x)Ta_2O_5{\cdot}xAl_2O_3$ solid solution series was re-investigated using synchrotron X-ray powder diffraction and neutron powder diffraction data. Structures were refined using the Rietveld method and an incommensurately modulated composite structure approach in superspace group $Xmmm(0\beta0)s00$ with a composition dependent modulation wave vector. Variable temperature synchrotron X-ray powder diffraction data were collected between 27 and 990 °C. The magnitude of the modulation wave vector was found to be constant over most of the temperature range.

25.1 Introduction

The low temperature form of tantalum pentoxide, L-Ta_2O_5, changes to the tetragonal high temperature modification, α-Ta_2O_5 (or H-Ta_2O_5), via a slow, reversible, phase transition at ~1360 °C [1, 2]. It was found that the addition of a range of other oxides (e.g. WO_3, GeO_2, Al_2O_3) [3], is possible, while maintaining the characteristics of the L-Ta_2O_5 structure. In addition, the temperature of the phase transition increases for these examples.

One example of the resulting phases, the $(1-x)Ta_2O_5{\cdot}xWO_3$, $0 \leq x \leq 0.267$ solid solution, has been described as a continuous incommensurate composite phase [4, 5]. The composite structure description considers two periodic substructures (M and O), with identical $\boldsymbol{a}$ and $\boldsymbol{c}$ subcell dimensions, and an incommensurate relationship in the $\boldsymbol{b}$-direction. The M-substructure is made up of the metal atoms and the oxygen atoms at apical positions, while the O-substructure consists of the remaining oxygen atoms (at $z=0$), as illustrated in Fig. 25.1.

Traditionally, the more strongly scattering M-substructure has been taken as the main substructure, but when neutron powder diffraction results are included this assignment is no longer conclusive. The $\boldsymbol{b}^*$ basis vectors of each substructure are at the same time the modulation vectors $\boldsymbol{q}$ of the other substructure, i.e. resulting in re-

D.T. Murphy · V. Fung · S. Schmid (✉)
School of Chemistry, The University of Sydney, Sydney NSW 2006, Australia
e-mail: siegbert.schmid@sydney.edu.au

S. Schmid et al. (eds.), *Aperiodic Crystals*, DOI 10.1007/978-94-007-6431-6_25,

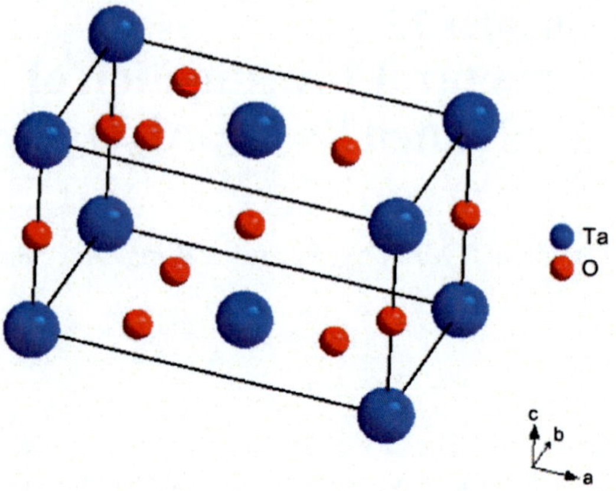

Fig. 25.1 Diagram of an average structure unit cell of L-Ta_2O_5 related structures. The metal atoms and the apical oxygen atoms belong to the M-substructure. The remaining oxygen atoms belong to the O-substructure

ciprocal basis vectors $\boldsymbol{a}^*, \boldsymbol{b}_M^*, \boldsymbol{c}^*, \boldsymbol{b}_O^*$, with $\boldsymbol{q}_M = \boldsymbol{b}_O^*$ and $\boldsymbol{q}_O = \boldsymbol{b}_M^*$. The modulation wave-vector $\boldsymbol{q}$ was found to be composition and temperature dependent.

We have recently studied the addition of Al_2O_3 to L-Ta_2O_5 and established many similarities to the $(1-x)Ta_2O_5{\cdot}xWO_3$ system [6]. In particular, it was found that the atomic modulation functions were similar for both systems. We report here on further synthesis and structural characterisation of the $(1-x)Ta_2O_5{\cdot}xAl_2O_3$ system with L-Ta_2O_5 type structure. Particular emphasis is placed on the $x = 0.06$ member of the solid solution and the use of room and variable temperature synchrotron X-ray powder diffraction and neutron powder diffraction data.

25.2 Experimental

The $(1-x)Ta_2O_5{\cdot}xAl_2O_3$ system was synthesised from $x = 0$ to 0.08 in steps of 0.01 via solid state reaction. Stoichiometric mixtures were heated at temperatures between 1000 and 1600 °C for between 10 and 40 hours to produce pure samples.

Synchrotron X-ray powder diffraction data were collected at the powder diffraction beamline, 10-BM-1, at the Australian Synchrotron. Due to the high absorption coefficient of tantalum, the samples were diluted with BN for data collection. Variable temperature data were collected using a hot air blower for temperatures up to 990 °C.

Neutron diffraction data were collected on Echidna at ANSTO. Data were collected at room temperature only, with a wavelength of 1.622(1) Å.

All refinements and structural analyses were carried out in JANA2006 [7]. Displacive modulation waves were refined up to fifth order for both subsystems, starting from lower order terms, which are generally associated with stronger satellite reflection intensities.

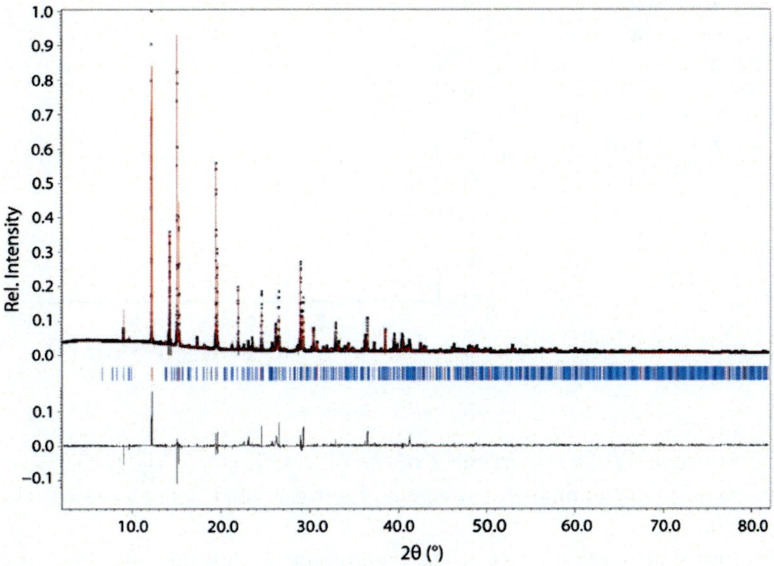

Fig. 25.2 Synchrotron X-ray powder diffraction pattern for $(1-x)Ta_2O_5{\cdot}xAl_2O_3$, $x = 0.06$ ($\lambda = 0.82555$ Å), showing the observed data (*black crosses*), calculated pattern (*red line*), and difference between the two (*black line* underneath). The *markers* show the locations of main reflections of both substructures (*red*), main reflections of M- and satellites of O-substructure (*black*), main reflections of O- and satellites of M-substructure (*orange*), and satellites of both substructures (*blue*)

25.3 Results and Discussion

Room temperature synchrotron X-ray diffraction patterns were collected for samples across the composition range. The $x = 0.06$ sample was diluted with BN for data collection, to reduce the overall absorption of the sample while still using a capillary of manageable diameter. The diffraction pattern for this sample is shown in Fig. 25.2, where the BN reflections were excluded from the Rietveld refinement as incommensurate composite structure. In contrast to an incommensurate modulated structure, which has main reflections and satellites, for an incommensurate composite structure four types of reflections can be distinguished. In the present case, these are reflections of type $(h0l0)$, i.e. main reflections of both substructures; $(hkl0)$, i.e. main reflections of the M- and satellites of the O-substructure; $(h0lm)$, i.e. satellites of the M- and main reflections of the O-substructure; $(hklm)$, i.e. satellites of both substructures (see Fig. 25.2). Refinements were carried out in the orthorhombic superspace group $Xmmm(0\beta0)s00$, where X indicates the non-standard centring $(1/2, 1/2, 0, 1/2)$ [8].

As the amount of aluminium in the structure increased, the magnitude of the modulation wave vector $\boldsymbol{q}$ decreased, leveling out at ∼6 % aluminium content (Fig. 25.3). This signals the end of the solid solution range, with $AlTaO_4$ forming as a second phase on further addition of Al_2O_3. The refinements for $x = 0.07$ and 0.08

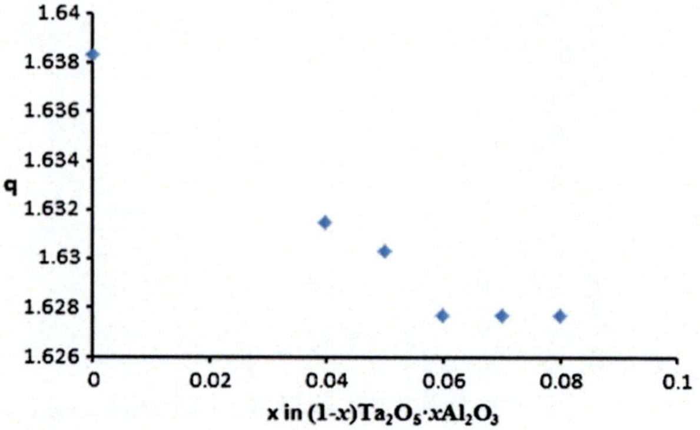

Fig. 25.3 Magnitude of q-vector versus Al_2O_3 content in the $(1-x)Ta_2O_5{\cdot}xAl_2O_3$ composition range (the values for $x \neq 0.06$ have been reported previously [6])

both correspond to the end member of the $(1-x)Ta_2O_5{\cdot}xAl_2O_3$, $0 \leq x <\sim 0.06$ solid solution. This composition range is slightly smaller than reported in the literature [3, 6].

Neutron diffraction data are particularly useful in the case of $(1-x)Ta_2O_5{\cdot}xAl_2O_3$. The partial diffraction pattern for the $x = 0.05$ composition, shown in Fig. 25.4, has many more strong satellite reflections (see Fig. 25.2), due to the relatively higher scattering power of oxygen when compared to X-ray diffraction data. The contribution of the O-substructure to the pattern is therefore enhanced. In addition, the lack of significant absorption effects is another advantage of neutron powder diffraction for this particular system. It is appropriate to extract the most reliable data from each of the X-ray or neutron refinements, e.g. parameters pertaining to the metal atoms may be better determined from X-ray data while those of the oxygen atoms are likely to benefit from neutron data. For example, the bond distances of interest in this structure are those between the metal atoms and the oxygen atoms, and in particular those between the different subsystems.

The distances between atoms in different subsystems are continuously changing throughout the incommensurate structure. Therefore, plots of atomic modulations functions (AMFs) are used to succinctly describe the structure (in addition to the average structure unit cell). AMFs are plotted as a function of the internal parameter t. For an incommensurate structure, $t = \boldsymbol{q}.\boldsymbol{T}$ modulo integer ($\boldsymbol{q}$—modulation wave vector; $\boldsymbol{T}$—translation vector of average subcell), can take all possible values between 0 and 1, but only once [9].

Metal–oxygen distances are plotted in Fig. 25.5 as a function of the internal parameter t. The blue lines represent a distance between a metal and an oxygen of the O-substructure, and show a continuous variation of distances from a minimum of ~1.9 Å. Bond distances between tantalum and oxygen of the M-substructure are depicted with the solid green line, and are approximately constant, with small variations around $t = 0$ and 0.5. The dotted red lines represent metal to O-substructure oxygen distances of the average structure (i.e. without the modulations). It can be seen that the modulations increase the minimum distance from ~1.6 Å to a more reasonable value of ~1.9 Å. The bond distance plot also highlights the variable

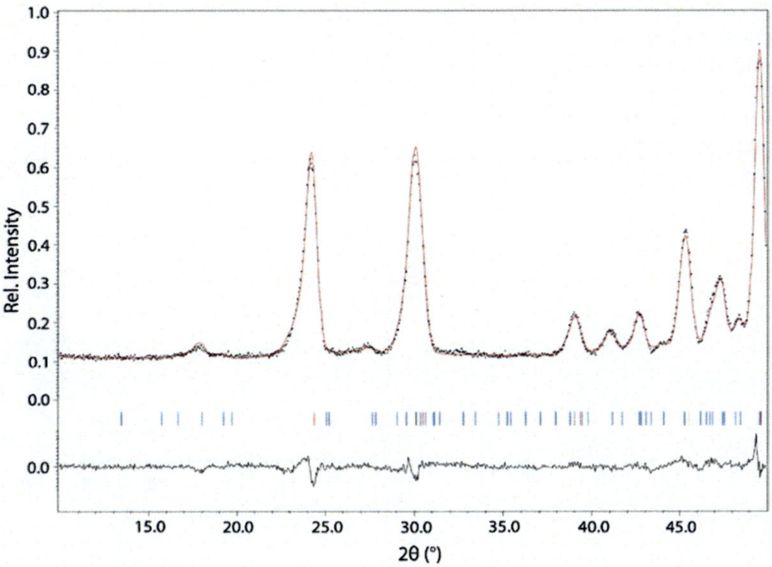

Fig. 25.4 Observed data (*black crosses*), calculated pattern (*red line*), and difference line (*black*) for the Rietveld refinement of $(1-x)Ta_2O_5{\cdot}xAl_2O_3$, $x = 0.05$, against neutron powder diffraction data (Echidna at ANSTO; $\lambda = 1.622(1)$ Å). Peak markers are colour coded as in Fig. 25.2. Note the strong reflection at about 45° 2θ, which corresponds to a weak feature in the synchrotron pattern

Fig. 25.5 Plot of M–O distances (Å) as a function of the internal phase parameter t for $(1-x)Ta_2O_5{\cdot}xAl_2O_3$, $x = 0.05$. *Dark blue lines* show distances between M and O_O. The *green line* represents distances between M and O_M. *Dashed red lines* indicate non-modulated distances between M and O_O. Regions showing $(4+2)$-fold coordination and $(5+2)$-fold coordination are labelled with the respective polyhedra

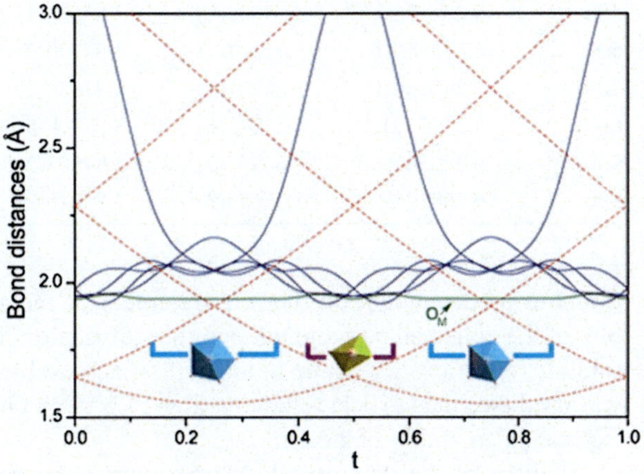

metal coordination environments from slightly distorted octahedra to pentagonal bipyramids (indicated in Fig. 25.5 with the respective polyhedra).

The determination of the solid solution field indicated that $x = 0.06$ was the end member of the $(1-x)Ta_2O_5{\cdot}xAl_2O_3$ solid solution. It was therefore considered important to collect variable temperature synchrotron X-ray diffraction data for this composition to establish the high temperature behaviour of the end member. In previously investigated L-Ta_2O_5 related systems, the modulation wave vector was

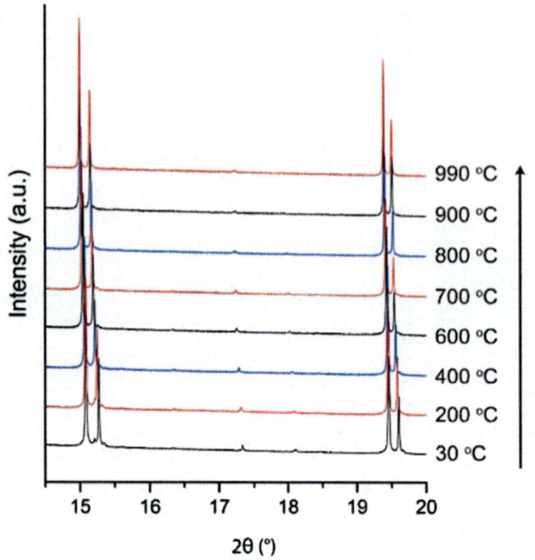

Fig. 25.6 Variable temperature synchrotron X-ray diffraction data collected at the powder diffraction beamline of the Australian Synchrotron ($\lambda = 0.82555$ Å)

Table 25.1 Cell dimensions for $(1 - x)Ta_2O_5{\cdot}xAl_2O_3$, $x = 0.06$

T [°C]	*a* (Å)	*b* (Å)	*c* (Å)	Vol. (Å^3)	*q*
27	6.21056(2)	3.64488(1)	3.87772(1)	87.7790(4)	1.62744(1)
200	6.22113(4)	3.64425(2)	3.87629(2)	87.8807(7)	1.62764(4)
400	6.23460(3)	3.64604(2)	3.87624(1)	88.1130(5)	1.62737(2)
600	6.24558(8)	3.65197(3)	3.87700(2)	88.4292(9)	1.62750(2)
700	6.24565(2)	3.65201(1)	3.87688(2)	88.4285(6)	1.62783(1)
800	6.2450 (2)	3.65235(6)	3.87708(5)	88.432(2)	1.62762(9)
900	6.26256(4)	3.66357(2)	3.87803(2)	88.975(1)	1.62913(1)
990	6.26819(5)	3.66659(3)	3.87815(3)	89.131(2)	1.63045(3)

found to be dependent on thermal treatment of the samples. Figure 25.6 shows a plot of the collected variable temperature diffraction patterns. It can be seen that the satellite reflections persist up to the highest achievable temperature in the set-up that was used (see the (210-1) reflection at ~17.3° 2θ). Otherwise the cell appears to be expanding smoothly on heating.

Analysis of the data revealed some interesting behaviour that warrants further investigation. Table 25.1 lists the cell dimensions from the Le Bail fits to the data. It can be seen that the cell volume increases slightly (~1.5 % overall), but it does that in small steps rather than continuously. In addition, the magnitude of the modulation wave vector stays virtually constant from 27 to 800 °C and only changes for the last 2 temperature steps. For an incommensurate structure the normal expectation is that the modulation wave vector would vary smoothly with changing temperature. This raises the question whether the $x = 0.06$ end member may have locked-in to the value of ~1.6275.

25.4 Conclusions

The $(1-x)Ta_2O_5 \cdot xAl_2O_3$ composition range was investigated and from the change in the magnitude of the modulation wave vector the solid solution range was determined to extend from $x = 0$ to 0.06. The variable temperature investigation of the $x = 0.06$ composition revealed that the volume was expanding stepwise rather than smoothly and the modulation wave vector was constant up to at least 800 °C. Given the small change of the volume overall and the change to the modulation wave vector at high temperatures further investigation of the thermal behaviour in the solid solution is warranted.

Acknowledgements A part of this work was carried out on the powder diffraction beamline at the Australian Synchrotron, Victoria, Australia, and the authors would like to thank Dr Qinfen Gu for his assistance. Neutron data were collected at ANSTO and the authors thank Dr James Hester for his assistance.

References

1. Lagergren S, Magneli A (1952) Acta Chem Scand 6:444–446
2. Grey IE, Mumme WG, Roth RS (2005) J Solid State Chem 178:3308–3314
3. Roth RS, Waring JL (1970) J Res Natl Bur Stand A 74:485–493
4. Schmid S, Withers RL, Thompson JG (1992) J Solid State Chem 99:226–242
5. Schmid S, Fütterer K, Thompson JG (1996) Acta Crystallogr, Sect B 52:223–231
6. Schmid S, Fung V (2012) Aust J Chem 65:851–859
7. Petříček V, Dušek M, Palatinus L (2006) Jana2006. The crystallographic computing system. Institute of Physics, Praha
8. Stokes HT, Campbell BJ, van Smaalen S (2011) Acta Crystallogr, Sect A 67:45–55
9. Perez-Mato JM, Madariaga G, Zúñiga FJ, Garcia Arribas A (1987) Acta Crystallogr, Sect A 43:216–226

Chapter 26
First-Principles Study for Phase Diagrams of Cd–Ca and Cd–Y Tsai-Type Approximants Under Pressure

K. Nozawa and Y. Ishii

Abstract We discuss the crystal structure and phase transitions in the Cd–Ca system by comparing the phase diagram and pressure dependence of the potential energy curves for the rotation of the tetrahedral cores. It turns out that some phase boundaries especially between the cubic phases seem to be well described by the rotation of the tetrahedral cores. Based on this analysis, we predict expected differences in the phase diagram in the Cd–Y system. We also point out the connection between the orientational configurations of the tetrahedral cores and valency of the second element. Furthermore, we also mention the possibility of the binary Cd–Y QC.

26.1 Introduction

Cd_6Ca and Cd_6Y are known as 1/1 cubic approximants of Tsai-type quasicrystals (QCs) $Cd_{5.7}Ca$ and $Cd_{5.7}Yb$ [1]. A structural building block of Tsai-type QCs and approximants is a five-layered atomic cluster referred to as Tsai-cluster. Tsai-clusters are ordered quasiperiodically in QCs, but periodically in approximants. As shown in Fig. 26.1, a Cd-tetrahedron is sitting at the center of the Tsai-cluster. Although there are still some ambiguities about the structure of the tetrahedral core, it inevitably breaks icosahedral symmetry of the outer shells.

Several phase transitions accompanied with changes in orientational configuration of the tetrahedral cores are found in Tsai-type 1/1 approximants. Tamura et al. found out that an order–disorder transition is universally observed in binary approximants at about 100 K [2]. It is also reported that the order–disorder transition disappears when even a small amount of impurities, or a third element, is added in the binary system [3]. First-principles density functional study for the Zn–Sc system shows that a Cu atom introduced as an impurity stabilizes a specific orientational configuration of the tetrahedral core, meaning the impurity increase the energy difference between different orientational configurations [4]. An increased

K. Nozawa (✉) · Y. Ishii
Department of Physics, Chuo University, 1-13-27, Kasuga, Tokyo 112-8551, Japan
e-mail: kazuki.nozawa@gmail.com

S. Schmid et al. (eds.), *Aperiodic Crystals*, DOI 10.1007/978-94-007-6431-6_26,

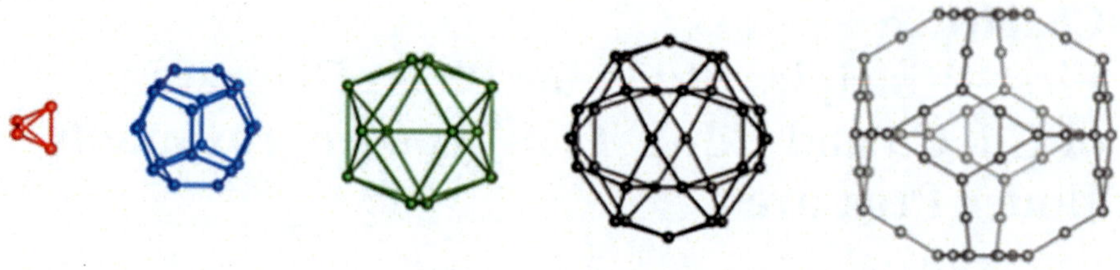

Fig. 26.1 Five-layered atomic shells of the Tsai-cluster. The innermost shell is a Cd tetrahedron

Fig. 26.2 Phase diagrams of Cd–Yb approximant [5]

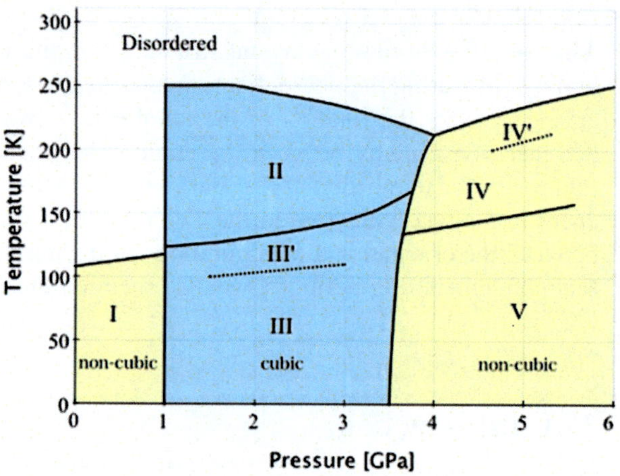

energy difference implies that it would prevent orientational changes of the tetrahedral cores as observed in the experiment. As for the other phase transition accompanied with orientational changes in the tetrahedral core, Watanuki et al. reported that 1/1 approximant phase exhibits various phase transitions under pressure as a phase diagram, which is reproduced in Fig. 26.2 [5]. In the phase diagram, ordered phases are labeled from I to V. As mentioned above, the order–disorder phase boundary is found at about 100 K. Phase I transforms into phase III at 1 GPa and this structure is kept until transforming the higher-pressure phase V at about 3.5 GPa. Depending on temperature, there are several phases in this pressure range, which are labeled II, III and III′. We investigated the phase transitions under pressures for an Cd–Ca approximant using the first-principles calculation within the local density approximation and found out that the tetrahedral core rotate around the twofold axis under high pressure [6]. The potential energy curve of the rotation around the twofold axis is of parabolic shape at the ambient pressure, but it turns to double-well shape at about 1 GPa, in good agreement with the transition pressure from I to III. In this paper, we present a more detailed comparison of the phase diagram and calculated results within the generalized gradient approximation (GGA). We also try to predict the phase diagram of Cd_6Y under pressure based on the analysis for Cd_6Ca.

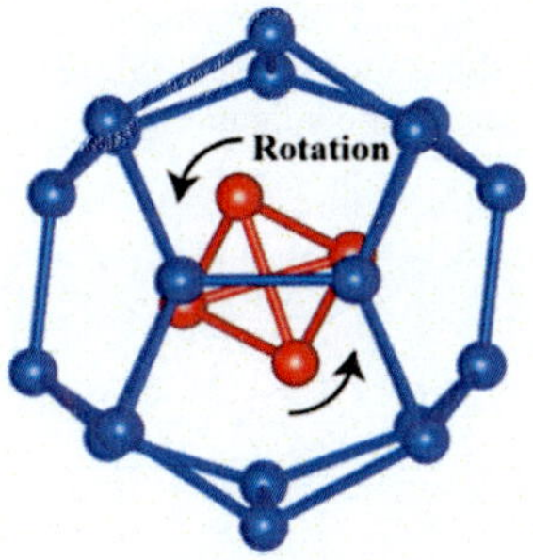

Fig. 26.3 Rotation around the twofold axis of the tetrahedral core

26.2 Calculation Methods

The density functional first-principles method is used to calculate the total energy of Cd_6Ca and Cd_6Y compounds. A cubic cell including 144 Cd atoms and 24 Ca or Y atoms is used as a unit cell. The unit cell contains two Tsai-clusters. An earlier study reported that the low-temperature phase at ambient pressure (Phase I) has a tetragonal unit cell including four Tsai-clusters [2]. Therefore, the cubic unit cell used here cannot represent structural configurations of Phase I completely, but some kind of tendency is probably derived from this calculation. Effective potentials of ionic cores are constructed using the projector-augmented wave method [7]. Also the GGA proposed by Perdew et al. is used for the exchange correlation term [8]. Calculations are done using the VASP code [9, 10], and standard settings for the plane-wave cut-off energy are adopted.

We previously reported a favorable structure of the tetrahedral core in the Cd_6Ca system. The four atoms of the tetrahedral core are located inside the pentagonal faces of outer dodecahedral shell [6]. Moreover, we found out that the tetrahedral cores are rotated around their twofold axis under pressure as shown in Fig. 26.3. In this paper, we concentrate on a rotation around the twofold axis. The optimal structure of the tetrahedral core is not the regular tetrahedron but the one having the C_{2v} symmetry. The four atoms of the tetrahedral core are placed inside the pentagonal faces of the outer dodecahedral shell. The twofold axis of the tetrahedral core can be aligned with the x, y or z axes, meaning the tetrahedral core can face six directions. Since we have Tsai-clusters at the vertex and body-center of the cubic unit cell, we can have three combinations as cluster configuration, and it can be described using the directions of the twofold axes as parallel: (x, x), anti-parallel: $(x, -x)$ and threefold: (x, y). However, we consider only parallel and anti-parallel configurations in this paper. We also consider rotational degrees of freedom around the twofold axis. Consequently, the structural configurations we argue here are four configurations shown in Fig. 26.4. Tetrahedral cores are anti-parallel and rotated different/same directions in IS2/ISI configurations, while those are parallel and rotated different/same directions in ISS/ISE configurations.

The equilibrium lattice constants are determined by calculating the total energy with varying the lattice constant, and evaluated as 15.88 Å for Cd–Ca and 15.72 Å for Cd–Y. Those are slightly larger than the experimental values 15.702 Å and

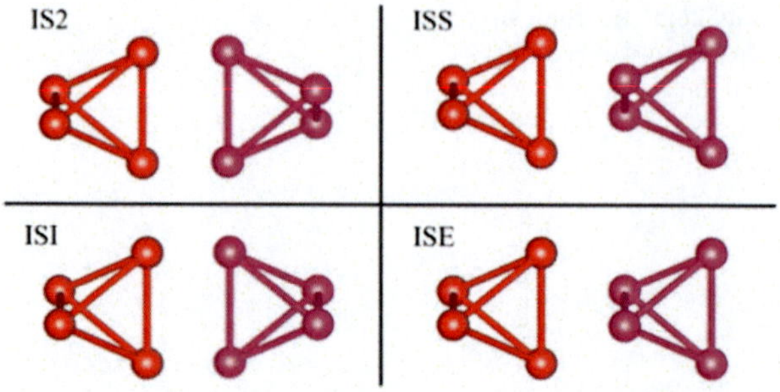

Fig. 26.4 Schematic illustrations of IS2, ISI, ISS and ISE configurations

15.482 Å [11], and the errors are only 1.1 % and 1.5 %, respectively. The differences can be explained as an error from the GGA. We evaluate pressure as a numerical differentiation of the total energy with respect to the volume. Strictly speaking, the calculated value depends on the configuration of the tetrahedral cores because the total energy depends on the configuration. However, it is confirmed that different configurations give almost same value of pressure.

26.3 Results and Discussions

Potential energy curves for rotation around the twofold axis at different pressures are shown in Fig. 26.5. We begin with comparing the results of the Cd–Ca system with the experimentally determined phase diagram for the Cd–Yb system (Fig. 26.2). Cd–Ca and Cd–Yb are considered to be isostructural. Up to 1 GPa, experiment shows that a non-cubic ordered phase exists below about 100 K. As previously mentioned, Phase I has a tetragonal unit cell containing four Tsai-clusters. We can find from calculated potential energy curves that anti-parallel configurations are favorable at the rotation angle $\theta = 0°$. The full vertical range for the potential energies in Fig. 26.5 is taken as 60 meV (600 K), which is compatible with the melting temperature of these systems. The parallel configurations (ISS and ISE) are not in the plotted energy range. Therefore, it is expected that the configurations of tetrahedral cores in Phase I are related to the anti-parallel configuration.

We notice that the potential energy curve for IS2 configuration turns to double-well shape at 2 GPa, meaning the ground state structure at this pressure is no longer the same as that of ambient pressure but the core is rotated about $\theta = 12°$. This is consistent with the experimental result that I–III phase boundary is located at 1 GPa. As indicated in the figure of Cd–Ca 2 GPa, the excitation from IS2 $\theta = \pm 12°$ configuration to IS2/ISI $\theta = 0°$ configuration needs 50 K. It is also 70 K from IS2 $\theta = \pm 12°$ configuration to ISI $\theta = \pm 12°$ configuration. Comparing the result of 2 GPa and 5 GPa, it is expected that these required excitation energies increase with pressure. They are consistent with the experimentally determined phase boundary from III to III′ (100 K) or III to II ($\sim$120 K) and with the increasing behavior of those.

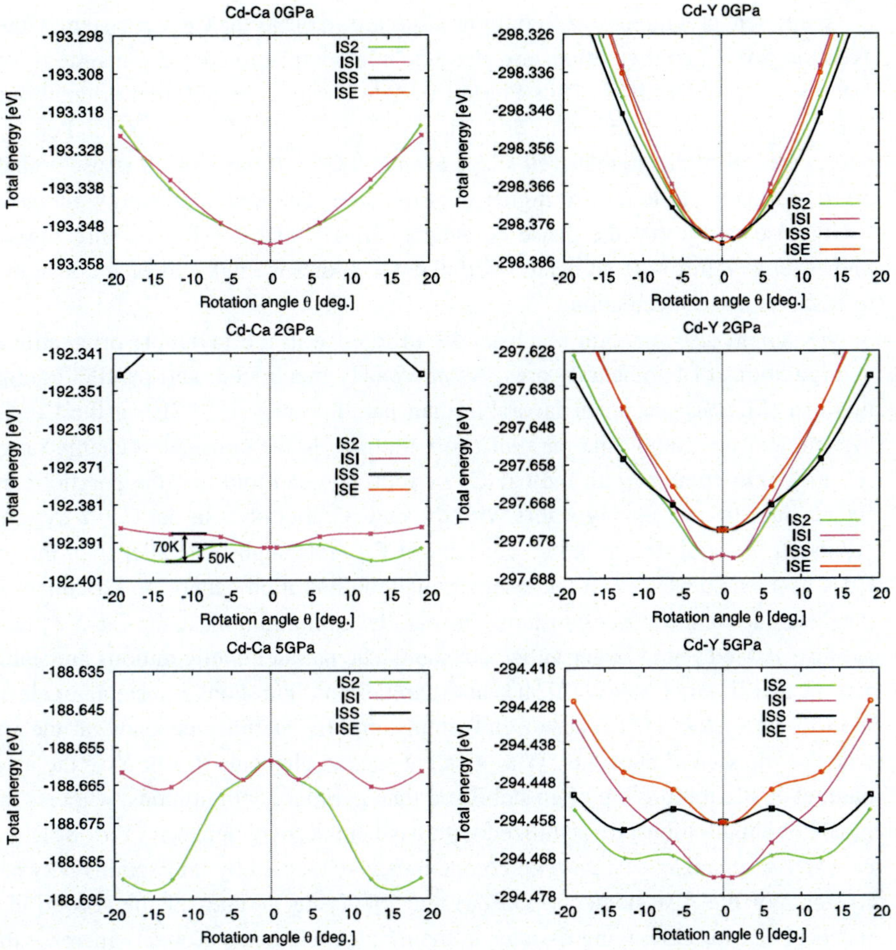

Fig. 26.5 The potential energy curves for the rotation around the twofold axis of the tetrahedral cores at 0, 2 and 5 GPa

At 5 GPa, the potential energy curves of the IS2 configuration has deep two minima at about $\theta = \pm 15^\circ$, and $\theta = 0^\circ$ is not even a metastable state. The excitation energy from IS2 to ISI seems not to correspond to the experimentally determined phase boundary. Since, however, Phase IV, IV′ and V are also assigned as a non-cubic structure, there may be other possible combinations, which can describe the thermal transition between these phases.

From the comparison of the pressure dependence of the potential energy curves and phase diagram, it turns out that our analysis based on the cubic cell seems to well describe the phase transition from Phase I to III and those in cubic phases (II, III, III′). Thus, we try to predict the phase diagram of Cd_6Y based on the analogy of the Cd–Ca case.

At ambient pressure, $\theta = 0°$ configurations are stable as the Cd–Ca system. However, the parallel configurations are also stable in contrast to the Cd–Ca system. We discuss this problem later. Thus Phase I of Cd–Y may have different orientational ordering of the tetrahedral cores or perhaps different unit cell. The potential energy curves are not of the double-well shape even at 5 GPa. It means I–III phase boundary presumably shifts toward higher-pressure side. The potential energy curves at 5 GPa also imply that the phase boundaries III–III′ and III′–II may shift toward higher-temperature side compared with Cd–Ca. Experimental studies are desirable to examine these speculations.

We should also mention a substantial difference in the favorable orientational configurations of tetrahedral cores. As previously mentioned, anti-parallel configurations (IS2/ISI) are much favorable than parallel ones (ISS/ISE) in the Cd–Ca system, and we cannot find parallel configurations in the considered energy range (60 meV). In contrast with the Cd–Ca system, one can find that the parallel configurations are comparably stable with the anti-parallel ones in the Cd–Y system. Although the equilibrium lattice constant of Cd–Y is slightly smaller than that of Cd–Ca, the tendency about the orientational configuration cannot be explained by the difference in the lattice constant. The parallel configurations in the Cd–Y system are also stable even at larger lattice constant. The parallel configurations are stable also in the Zn–Sc system [12], but anti-parallel configurations are stable again in the Cd–Yb system [13]. These trends imply that the stability depends on the valency of the second elements. With divalent second elements (Ca or Yb), the anti-parallel configurations are more stabilized than parallel configurations, whereas the parallel configurations are stabilized with trivalent second elements (Y or Sc). The cohesive mechanism of Tsai-type compounds is explained by interactions between *sp*-states of the first elements (Cd/Zn) and low-lying *d*-states of the second elements [14]. More low-lying *d*-states in the trivalent elements probably increase the cohesive energy as shown in the values in Fig. 26.5, and also can affect on the orientational configuration. On the other hand, differences in the atomic radius also may contribute to determine the stable configuration. The atomic radius is closely related to the valency, and the trivalent elements (Sc:1.62 Å, Y:1.80 Å) has a smaller atomic radius than Ca (1.98 Å). It is not clear what determines the stability of the orientational configurations so far, but some connections have emerged from these results. This may concern with the stability of the binary quasicrystalline phase. It is known that Cd-based binary QCs form at $Cd_{5.7}M$ (M = Ca/Yb), but in the Zn–Sc system QC forms at far Zn-richer composition $Zn_{88}Sc_{12}$ [15]. Although the role of the second element for the formation of the binary QC has also not been completely clarified yet, the series of tendencies between the orientational configurations and second elements may imply existence of the Cd–Y binary QC at different, maybe Cd-richer, compositions compared with the Cd–Ca and Cd–Yb systems.

26.4 Summary

We discussed the crystal structure and phase transitions in the Cd–Ca system by comparing the phase diagram and pressure dependence of the potential energy curves for the rotation of the tetrahedral cores. We found out that some phase boundaries, especially related to the cubic phase, seem to be well described by the rotation of the tetrahedral cores. Based on this analysis, we predicted some expected differences in the phase diagram in the Cd–Y system. Furthermore, we pointed out the connection between the orientational configurations of the tetrahedral cores and valency of the second element. We also mentioned about the possibility of the binary Cd–Y QC.

Acknowledgements Calculations are carried out using supercomputer facilities at ISSP, the University of Tokyo and Yukawa Institute, Kyoto University.

References

1. Tsai AP, Guo JQ, Abe E, Takakura H, Sato TJ (2000) Nature (London) 408:537
2. Tamura R, Nishimoto K, Takeuchi S, Edagawa K, Isobe M, Ueda Y (2005) Phys Rev B 71:092203
3. Yamada T, Tamura R (2009) Z Kristallogr 224:101
4. Nozawa K, Ishii Y, Makoshi K (2010) J Phys Conf Ser 226:012031
5. Watanuki T, Machida A, Ikeda T, Aoki K, Kaneko H, Shobu T, Sato TJ, Tsai AP (2006) Phys Rev Lett 96:105702
6. Nozawa K, Ishii Y (2008) J Phys Condens Matter 31:315206
7. Kresse G, Jourbert D (1999) Phys Rev B 59:1758
8. Perdew JP, Chevary JA, Vosko SH, Jackson KA, Pederson MR, Singh DJ, Fiolhais C (1992) Phys Rev B 46:6671
9. Kresse G, Hafner J (1993) Phys Rev B 47:RC558
10. Kresse G, Furthmüller J (1996) Phys Rev B 54:11169
11. Gomez CP, Lidin S (2003) Phys Rev B 68:024203
12. Hatakeyama T, Nozawa K, Ishii Y (2008) Z Kristallogr 223:830
13. Nozawa K, unpublished
14. Ishii Y, Fujiwara T (2001) Phys Rev Lett 87:206408
15. Canfield PC, Caudle ML, Ho C-S, Kreyssig A, Nandi S, Kim MG, Lin X, Kracher A, Dennis KW, McCallum RW, Goldman AI (2010) Phys Rev B 81:020201(R)

Chapter 27
The Choice of Vector Basis for Ammann Tiling in a Context of the Average Unit Cell

R. Strzalka, J. Wolny, and P. Kuczera

Abstract For the construction of the average unit cell (AUC) within the statistical approach, the use of a so-called reference lattice is needed. The choice of the lattice constants is in general arbitrary. However, it is convenient to bind them with the reciprocal space vectors k and q (main and modulation vector, $q = k/\tau$) which we use for indexing the diffraction pattern, $\lambda_k = 2\pi/k$, $\lambda_q = 2\pi/q$. AUC is a distribution of projections of atomic positions in real space on the reference lattice. With the choice of lattice as above, the shape of the AUC is related to the shape of the atomic surface (AS), used in the higher-dimensional approach. In this paper, the discussion on the choice of the set of wave vectors k and q is provided in terms of different geometrical bases used for a construction of Ammann–Kramer–Neri tiling (simply called Ammann tiling—AT, a model for icosahedral quasicrystal) and relation of the AUC and AS shapes. The dependence of the AUC shape on the choice of wave vectors is also demonstrated. Additionally, it is proved that the diffraction pattern does not depend on the basis chosen.

27.1 Introduction

Ammann tiling is one of the ideas used for the structural modelling of icosahedral quasicrystal [1, 2]. For the statistical approach, where the numerical calculations are crucial, AT seems to be most promising [8]. The AT is a kind of space covering with the points reflecting the icosahedral symmetry (the 5-fold axes in particular). It is the fully aperiodic covering, which means that atoms put in the positions of AT build a quasiperiodic system. In the higher-dimensional mathematical description, one can obtain the real-space AT points as projections of points in 6D hyperspace with the cubic arrangement. The 3D window through which one projects the hypercube is defined by the geometrical basis of a system. After [8], at least two settings of different bases can be used for the description of AT. For these two sets, the discussion on the AUC approach will be held.

R. Strzalka (✉) · J. Wolny · P. Kuczera
Faculty of Physics and Applied Computer Science, AGH University of Science and Technology, Krakow, Poland
e-mail: Radoslaw.Strzalka@fis.agh.edu.pl

S. Schmid et al. (eds.), *Aperiodic Crystals*, DOI 10.1007/978-94-007-6431-6_27,

The AUC is built as a projection of the AT points (in real space) on the reference lattice. This is a numerical derivation. The reference lattice is a 3D periodic grid with parameters simply related to three wave vectors $\mathbf{k}_i$ in reciprocal space:

$$|\mathbf{k}_i| = \frac{2\pi}{\lambda_{\mathbf{k}i}}, \quad i = x, y, z, \tag{27.1}$$

λ_i are the lattice constants of the three directions. Vectors $\mathbf{k}_i$ are used for indexing the diffraction pattern. Relation (27.1) is just a choice. However, it is approved to be used within the statistical approach [10]. The atomic positions u_i in AUC are the distances from the projections to the nodes of the reference lattice and form the continuous distribution $P(u)$.

As it was posted in many works [3, 4, 7], because of the aperiodicity of a system, the second reference lattice is needed. The parameters of this new lattice are τ^- times larger, and the reciprocal space vectors $\mathbf{q}_i$ connected with that lattice are τ^- times smaller and called modulation vectors. The choice of vectors $\mathbf{k}_i$ and $\mathbf{q}_i$ is in general arbitrary, but the shape of the AUC depends on the chosen reference lattice.

The important property of the AUC shape is its linear relation to the shape of the atomic surface (AS), which is the perpendicular component of the AT in the direct space [9]. AUC is an oblique projection of the AS along the certain direction (for details, see [5, 6]). The coordinates of the positions u_i in the AUC and $i^\perp$ in the perpendicular space are related as follows [8]:

$$u_i = -\frac{1}{\tau} i^\perp, \quad i = x, y, z. \tag{27.2}$$

However, it is more convenient to get the "simple" shape of the AUC (as a linear transformation of the AS) in the context of analytical calculations. This is dictated by the basis $\mathbf{k}_i$, $\mathbf{q}_i$ chosen for derivation of the AUC. In this paper, such a choice and its consequences will be discussed.

27.2 Icosahedral Basis (*Setting 1*)

The reciprocal space basis $\mathbf{V}^*$ within the *setting 1* is related directly to the icosahedral symmetry (the so-called "icosahedral basis"). The 6D basis vectors $\mathbf{d}_i^*$ are defined as follows:

$$\mathbf{d}_1^* = a^* \begin{pmatrix} 0 \\ 0 \\ 1 \\ 0 \\ 0 \\ 1 \end{pmatrix}_{V^*}, \quad \mathbf{d}_i^* = a^* \begin{pmatrix} \sin\theta \cos\frac{2\pi i}{5} \\ \sin\theta \sin\frac{2\pi i}{5} \\ \cos\theta \\ -\sin\theta\cos\frac{4\pi i}{5} \\ -\sin\theta\sin\frac{4\pi i}{5} \\ -\cos\theta \end{pmatrix}_{V^*}, \quad i = 2, \ldots, 6, \ \theta = 2\,\mathrm{arctg}\frac{1}{\tau} \tag{27.3}$$

a^* is a constant (arbitrarily set to 1).

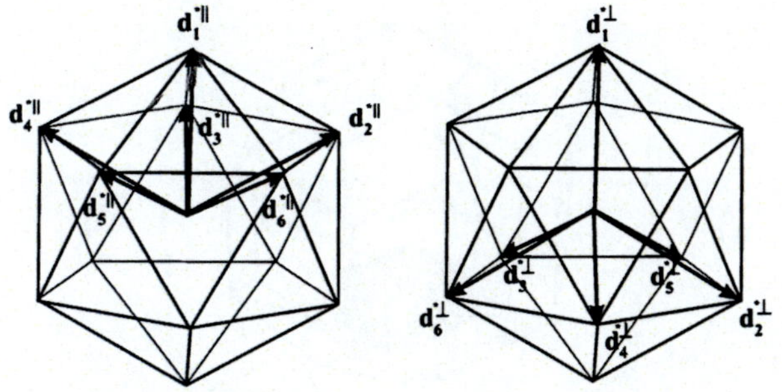

Fig. 27.1 Real-space (*left*) and perpendicular-space (*right*) components of the reciprocal space vectors $\mathbf{d}_i^*$ (*setting 1*)

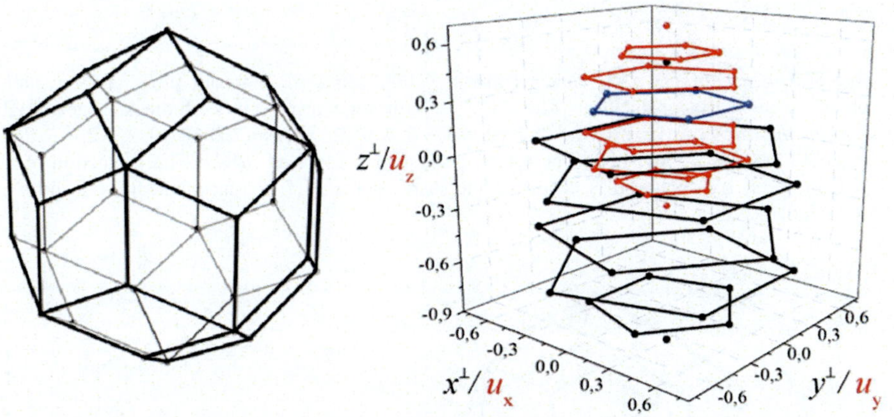

Fig. 27.2 *Left*: rhombic triacontahedron. *Right*: atomic surface (*black*) and average unit cell (*red*). The *blue layer* is common both for AS and AUC. The shape of AS and AUC is presented with 6 pentagons and 2 points building the rhombic triacontahedron

The 3D components of the basis V^* (three upper—real-space, and three lower—perpendicular-space components) can be defined geometrically as vectors spanned by an icosahedron (see Fig. 27.1).

The set of vectors $\mathbf{d}_i^*$ can be used as an indexing scheme for the diffraction pattern of the icosahedral quasicrystals. The 6D wave vector $\mathbf{K}$ can be expressed in basis V^* as $\mathbb{V}^*\mathbf{H}$, where $\mathbb{V}^*$ is a matrix with vectors $\mathbf{d}_i^*$ set in columns and $\mathbf{H}$ is column vectors of integer indices $(h_1, h_2, h_3, h_4, h_5, h_6)$. The wave vector of each diffraction peak is given by the real space component $\boldsymbol{\kappa}$ of a vector $\mathbf{K}$.

The direct space basis V (vectors $\mathbf{d}_i$) is just orthogonal to V^* (basis vectors look like (27.3) with opposite signs and different unit). The projection of the 6D cubic set of points via perpendicular-space components of the basis vectors $\mathbf{d}_i$ gives the AS which has the shape of the rhombic triacontahedron (see Fig. 27.2, *left*). The AUC shape is related to the shape of AS by formula (27.2). It also has the shape of rhombic triacontahedron. This is an exact result obtained purely from analytical derivation (see Fig. 27.2, *right*).

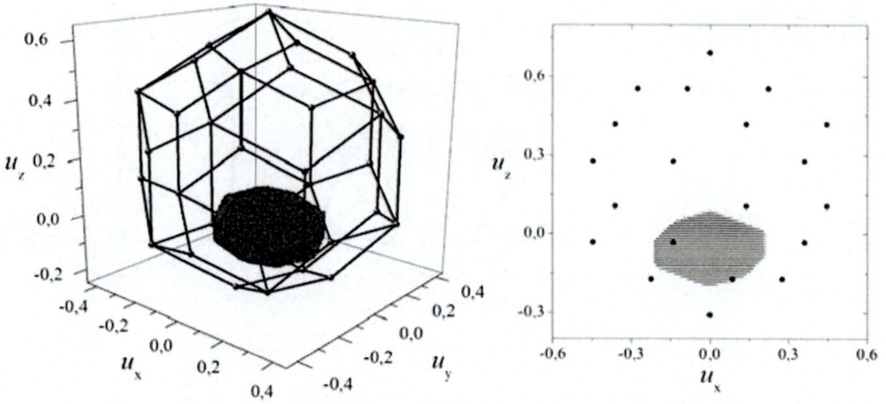

Fig. 27.3 *Left*: Average Unit Cell for about 77 000 points of Ammann tiling and the skeleton of rhombic triacontahedron. The AUC is built for wave vectors of the strongest peaks (*indices in brackets*): $|\boldsymbol{\kappa}_x| = 13.7082$ $(0, -5, -5, 2, 6, 2)$; $|\boldsymbol{\kappa}_y| = 7.2068$ $(0, 2, -2, -3, 0, 3)$; $|\boldsymbol{\kappa}_z| = 17.9443$ $(9, 4, 4, 4, 4, 4)$. The reference system (3) is used. *Right*: The projection of the AUC on the $u_x u_z$—plane. It is clear that the shape of the AUC is triacontahedral but distorted (re-scaled)

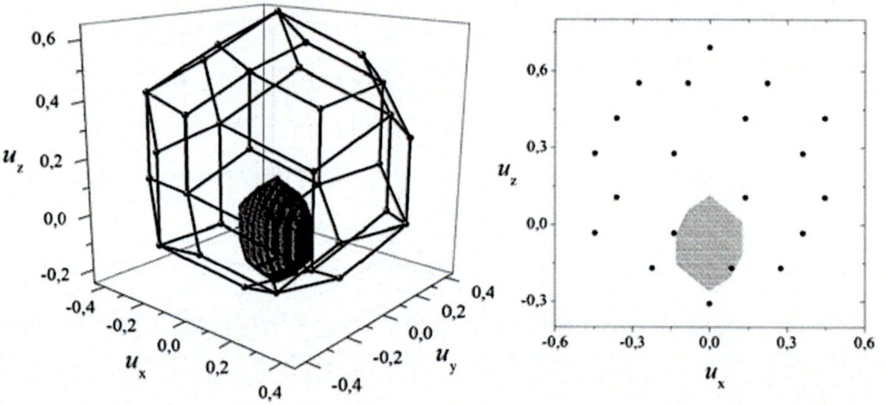

Fig. 27.4 Same graphs as in Fig. 27.3, but for other wave vectors: $|\mathbf{k}_x| = 22.1803$ $(0, -8, -8, 3, 10, 3)$; $|\mathbf{k}_y| = 11.6609$ $(0, -3, 3, 5, 0, -5)$; $|\mathbf{k}_z| = 4.2361$ $(2, 1, 1, 1, 1, 1)$. The triacontahedral shape is irregular again

As mentioned, to obtain the statistical distribution of AT points with respect to the reference lattice, some choice of the reciprocal space basis is needed. One of the possibilities is to take the real-space components of the 6D vectors of "icosahedral basis" for AT. However, there is some inconvenience. The reference lattice is 3D, so according to formula (27.1), one needs only 3 vectors $\mathbf{k}_i$ (and 3 vectors q_i for the second reference lattice, as well). One could also choose any 3 of projections of the 6D Cartesian unit vectors spanning the 6D cube on the real space (using *setting 1*).

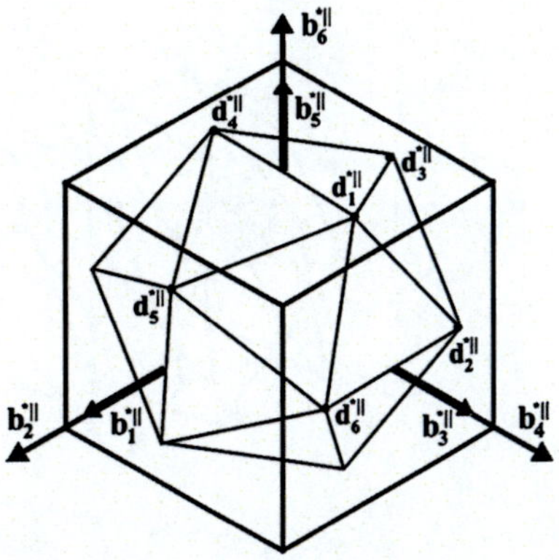

Fig. 27.5 The relative orientation of two bases $\boldsymbol{B}^*$ and $\boldsymbol{V}^*$ in real space. For vectors $\mathbf{d}_i^*$ only the ends were marked (in the vertices of an icosahedron). The shape of AS in basis $\boldsymbol{B}^*$ it is also given by a rhombic triacontahedron but rotated relative to that from Fig. 27.2

Such a choice gives, however, the oblique reference system. It could be expected in this case that the distribution $P(u)$ will have a distorted shape. As it is shown in Figs. 27.3 and 27.4, it is still a triacontahedron, but not regular anymore. These figures prove additionally that the shape of the AUC is strongly dependent on the choice of wave vectors $\mathbf{k}_i$ and $\mathbf{q}_i$.

For analytical calculation (like integrating to obtain the diffraction pattern), it is inconvenient to handle with irregular shapes. The integral over the AS (in higher-dimensional approach) or the AUC (in statistical method) gives the diffraction pattern. The integration over platonic solids (like rhombic triacontahedron) is much more preferable.

27.3 Cubic Basis (*Setting 2*)

There is an alternative reciprocal space basis $\boldsymbol{B}^*$ which is cubic and easily related to the "icosahedral basis" $\boldsymbol{V}^*$ by the following formula:

$$\begin{pmatrix} \mathbf{b}_1^* \\ \mathbf{b}_2^* \\ \mathbf{b}_3^* \\ \mathbf{b}_4^* \\ \mathbf{b}_5^* \\ \mathbf{b}_6^* \end{pmatrix}_{\boldsymbol{B}^*} = \frac{1}{2} \begin{pmatrix} 0 & \bar{1} & 0 & 0 & 0 & 1 \\ 0 & 0 & \bar{1} & 0 & 1 & 0 \\ 1 & 0 & 0 & \bar{1} & 0 & 0 \\ 0 & 1 & 0 & 0 & 0 & 1 \\ 0 & 0 & 1 & 0 & 1 & 0 \\ 1 & 0 & 0 & 1 & 0 & 0 \end{pmatrix} \begin{pmatrix} \mathbf{d}_1^* \\ \mathbf{d}_2^* \\ \mathbf{d}_3^* \\ \mathbf{d}_4^* \\ \mathbf{d}_5^* \\ \mathbf{d}_6^* \end{pmatrix}_{\boldsymbol{V}^*} = \frac{1}{2} \begin{pmatrix} \mathbf{d}_6^* - \mathbf{d}_2^* \\ \mathbf{d}_5^* - \mathbf{d}_3^* \\ \mathbf{d}_1^* - \mathbf{d}_4^* \\ \mathbf{d}_6^* + \mathbf{d}_2^* \\ \mathbf{d}_5^* + \mathbf{d}_3^* \\ \mathbf{d}_1^* + \mathbf{d}_4^* \end{pmatrix}_{\boldsymbol{V}^*} . \tag{27.4}$$

Basis $\boldsymbol{B}^*$ is called after [8] *setting 2*. The orientation of basis $\boldsymbol{B}^*$ versus basis $\boldsymbol{V}^*$ in real space is shown in Fig. 27.5. Interestingly, the vectors $\{\mathbf{b}_1^*, \mathbf{b}_2^*\}$, $\{\mathbf{b}_3^*, \mathbf{b}_4^*\}$, $\{\mathbf{b}_5^*, \mathbf{b}_6^*\}$ are parallel in pairs and the lengths of vectors with even indices are τ-times larger. This is the reason why basis $\boldsymbol{B}^*$ comes out naturally as a choice for sets of main and modulation vectors $\mathbf{k}_i$ and $\mathbf{q}_i$.

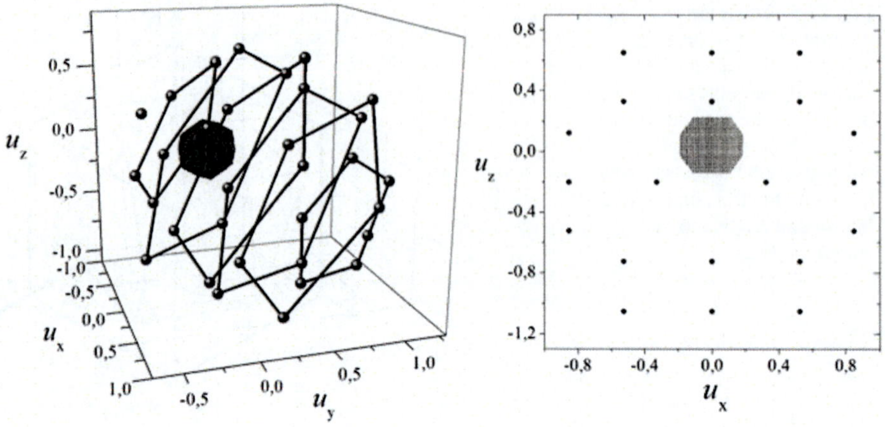

Fig. 27.6 *Left*: Average Unit Cell for about 77 000 points of Ammann tiling and the skeleton of rhombic triacontahedron constructed in cubic basis $\boldsymbol{B}^*$ (the reference system (27.4)). The AUC is built for basis wave vectors from Eq. (27.5). *Right*: The projection of the AUC on the $u_x u_z$-plane. It is clear that the shape of the AUC is given by rhombic triacontahedron without any distortion or scaling

Using the relations of the golden ratio τ and the cosine function of 36° ($\tau = 2\cos\frac{\pi}{5}$), one can easily get wave vector $\mathbf{K}$ in the cubic reference system (*setting* 2):

$$\mathbf{K} = h\mathbf{Q}_x + h'\mathbf{K}_x + k\mathbf{Q}_y + k'\mathbf{K}_y + l\mathbf{Q}_z + l'\mathbf{K}_z,$$
$$\boldsymbol{\kappa} = h\mathbf{q}_x + h'\mathbf{k}_x + k\mathbf{q}_y + k'\mathbf{k}_y + l\mathbf{q}_z + l'\mathbf{k}_z,$$
$$\mathbf{q}_x = \left[\frac{1}{\sqrt{\tau+2}}, 0, 0\right]; \quad \mathbf{q}_y = \left[0, \frac{1}{\sqrt{\tau+2}}, 0\right]; \quad \mathbf{q}_z = \left[0, 0, \frac{1}{\sqrt{\tau+2}}\right],$$
$$\mathbf{k}_x = \left[\frac{\tau}{\sqrt{\tau+2}}, 0, 0\right]; \quad \mathbf{k}_y = \left[0, \frac{\tau}{\sqrt{\tau+2}}, 0\right]; \quad \mathbf{k}_z = \left[0, 0, \frac{\tau}{\sqrt{\tau+2}}\right],$$
(27.5)

where $\frac{2}{\sqrt{\tau+2}} \approx 0.5257$ and $\frac{2\tau}{\sqrt{\tau+2}} \approx 0.8507$. Integers h–l' are the peak indices. $\mathbf{Q}_i$, $\mathbf{K}_i$ are the basic and modulation wave vectors in 6D and $\mathbf{q}_i$, $\mathbf{k}_i$ are their real-space components, $i = \{x, y, z\}$. $\boldsymbol{\kappa}$ is the real-space wave vector.

Now for the construction of the AUC, the reference system (27.4) may be used. Every diffraction peak can now be indexed with vectors $\mathbf{q}_i$, $\mathbf{k}_i$ from Eq. (27.5). Indices are transformed from the set h_1–h_6 to h–l' according to the transformation matrix from Eq. (27.4).

Figure 27.6 presents the shape of the AUC generated for the basis vectors $\mathbf{q}_i$, $\mathbf{k}_i$ from Eq. (27.5). One can see that the shape of the AUC in this picture is regularly triacontahedral.

Finally, it has to be pointed out that the diffraction pattern derived as the Fourier transform of the atomic positions in the AUC (numerical calculation) is the same

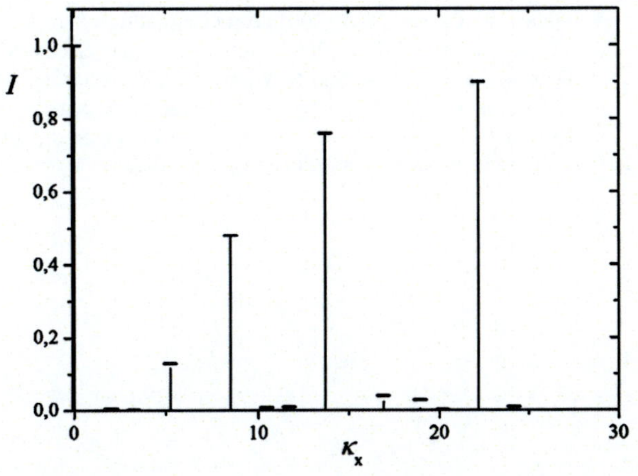

Fig. 27.7 Diffraction pattern for AT (empty lattice, with no decoration) in the direction κ_x of a diffraction wave vector calculated as the numerical Fourier transform for two AUCs: icosahedral basis *setting 1* (*vertical lines*)—data from Fig. 27.3; cubic basis *setting 2* (*dashes over peaks*)—data from Fig. 27.6. Patterns agree with little deviation for small peaks

for both bases used to build the Ammann tiling points and the average unit cell. The comparison of diffraction intensities in direction κ_x for both considered cases (i.e. for the AUCs from Figs. 27.3 and 27.6) is shown in Fig. 27.7. The intensity of a peak is understood as a square of the structure factor for empty Ammann tiling (with no decorative atoms). The structure factor is simply the Fourier transform of atomic positions (in the AUC).

27.4 Conclusions

This paper is related to Ammann tiling (also called Ammann–Kramer–Neri tiling) as a model for an icosahedral quasicrystal. The statistical approach using the concept of an average unit cell is applied for consideration. It is shown that the AUC shape as a distribution of atomic positions projected on the reference grid is strongly dependent on the vectors **k** and **q** chosen for building a grid. The two complementary geometrical bases ("icosahedral" and cubic) were used and the argumentation for the choice of the cubic one for the AUC method is delivered. Finally, the numerical diffraction pattern (over points in the AUC) is calculated to show that the diffraction pattern does not depend on the choice of reference system.

Acknowledgements Partially supported by the Polish Ministry of Science and Higher Education and its grants for Scientific Research: 3264/B/H03/2011/40 and DEC-2011/01/N/ST3/02250.

References

1. Kramer P, Neri R (1984) Acta Crystallogr, Sect A 40:580–587
2. Levine D, Steinhardt PJ (1984) Phys Rev Lett 53:2477
3. Wolny J, Pytlik L, Lebech B (1990) J Phys Condens Matter 2:785–795
4. Wolny J (1998) Philos Mag A 77:395–412

5. Steurer W, Cervellino A (2001) Acta Crystallogr, Sect A 57:333
6. Wolny J, Kozakowski B (2003) Acta Crystallogr, Sect A 59:54
7. Dabrowska A, Kozakowski B, Wolny J (2005) Acta Crystallogr, Sect A 61:350
8. Steurer W, Deloudi S (2009) Crystallography of Quasicrystals Springer, Berlin, pp 166–170
9. Kozakowski B, Wolny J (2010) Acta Crystallogr, Sect A 66:489–498
10. Wolny J et al. (2011) Isr J Chem 51:1275–1291

Chapter 28
Real Space Structure Factor and Scaling for Quasicrystals

J. Wolny, B. Kozakowski, P. Kuczera, L. Pytlik, and R. Strzalka

Abstract This paper describes the average unit cell (AUC) approach and its application to a derivation of the structure factor for quasicrystals. Scaling plays a special role in this approach. The TAU2-scaling in AUC simplifies the formulae for the structure factor. The TAU-scaling of peak positions distinguishes quasicrystals from twins, indicating a phase transition at some critical concentration of atoms.

28.1 Introduction

Quasicrystals are aperiodic structures. Consequently, their structure cannot be characterized by a single periodic unit cell defined in the physical space. Recently, it was shown that the structure of quasicrystals can be successfully described in the physical space by the statistical approach where a regular unit cell is replaced by an average unit cell (AUC) [3–5, 12–14] in which the atomic positions are occupied by atoms with some probability. The knowledge of the probability distribution allows calculating the structure factor and using it during a structure refinement [7, 8, 14]. An advantage of the AUC approach is that the structure refinement can be carried out in physical space only. Moreover, with this method one can go beyond quasicrystals, even to more complicated structures with singular continuous diffraction patterns [17].

In most cases, the structure of a quasicrystal is analyzed by means of the "cut and project" method which works in a high-dimensional space [2, 6, 9, 10]. According to this method, the structure is represented by its atomic surface which provides statistical information of high-dimensional atomic coordinates. One disadvantage of the method is that it is not easy to incorporate dynamic structural properties into a structure model. Randomness or structural defects is another issue which is not easily accessible by this method [1]. It was shown that an AUC and an atomic surface are related by an oblique projection onto the physical space [5, 11].

J. Wolny (✉) · B. Kozakowski · P. Kuczera · L. Pytlik · R. Strzalka
Faculty of Physics and Applied Computer Science, AGH University of Science and Technology, Krakow, Poland
e-mail: Janusz.Wolny@fis.agh.edu.pl

S. Schmid et al. (eds.), *Aperiodic Crystals*, DOI 10.1007/978-94-007-6431-6_28,

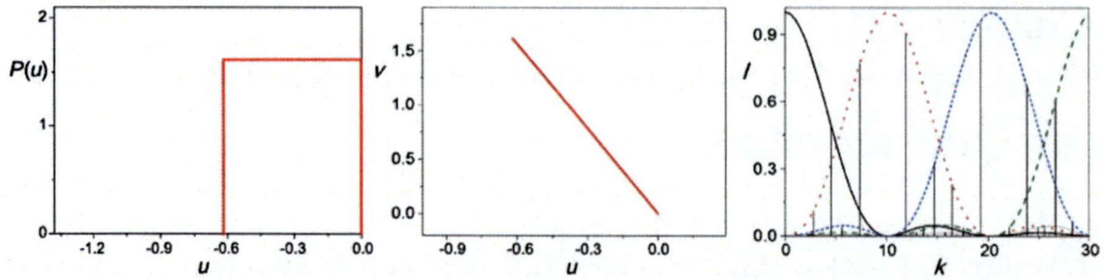

Fig. 28.1 (*Left*) $P(u)$ distribution for the Fibonacci chain calculated for the scattering vector $k_0 = 2\pi/a_0 \approx 4.55$, where a_0 corresponds to a mean distance of atoms equal to $3 - \tau \approx 1.38$. (*Middle*) Trace of $P(u, v)$ distribution calculated for scattering vector k_0 and $q_0 = k_0/\tau$. The distribution is monotonic and nonzero only along a line segment described by formulae: $v = -\tau^2 u$, for $u \in \langle 1 - \tau, 0)$. A Fourier transform of this AUC leads to the envelope functions connecting periodic series of peaks as is shown in the *right figure*

For all crystalline structures (including quasicrystals) one can define an appropriate scaling factor. The distances scaled by such a factor result in a self-similar structure. The same holds for the reciprocal space. When the scattering vector of any diffraction peak is multiplied by the scaling factor, another diffraction peak is reached. For the model quasicrystals, like the Fibonacci chain, the Penrose tiling or the 3D Amman–Kramer–Neri tiling, the scaling factor is related to the value of $\tau \approx 1.618$. The fact that this number is irrational is a source of quasicrystal aperiodicity. It is also possible to express this scaling factor as a rational number. In such a situation, we observe structural approximants.

28.2 Average Unit Cell

The concept of the reference lattice was presented in details in [5, 12, 16]. Suppose we have a one-dimensional set $\{r_n\}$ of N points (atoms) with scattering intensities equal to one. With some appropriate assumptions about the sequence $\{r_n\}$, we get the following expression for the structure factor for the scattering vector k:

$$F(k) = N^{-1} \sum_{n=1}^{N} \exp(ikr_n) = \int_{-\lambda/2}^{\lambda/2} P(u) \exp(iku)\, du \qquad (28.1)$$

where $P(u)$ is a probability distribution of distances u_n measured from the atomic positions to the nearest points of the reference lattice points $m\lambda$, $\lambda = 2\pi/k$, m is an integer number. The function $P(u)$ may be viewed as a probability distribution within an average unit cell. The structure factor for the scattering vector k is just the first Fourier mode of this distribution. A single average unit cell is sufficient to analyze structures in situations when scattering occurs at multiples of a fixed scattering vector k_0. For the Fibonacci chain, the AUC has a rectangular shape as shown in Fig. 28.1 (*left*).

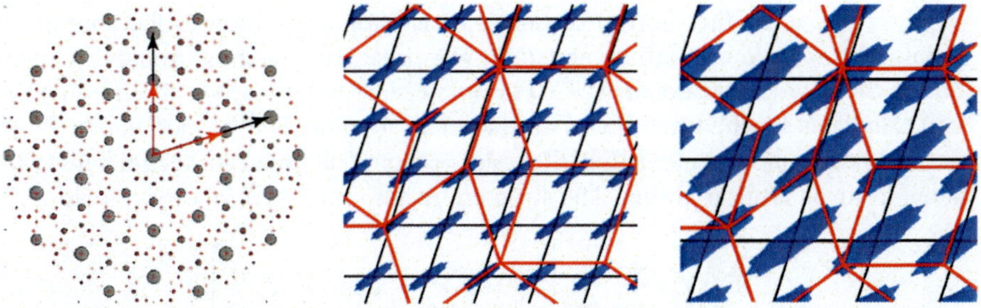

Fig. 28.2 Diffraction patter of the Penrose tiling (*left*) with two sets of scattering vectors $\mathbf{k}_1$, $\mathbf{k}_2$ and τ-times smaller vectors $\mathbf{q}_1$, $\mathbf{q}_2$, used for further calculations of the AUC. The AUC obtained for the first set of scattering vectors is shown in the *middle part of the figure*. The AUC obtained for vectors $\mathbf{q}_1$, $\mathbf{q}_2$ is shown on the *right*; $k_1 = k_2 = 20.25$, $q_1 = q_2 = 12.52$

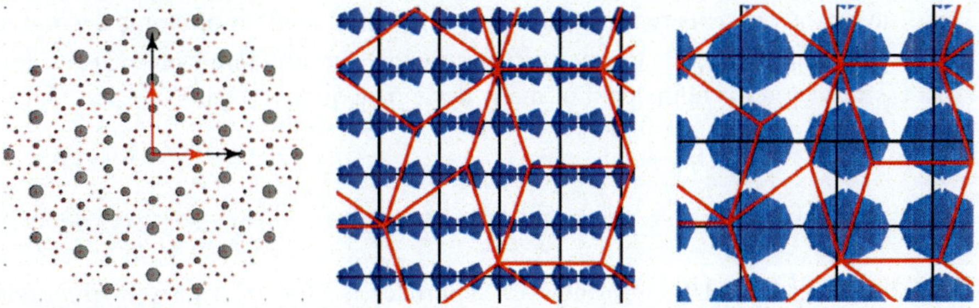

Fig. 28.3 The same as in Fig. 28.2 but for a different set of scattering vectors (orthogonal frame) as marked in the diffraction pattern (*left*): $k_x = 17.23$, $q_x = 10.65$; $k_y = 20.25$, $q_y = 12.52$

In case of 2D quasiperiodic structures, like the Penrose tiling, to obtain a probability distribution a pair of scattering vectors is required. As an example, histograms for a set of 10^5 points of the Penrose tiling are presented in Figs. 28.2 and 28.3. In the *left* figures, the scattering vectors that were used are marked. Two nonlinear vectors, $\mathbf{k}_1$ and $\mathbf{k}_2$, were used to construct the histogram of the probability distribution shown in the *middle* figures. In the *right* figures, similar distributions are shown but for scattering vectors τ-times shorter: $\mathbf{q}_1 = \mathbf{k}_1/\tau$ and $\mathbf{q}_2 = \mathbf{k}_2/\tau$. The AUC of the Penrose tiling consists of four pentagons resembling the atomic surface used in the 5D analysis. The periodic arrangements of AUCs superimposed on the Penrose tiling are shown in those figures as well.

In order to describe modulated structures (including quasicrystals), there is usually a need for two periods, a and b, which may be incommensurate. With two reference lattices, the first one of periodicity a and the second one of periodicity b, the structure factor can properly describe peaks positioned at a linear combination of $k_0 = 2\pi/a$ and $q_0 = 2\pi/b$ ($k = nk_0 + mq_0$), and can be expressed as:

$$F(k) = \iint_{\mathrm{AUC}} P(u, v) \exp\big(i(nk_0 u + mq_0 v)\big)\, \mathrm{d}u\, \mathrm{d}v \qquad (28.2)$$

where u and v are the shortest distances of an atomic position to the appropriate points of two reference lattices and $P(u, v)$ is the corresponding probability distribution of a two dimensional AUC. Formula (28.2) indicates that the average unit cell calculated for the wave vectors of the main structure and its modulation can be used to determine the intensities of peaks of any of the main reflections and their satellites in an arbitrary order. The structure factor in the AUC approach reads:

$$F(\mathbf{k}) = \sum_t \sum_\alpha \left[F_t^\alpha(\mathbf{k}) \sum_{j=1}^{n_t} f_j^t \exp\left(i\mathbf{k} \cdot \mathbf{r}_j^{t,\alpha}\right) \right] \tag{28.3}$$

where $F_t^\alpha(\mathbf{k})$ is the Fourier transform of a triangular probability distribution associated with a given structure unit t of the Penrose tiling and directed at angle α. The first sum (t) is over all independent structure units and the second (α) over all possible orientations of each structure unit. The position of an atom j in a given structure unit t at orientation α is represented by $\mathbf{r}_j^{\alpha,t}$, n_t is the number of atoms decorating the structure unit and f_j^t is the atomic form-factor. The structure factor (3) has been successfully used for refinement of many decagonal quasicrystals [7, 8, 14, 15].

28.3 Scaling

TAU2-Scaling for AUC In the AUC approach, scaling plays a very important role. It is observed for quasicrystals in the form of the probability distribution of atomic positions being nonzero only along a line, as shown in Fig. 28.1 (*right*). Such a linear relationship has an essential effect on the calculation of the corresponding structure factor. Its general form is $v = -\tau^2 u + b$, and we call such a relationship "TAU2-scaling". If the scaling is expressed as a rational approximant of τ, the distribution becomes broader and this affects the intensity of diffraction peaks at the inflated positions of the scattering vector. The size of the broadening is an order parameter. It can be used to describe the critical behavior of a diffraction pattern when the value of the scaling factor approaches the golden mean value.

Figure 28.4 (*left*) shows the distribution $P(u, v)$ which was calculated for the Fibonacci cluster consisting of 10^3 points (atoms). The chosen wave vector is $k_0 = 2\pi/a_0$ and the modulation vector is $q = k_0/\kappa$, where κ is some rational approximant of τ ($\kappa = 1.6177$ in the figure). Similarly to the previous results, the obtained distribution $P(u, v)$ is nonzero along a line segment ($v = -\tau^2 u$), but this time the line is broadened and the width is marked as Δv. The width depends linearly on κ and reaches zero for the critical value equal to τ (*middle* figure). The size of the width depends not only on the value of the ratio k_0/q but also on the number of atoms in the cluster. The linear dependence of the width on the number of atoms is shown in Fig. 28.4 (*right*). Only at the critical point, the width decreases to zero, leading to the TAU2-scaling.

The TAU2-scaling observed in the Penrose tiling reduced to the AUCs, which are shown in Figs. 28.2 and 28.3, is presented in Fig. 28.5. In this case, the linear

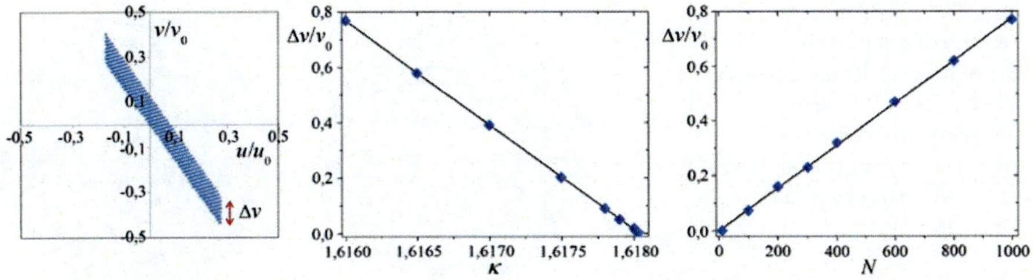

Fig. 28.4 TAU2-scaling is a typical behavior of AUC at the critical ratio of the two scattering vectors. In the *left figure*, the $P(u, v)$ was calculated for a finite cluster ($N = 10^3$) of the Fibonacci chain and a ratio of scattering vectors given by an rational approximant $\kappa = 1.6177$ (very close to the irrational value of τ). The observed broadening of the distribution is analyzed versus κ for $N = 10^3$ (*middle*) and number of N atoms for $\kappa = 1.616$ (*right*)

Fig. 28.5 TAU2-scaling for AUC of the Penrose tiling shown in Fig. 28.2 along two directions: x (*left*) and y (*right*). The AUC distributions from Fig. 28.3 for both directions, $\mathbf{k}_1$ and $\mathbf{k}_2$, scales identically as shown in the *right figure*

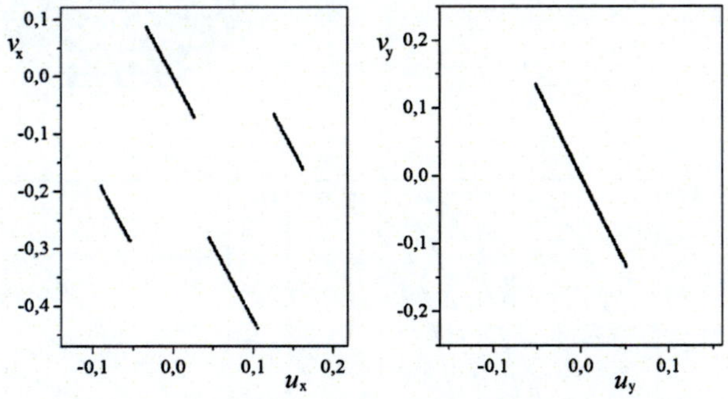

relationship between u and v variables is observed as well. This time, however, the relationship is not monotonic as it was in the case of the Fibonacci chain. The shape of the obtained distributions resembles the pentagonal distributions of the perfect Penrose tiling. In many cases, such a distribution can be successfully approximated by a Gaussian, which essentially simplifies calculations of the structure factor.

By using the TAU2-scaling, one can reduce the expression for the structure factor to the equation:

$$F(nk_0 + mq_0) = e^{imq_0 b} \int_{-u_1}^{+u_1} P(u) e^{ik_0(n-m\tau)u} \, du = A_{nm} e^{im\vartheta} \tag{28.4}$$

where A_{nm} is an amplitude, $\vartheta = q_0 b$, and the probability distribution $P(u)$ is assumed to be uniform for the Fibonacci chain. After its symmetrization $\vartheta = \pi$, the final result is:

$$F(k) = e^{im\pi} \frac{\sin(w)}{w} \tag{28.5}$$

where: $w \equiv \frac{1}{2\tau}(k - mk_0\sqrt{5})$ and k is a running scattering vector, $k = nk_0 + mq_0$. The diffraction intensity is shown in Fig. 28.1 (*right*). For continuous k one gets the so-called envelope function, also marked in the figure.

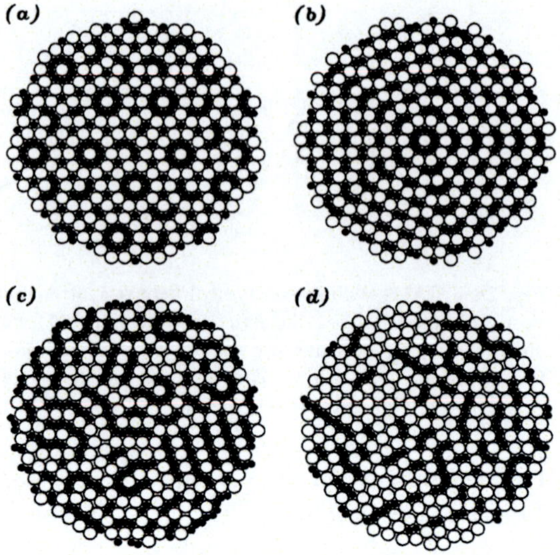

Fig. 28.6 Different structures obtained by decoration of 2D arrangement of decorated Robinson triangles: (**a**) The Penrose tiling, (**b**) twins, (**c**) random structure, (**d**) precipitated structure

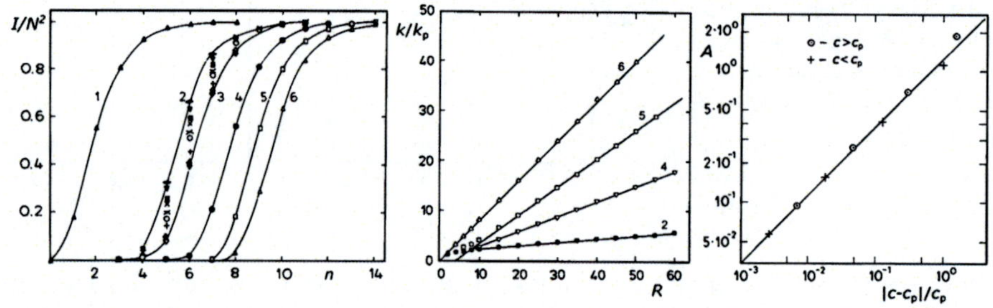

Fig. 28.7 (*Left*) Series of diffraction peaks for different structures: *1*—the Penrose tiling (c_p), *2*, *3*—random structures ($c \approx 0.37$), *4*—twins ($c = 0.5$), *5*—precipitated structure ($c \approx 0.8$), and *6*—twins ($c = 1$). (*Middle*) Linear scaling of scattering vectors versus the radius of the cluster. (*Right*) Critical behavior of the linear scaling coefficient near the Penrose concentration

As already mentioned, the distributions of the Penrose tiling atomic positions can be approximated by Gaussians. In such a situation, the resulting structure factor is also a Gaussian. Moreover, the integral in (28.4) is a real function for any symmetrical $P(u)$. Consequently, the phase value of each component of the structure factor must be equal to $\vartheta = q_0 b$.

Scaling for Peak Positions We used the TAU-scaling to differentiate between different 2D structures shown in Fig. 28.6. All these structures have been constructed by simple decoration of Robinson triangles [16]. Four different classes of structures have been used: (a) Penrose tiling ($c = c_\mathrm{p} = 1/\tau^2 \approx 0.382$), (b) twins ($c = 0.5$), (c) random structure ($c \approx 0.37$), and (d) precipitated structure ($c \approx 0.8$). The concentration of small Robinson triangles is given in parentheses.

Diffraction patterns of all the structures discussed consist of peak series positioned at $k = k_0\tau^n$ where n is an integer; it is shown in Fig. 28.7 (*left*). The tops of the diffraction peaks form the shape of the envelope function. The envelope obtained for the Penrose tiling can be used to describe other structures as well. To achieve that, it must be shifted along the scattering vector. This shift allows us to define the similarity parameter as a ratio of the position of a scattering vector $\boldsymbol{k}$ for a chosen structure to the position of the corresponding scattering vector for the Penrose tiling k_{P}, that is, (k/k_{P}), conditioned on $F(k) = F(k_{\mathrm{P}})$. For all the structures discussed, the similarity parameter depends linearly on the size of the cluster, as shown in Fig. 28.7 (*middle*). The linear coefficient scales critically with the concentration of atoms, as shown in Fig. 28.7 (*right*).

28.4 Conclusions

The AUC approach operates in physical space only. For each scattering vector one can construct a grid of planes and calculate an atomic position with respect to the grid. A probability distribution of these new positions defines the AUC and can be used in determination of periodic series of peaks within a diffraction pattern. To fully reconstruct a diffraction pattern of a quasicrystal (or a modulated structure), one has to use another wave-vector, a modulation vector, and calculate another AUC. The correlation between the atomic coordinates expressed in two reference lattices leads to a linear relationship $v = -\tau^2 u + b$ which we call the "TAU2-scaling". In case of the centrosymmetric distributions, the structure factor is given by a simpler formula (4) where the amplitude is a Fourier transform of the distribution, and all phases can be easily calculated as a product of the length of the modulation vector (q_0), parameter b, and satellite peak's index (m).

Another scaling property, the TAU-scaling of peak positions, is used to define the similarity parameter which distinguishes between different structures, including twins and random structures. This similarity parameter behaves critically at the critical concentration, indicating a continuous phase transition.

Acknowledgements Partially supported by the Polish Ministry of Science and Higher Education and its grants for Scientific Research: 3264/B/H03/2011/40 and DEC-2011/01/N/ST3/02250.

References

1. Baake M, Moody RV, Richard C, Sing B (2003) In: Trebin H-R (ed) Quasicrystals: structure and physical properties. Wiley-VCH, Berlin, pp 188–207
2. de Bruijn NG (1981) Proc K Ned Akad Wet, Ser A, Indag Math 43:38–66
3. Buczek P, Sadun L, Wolny J (2005) Acta Phys Pol B 36:919
4. Dabrowska A, Kozakowski B, Wolny J (2005) Acta Crystallogr, Sect A 61:350–357
5. Kozakowski B, Wolny J (2010) Acta Crystallogr, Sect A 66:489–498
6. Kramer P, Neri R (1984) Acta Crystallogr, Sect A 40:580

7. Kuczera P, Kozakowski B, Wolny J, Steurer W (2010) J Phys Conf Ser 226:012001
8. Kuczera P, Wolny J, Fleischer F, Steurer W (2011) Philos Mag 91:2500–2509
9. Levine D, Steinhardt PJ (1984) Phys Rev Lett 53:2477
10. Senechal M (1995) Quasicrystals and geometry. Cambridge University Press, Cambridge
11. Steurer W, Cervellino A (2001) Acta Crystallogr, Sect A 57:333
12. Wolny J (1998) Philos Mag A 77:395–412
13. Wolny J, Kozakowski B (2003) Acta Crystallogr, Sect A 59:54
14. Wolny J et al. (2011) Isr J Chem 51:1275–1291
15. Wolny J, Kozakowski B, Kuczera P, Takakura H (2008) Z Kristallogr 223:847–850
16. Wolny J, Pytlik L, Lebech B (1990) J Phys Condens Matter 2:785–795
17. Wolny J, Wnek A, Verger-Gaugry JL (2000) J Comput Phys 163:313–327

Chapter 29
Direct Observations of Aperiodic Arrangements of Transition-Metal Atoms in Al–Co–Ni Decagonal Quasicrystals by *Cs*-Corrected HAADF-STEM

A. Yasuhara, K. Saito, and K. Hiraga

Abstract HAADF (high-angle annular detector dark-field) images with *Cs*-corrected scanning transmission electron microscopy (STEM) have been observed for a W-(AlCoNi) crystalline phase and two-types of decagonal quasicrystals in $Al_{71.5}Co_{25.5}Ni_3$ and $Al_{72.5}Co_{17.5}Ni_{10}$ alloys. We have secured positive evidence that three-dimensional arrangements of transition-metal (TM) atoms of decagonal quasicrystals can be directly derived from the arrangements of bright dots in HAADF-STEM images, which correspond to individual TM atoms, by reference to results on HAADF-STEM observation and the structure of W-(AlNiCo) determined by X-ray diffraction analysis. We could conclude that pure TM atomic sites and mixed TM sites (with Al atoms) on A and B planes stacking along the periodic axis are located at the lattice points of a Penrose lattice with a bond length of 0.25 nm. In both planes atomic sites form pentagonal tilings with bond lengths of 0.47 nm and 0.76 ($=0.47 \cdot \tau$) nm, respectively, in both the $Al_{71.5}Co_{25.5}Ni_3$ and $Al_{72.5}Co_{17.5}Ni_{10}$ decagonal quasicrystals, whose structures were formally characterized as rhombic and pentagonal tilings of atom columnar clusters with a bond length of 2 nm.

29.1 Introduction

Several types of decagonal quasicrystals have been found in Al–Co–Ni alloys [1], and their structures have been characterized as two-dimensional aperiodic arrangements of decagonal columnar atom clusters whose size is 2 nm in diameter, with the so-called bond-orientational order (BOO) that is one of important structural features of quasicrystals [2, 3]. On the other hand, the structures of W-(AlCoNi) [4] and τ^2-$Al_{13}Co_4$ [5] crystalline phases show that TM atoms are located at lattice points of

A. Yasuhara (✉)
EM Application Group, EM Business Unit, JEOL Ltd., Tokyo 196-8558, Japan
e-mail: ayasuhar@jeol.co.jp

K. Saito
Department of Materials Science and Engineering, Akita University, Akita 010-8502, Japan

K. Hiraga
Institute for Materials Research, Tohoku University, Sendai 980-8577, Japan

S. Schmid et al. (eds.), *Aperiodic Crystals*, DOI 10.1007/978-94-007-6431-6_29,

a Penrose lattice with a bond length of 0.25 nm, and are arranged with BOO. Recent *Cs*-corrected STEM has successfully reproduced individual TM atoms in decagonal quasicrystals projected along the periodic axes [6]. However, in a previous report on *Cs*-corrected HAADF-STEM observations, it has still been used for examining the structure of 2 nm atom clusters, and the structures in wide regions including the outside of the atom clusters, i.e., glue regions, have never been discussed. In this paper, we aim to derive the structures of the Al–Co–Ni decagonal quasicrystals from direct observations of individual TM atoms by *Cs*-corrected HAADF-STEM with reference to the result, by our HAADF observation of the W-(AlCoNi) crystalline phase and its structure determined by X-ray structure analysis.

29.2 Experiment

Two alloys of nominal compositions of $Al_{71.5}Co_{25.5}Ni_3$ and $Al_{72.5}Co_{17.5}Ni_{10}$ were prepared from molten high-purity (99.99 %) Al, Co, and Ni metals, which were heated and mixed in an arc furnace in Ar atmosphere. The alloy ingots of $Al_{71.5}Co_{25.5}Ni_3$ and $Al_{72.5}Co_{17.5}Ni_{10}$ sealed in evacuated quartz tubes were annealed at 1160 °C for 3 h and at 900 °C for 40 h, respectively. HAADF-STEM images of those samples were observed by using a *Cs*-corrected electron microscope (JEM-ARM200F) with the incident beam along the periodic axis. And an HAADF-STEM image of the W-(AlCoNi) crystalline phase in the $Al_{72.5}Co_{20}Ni_{7.5}$ alloy annealed at 900 °C for 280 h was also observed with an incident beam along the *b*-axis as a reference.

29.3 Results and Discussion

The arrangements of TM atomic sites and mixed sites of TM and Al atoms on the *A* and *B* planes of W-(AlNiCo), determined by the single-crystal X-ray analysis [4], are shown in Figs. 29.1(a) and (b). The structure of the W-(AlCoNi) crystalline phase has a stacking of $ABA'B$ along the *b*-axis. Most of TM atoms on the *A* and A' planes have the same arrangement, except for TM atoms indicated by the red circle (on the *A* plane) and the green circle (on the A' plane) on the deformed hexagonal tiles. It should be noted here that the TM atoms of red and green circles have a phase flip relation, as indicated by red and green lines. The TM atomic site and mixed sites in the structure are exactly located at lattice points of a Penrose lattice with a bond length of 0.25 nm, which is indicated by thin dashed lines in Figs. 29.1(a) and (b), and they form pentagonal lattices with a bond length of 0.47 nm on the *A* plane and 0.76 ($=\tau \cdot 0.47$) nm on the *B* plane. On the other hand, Al atoms are shifted from lattice points of a Penrose lattice, and they prefer to be located at energetically stable positions between the definite arrangements of TM atomic sites and mixed sites with BOO [4]. This result indicates that the structures of decagonal quasicrystals could be characterized as arrangements of TM atomic sites and mixed sites.

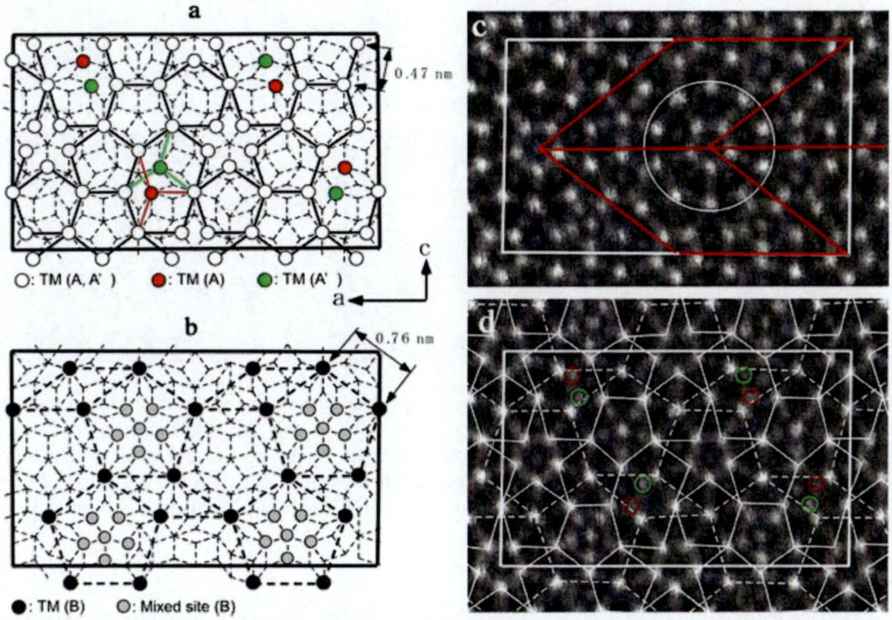

Fig. 29.1 (**a**)–(**b**) arrangements of TM atoms on the *A* and *B* planes, respectively, in the W-(AlCoNi) structure, and (**c**)–(**d**) HAADF-STEM images of the W-(AlCoNi) phase observed with the incident beam parallel to the *b*-axis. Note that almost all of *bright dots* in (**d**) are connected by pentagonal tilings with bond lengths of 0.47 nm and 0.76 nm, and that mixed sites of TM and Al atoms are represented as *weak dots* in *downward pentagons* indicated by *solid lines*

Figure 29.1(c) is an observed HAADF-STEM image. In the image, one can see characteristic clusters, formed in a ring of ten bright dots, which surrounds a downward pentagonal bright dot ring, indicated by a white circle. And one can say that the clusters are arranged within a thin rhombic frame with a side-length of 2 nm, as indicated by red lines. Almost all of the bright dots in the HAADF-STEM image of Fig. 29.1(c) can be connected by lines with bond lengths of 0.47 nm and 0.76 nm, which are indicated by white solid and dashed lines, respectively, as shown in Fig. 29.1(d). One can recognize that the bright dots at lattice points of pentagonal lattices with bond lengths of 0.47 nm and 0.76 nm in Fig. 29.1(d) correspond to TM atoms in the *A* and *B* planes of Figs. 29.1(a) and (b), respectively.

In addition to the bright dots corresponding to TM atoms, mixed sites of Al and TM in the B plane appeared as relatively weak bright dots and located in smaller downward pentagonal tiles indicated by solid lines in Fig. 29.1(d). Judging from the structure of the W-(AlNiCo), the TM atoms indicated by red and green circles in Fig. 29.1(a) correspond to single atomic TM sites along the *b*-axis in the unit cell. The other TM sites have two atoms in the unit cell. Therefore, the brightness of atomic sites corresponding to red and green circles in Fig. 29.1(d) should be of 1/4 weak brightness compared with the other bright dots in the HAADF-STEM image formed by Z^2 ($Z =$ atomic number) contrast. Therefore, the TM atomic sites,

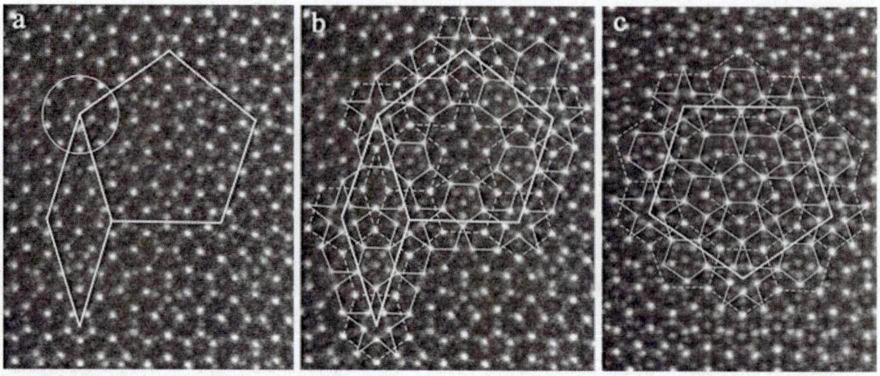

Fig. 29.2 HAADF-STEM images of the $Al_{72.5}Co_{17.5}Ni_{10}$ decagonal quasicrystal taken with the incident beam parallel to the periodic axis. (**a**) Shows that atom clusters indicated by *circles* are arranged in a pentagonal framework with a bond length of 2 nm, and (**b**) and (**c**) show that *bright dots* in the images can be connected by *solid* and *dashed lines* with bond lengths of 0.47 and 0.76 nm. Note that the *weak dots* in *downward pentagons of solid lines* correspond to mixed sites of TM and Al atoms

except for the sites indicated by red and green circles, are considered to be located in both atomic planes of A and B. With this consideration, we performed the structure analysis of two-types of quasicrystals ($Al_{71.5}Co_{25.5}Ni_3$ and $Al_{72.5}Co_{17.5}Ni_{10}$) with two atomic planes of A and B.

Figure 29.2(a) is an observed HAADF-STEM image of the decagonal quasicrystal in $Al_{72.5}Co_{17.5}Ni_{10}$. This structure of the decagonal quasicrystal was called as a Co-rich basic structure in [1], and its structure was formally characterized by a pentagonal tiling of 2 nm atom clusters [2, 7]. In the image, one can see that the clusters, indicated by a circle, are arranged by a pentagonal network with a side length of 2 nm, as indicated by thick solid lines. Almost all of bright dots in the HAADF-STEM image can be connected by solid and dashed lines with bond lengths of 0.47 and 0.76 nm, respectively, as shown in Figs. 29.2(b) and (c). Bright dots connected by solid lines and dashed lines correspond to TM atoms in the A and B planes, respectively, and the weak bright dots corresponding to mixed sites of TM and Al atoms are located in downward pentagonal tiles indicated by solid lines, except for an upward pentagon located at the center of a pentagonal network in Fig. 29.2(c). From the HAADF-STEM images of Figs. 29.2(b) and (c), arrangements of TM atoms and mixed sites in the A and B planes can be directly derived, as shown in Fig. 29.3. The actual structure of this decagonal quasicrystal is considered to be made of four atomic planes, with the $ABA'B$ stacking along the periodic axis, and the order of four atomic planes is a result of atoms indicated by red and green circles in Fig. 29.3(a), which are located in the A and A' planes, respectively.

Figure 29.4(a) is an observed HAADF-STEM image of the $Al_{71.5}Co_{25.5}Ni_3$ decagonal quasicrystal. The decagonal quasicrystal was called a Pentagonal superstructure in [1], and its structure was formally characterized as a rhombic tiling of 2 nm atom clusters [2, 8]. Bright dots in the image can be connected by lines

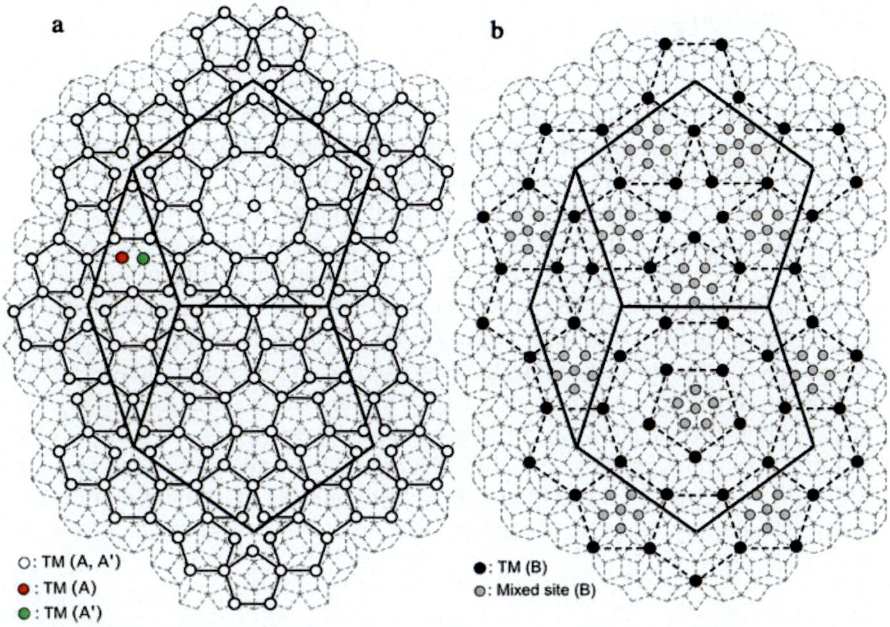

Fig. 29.3 Arrangements of TM atoms and mixed sites in *A* (**a**) and *B* (**b**) planes, determined from the observed HAADF-STEM images of Fig. 29.2

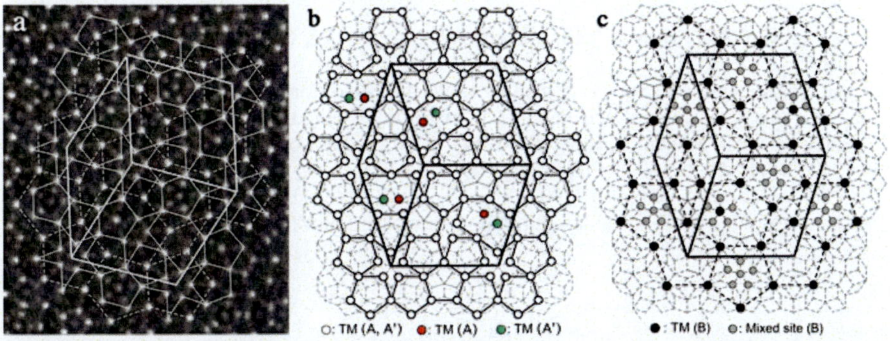

Fig. 29.4 (**a**) HAADF-STEM image of the $Al_{71.5}Co_{25.5}Ni_3$ decagonal quasicrystal, and (**b**)–(**c**) arrangements of TM atoms and mixed sites in *A* (**b**) and *B* (**c**) planes, respectively, determined from (**a**)

with bond lengths of 0.47 and 0.76 nm, as indicated by solid and dashed lines in Fig. 29.4(a), respectively. From Fig. 29.4(a), arrangements of TM atoms and mixed sites in the *A* and *B* planes can be directly derived as shown in Figs. 29.4(b) and (c), respectively. The structure of this quasicrystal is made of four atomic planes, with the order of $ABA'B$. The structure is considered to be a result of the atoms indicated by red and green circles in Fig. 29.4(b) in analogy with the W-(AlCoNi) structure.

29.4 Summary

The structures of two-types of decagonal quasicrystals in $Al_{71.5}Co_{25.5}Ni_3$ and $Al_{72.5}Co_{17.5}Ni_{10}$ alloys have been studied by *Cs*-corrected HAADF-STEM observations, on the basis of the close examination between an observed HAADF-STEM image of the W-(AlCoNi) crystalline phase and its structure determined by X-ray single crystal analysis. Individual TM atoms and mixed sites of TM and Al atoms appeared as bright dots in HAADF-STEM images observed with the incident beam parallel to the periodic axis. We can conclude that the three-dimensional arrangements of TM atoms and mixed sites in the quasicrystals can be directly determined from *Cs*-corrected HAADF-STEM images. The TM atoms and mixed sites in the two atomic planes stacking along the periodic axis are located at lattice points in a Penrose lattice of a bond length of 0.25 nm and form pentagonal tilings with bond lengths of 0.47 and 0.76 nm. This result suggests that the structures of Al–TM decagonal quasicrystals should be characterized as aperiodic arrangements of TM atoms with BOO, instead of previous cluster models in which the structures of decagonal quasicrystals were characterized by aperiodic arrangements of 2 nm atom clusters.

References

1. Ritsch S, Beeli C, Nissen H-U, Godecke T, Scheffe M, Lück R (1998) Philos Mag Lett 78:67–75
2. Hiraga K (2002) Advances in imaging and electron physics, vol 122. Elsevier Science, Amsterdam
3. Hiraga K, Ohsuna T, Sun W, Sugiyama K (2001) Mater Trans 42:2354–2367
4. Sugiyama K, Nishimura S, Hiraga K (2002) J Alloys Compd 342:65–71
5. Sugiyama K, Yasuhara A, Hiraga K (2012) In: Proceedings of aperiodic 2012
6. Taniguch S, Abe E (2008) Philos Mag 88:1949–1958
7. Hiraga K, Sun W, Ohsuna T (2001) Mater Trans 42:1146–1148
8. Hiraga K, Ohsuna T, Nishimura S (2001) Philos Mag Lett 81:123–127

Chapter 30
Arrangement of Transition-Metal Atoms in an Approximant Crystal Related to Al–Cu–Co Decagonal Quasicrystals Studied by *C*s-Corrected HAADF-STEM

K. Yubuta, A. Yasuhara, and K. Hiraga

Abstract A crystalline approximant, which is related to Al–Cu–Co decagonal quasicrystals with two aperiodic planes stacking along the periodic axis, in an $Al_{66}Cu_{15}Co_{19}$ alloy annealed at 900 °C for 36 h has been studied by high-angle annular detector dark-field (HAADF) observations with *C*s-corrected scanning electron microscopy (STEM). Observed HAADF-STEM images represent individual transition-metal (TM) atoms as bright dots, and so a three-dimensional arrangement of TM atoms in the approximant can be derived from the arrangement of bright dots. The structure has an orthorhombic unit cell with $a_0 = 10.1$ nm, $b_0 = 0.4$ nm and $c_0 = 6.7$ nm, formed by an ordered arrangement of two types of atom columnar clusters in a τ^3-inflated monoclinic $Al_{13}Co_4$ structure formed by a network of pentagons with an edge-length of 2 nm. The TM atoms in the two planes stacking along the *b*-axis are located at lattice points of a Penrose lattice with a bond length of 0.25 nm and pentagonal tilings with bond lengths of 0.47 and 0.76 nm.

30.1 Introduction

A series of decagonal quasicrystals with two or four aperiodic planes along the periodic axis have been found in Al–TM alloys, and also a number of crystalline approximants related to the decagonal quasicrystals are formed in alloys around the decagonal quasicrystals [1–3]. Almost all of those structures have been examined by HAADF-STEM, forming *Z* (atomic number) contrast images, and have been characterized by arrangements of columnar atom clusters with 2 nm in diameter, with a bond-orientational order, which is one of structural features of quasicrystals [2, 4].

K. Yubuta (✉) · K. Hiraga
Institute for Materials Research, Tohoku University, Sendai 980-8577, Japan
e-mail: yubuta@imr.tohoku.ac.jp

K. Hiraga
e-mail: hiraga@imr.tohoku.ac.jp

A. Yasuhara
EM Application Group, JEOL Ltd., Tokyo 196-8558, Japan
e-mail: ayasuhar@jeol.co.jp

S. Schmid et al. (eds.), *Aperiodic Crystals*, DOI 10.1007/978-94-007-6431-6_30,

In the present paper, we have examined the structure of a crystalline approximant in an $Al_{66}Cu_{15}Co_{19}$ alloy, which has a composition near that of $Al_{64}Cu_{22}Co_{14}$ for an Al–Cu–Co decagonal phase [5], by Cs-corrected HAADF-STEM observations. Modern Cs-corrected STEM has enough resolution to reproduce individual TM atoms as bright dots in observed images taken with the incident beam parallel to the periodic axis of the decagonal quasicrystals [5]. Older Cs-corrected HAADF-STEM is still being used for examining the structure of 2 nm atom clusters; however, it has never been employed for discussing the structures in wide regions, including the outside of the atom clusters, i.e., glue regions [5]. The purpose of the present paper is to discuss the structure of the approximant using an arrangement of TM atoms.

30.2 Experimental Procedures

An alloy of a nominal composition of $Al_{66}Cu_{15}Co_{19}$ was prepared by melting high-purity (99.99 %) Al, Cu, and Co metals in an arc furnace in Ar atmosphere. The ingot was sealed in an evacuated quartz tube and annealed at 900 °C for 36 h. HAADF-STEM images were taken by using a Cs-corrected electron microscope (JEM-ARM200F).

30.3 Results and Discussion

Figure 30.1 shows electron diffraction patterns of the $Al_{66}Cu_{15}Co_{19}$ approximant, taken with the incident beam parallel to three primary axes. Figure 30.1(a) shows the pseudo-tenfold symmetry, and Figs. 30.1(b) and (c) taken along the two directions of t and s indicated in Fig. 30.1(a) show a structure formed with two layers along the pseudo-tenfold axis, if diffuse scattering in Fig. 30.1(b) is ignored.

Figure 30.2(a) is an HAADF-STEM image of the $Al_{66}Cu_{15}Co_{19}$ approximant taken with the incident beam parallel to the pseudo-tenfold axis. In the image, one can see a characteristic cluster with a contrast distribution formed by ten bright dots surrounding a pentagonal arrangement of bright dots, as indicated by circles of solid and dotted lines, and one can also recognize that the clusters are arranged with a pentagonal framework of an edge length of 2 nm, as shown in Fig. 30.2(b). Hereafter, the cluster with a diameter of 1.2 nm is called a 1.2 nm decagonal cluster. The orthorhombic unit cell of $a_0 = 10.1$ nm and $c_0 = 6.7$ nm is formed by an ordered arrangement of two types of the 1.2 nm decagonal clusters, which have different directions of centered pentagons of bright dots, indicated by circles of solid and dotted lines in Fig. 30.2(b), in a τ^3-inflated monoclinic $Al_{13}Co_4$ unit cell indicated with red lines in Fig. 30.2(b) [6].

Almost all of bright dots corresponding to a TM atom in the HAADF-STEM image can be connected by two pentagonal tilings with bond lengths of 0.47 and

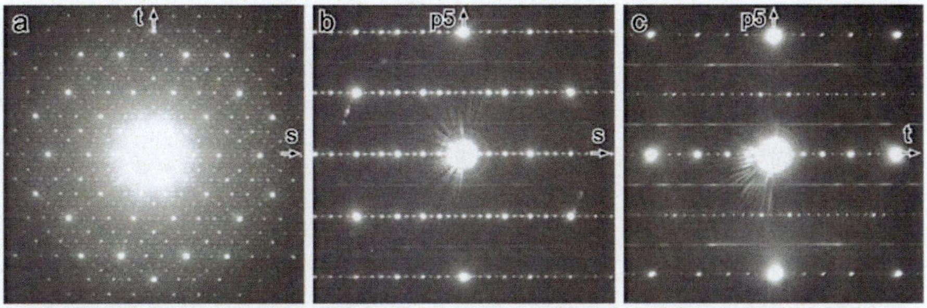

Fig. 30.1 Electron diffraction patterns of the $Al_{66}Cu_{15}Co_{19}$ approximant, taken with the incident beam parallel to the three primary axes. (**a**) shows a pseudo-tenfold symmetry, and (**b**) and (**c**) were respectively taken along the two directions of *t* and *s* in (**a**)

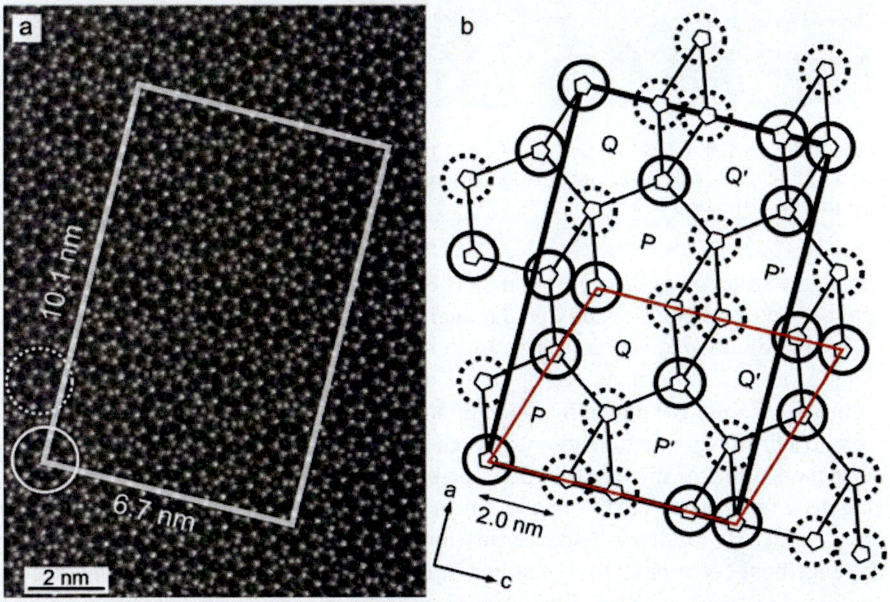

Fig. 30.2 (**a**) HAADF-STEM image of the $Al_{66}Cu_{15}Co_{19}$ approximant taken with the incident beam parallel to the pseudo-tenfold axis, and (**b**) ordered arrangement of two-types of 1.2 nm atom clusters in the unit cell. A τ^3-inflated monoclinic $Al_{13}Co_4$ unit cell is indicated by *red lines*

0.76 nm, as shown by solid and dashed lines in Figs. 30.3(a) and (b), which correspond to those in the *A* and *B* planes stacking along the *b*-axis. The pentagonal tilings with bond lengths of 0.47 and 0.76 nm exist in the *A* and *B* planes of a W-(AlCoNi) crystalline phase, respectively [7], which is an important approximant related to Al–TM decagonal quasicrystals. The structure of the W-(AlCoNi) phase is formed by one type of the 1.2 nm decagonal clusters connected with an interval of 2 nm, and consequently the pentagonal tilings with bond lengths of 0.47 nm and 0.76 nm are distinguished as those in different planes of *A* and *B* [7, 8]. On the other

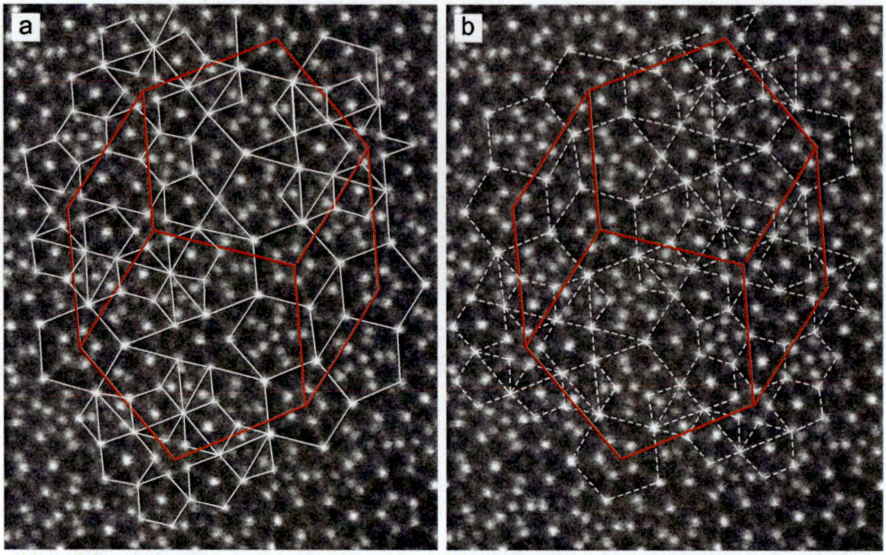

Fig. 30.3 HAADF-STEM images of the $Al_{66}Cu_{15}Co_{19}$ approximant and tilings formed by connecting bright dots corresponding to TM atoms in the *A* (**a**) and *B* (**b**) atomic planes stacking along the pseudo-tenfold axis

hand, the structure of this approximant is formed by two kinds of the 1.2 nm decagonal clusters, and consequently pentagonal tilings with bond lengths of 0.47 nm and 0.76 nm simultaneously appear in each of the planes, and the tiling with a bond length of 0.47 nm around the one type of decagonal clusters is connected to the tiling of 0.76 nm bond length around another type of decagonal clusters, and sometimes their tilings interpenetrate each other. In the images of Fig. 30.3, one can see that the starfish-shaped distributions of bright dots are located in downward pentagons with a bond length of 0.76 nm in Fig. 30.3(a) and in upward pentagons in Fig. 30.3(b). In the starfish-shaped distribution, most outside dots with a pentagonal arrangement correspond to TM atoms in the planes forming the pentagonal tilings, but inside pentagonal dots with a central one can be considered as corresponding to mixed sites of TM and Al atoms in different atomic planes. The above characteristic arrangements of TM and mixed sites can be seen in the structure of the W-(AlCoNi) approximant [7, 8].

From arrangements of bright dots in Fig. 30.3, we can directly derive the arrangements of TM and mixed sites in the *A* and *B* planes stacking along the *b*-axis, as shown in Fig. 30.4. In the model, the TM atoms and mixed sites of TM and Al atoms are distinguished from each other, in analogy with the W-(AlCoNi) structure [7, 8], but it should be noted that the difference between TM atoms and mixed sites is not clear from the observed HAADF-STEM image.

As can be seen in Fig. 30.4, the TM atoms and mixed sites are located at lattice points of a Penrose lattice having a bond length of 0.25 nm, and so they are arranged in a bond-orientational order. It should be noticed here that the atomic arrangement

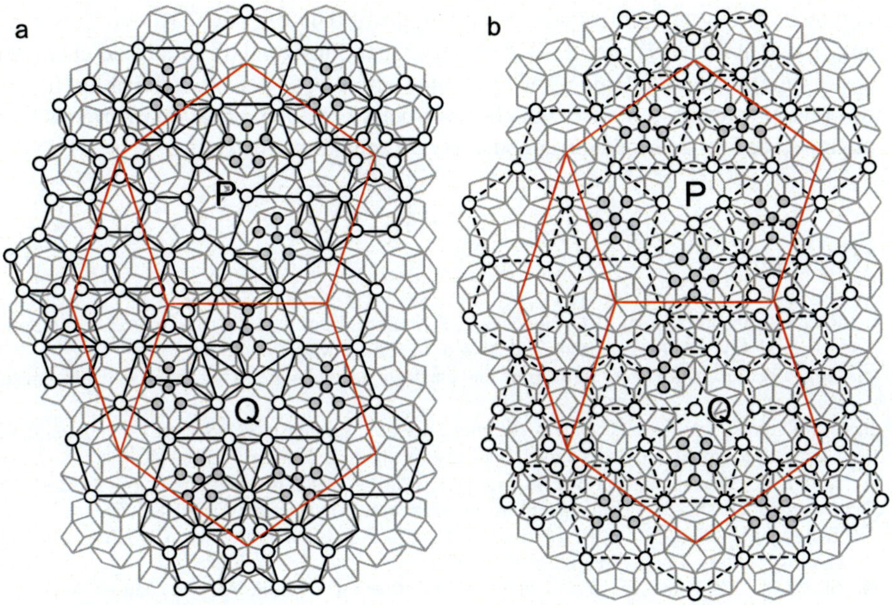

Fig. 30.4 Arrangements of TM atoms (*open circles*) and mixed sites (*gray circles*), and pentagonal tilings in the *A* (**a**) and *B* (**b**) planes. Note that TM atoms and mixed sites are located at lattice points of a Penrose lattice with a bond length of 0.25 nm, indicated by *thin lines*

in the pentagon frame of Q with an edge-length of 2 nm is formed from that in the P pentagon by two-fold screw operation of a rotation by 180 degrees and a glide of $b_0/2$. Also atomic arrangements in P' and Q' pentagons are formed from those of the P and Q pentagons, respectively, by the reflection operation against the vertical mirror plane. The structure of the crystalline approximant is formed by an ordered arrangement of those pentagons, as indicated in Fig. 30.2(b).

30.4 Summary

An $Al_{66}Cu_{15}Co_{19}$ crystalline approximant, which is related to Al–Cu–Co decagonal quasicrystals with two aperiodic planes stacking along the periodic axis, has been studied by Cs-corrected HAADF-STEM observations. The structure has an orthorhombic unit cell with $a_0 = 10.1$ nm, $b_0 = 0.4$ nm and $c_0 = 6.7$ nm, which is formed by an ordered arrangement of two types of atom clusters in a τ^3-inflated monoclinic $Al_{13}Co_4$ structure formed by a network of pentagons with an edge-length of 2 nm. Individual TM atoms and mixed sites of TM and Al atoms are represented as bright dots in HAADF-STEM images, and so a three-dimensional arrangement of TM atoms and mixed sites in the approximant can be directly determined. The TM atoms and mixed sites in the two atomic planes stacking along the b-axis are located at lattice points of a Penrose lattice with a bond length of 0.25 nm and form pentagonal tilings with bond lengths of 0.47 nm and 0.76 nm. This result suggests that the structure of Al–TM decagonal quasicrystals should be characterized as aperiodic arrangements of TM atoms with a bond-orientational order, instead of previous cluster models in which the structures of decagonal quasicrystals were characterized by aperiodic arrangements of 2 nm atom clusters [2, 4].

References

1. Ritsch S, Beeli C, Nissen H-U, Godecke T, Scheffe M, Lück R (1998) The existence regions of structural modifications in decagonal Al–Co–Ni. Philos Mag Lett 78:67–75
2. Hiraga K (2002) Advances in imaging and electron physics, vol 122. Elsevier Science, Amsterdam
3. Steurer W (2004) Twenty years of structure research on quasicrystals, part I: pentagonal, octagonal, decagonal and dodecagonal quasicrystals. Z Kristallogr 219:391–446
4. Hiraga K, Ohsuna T, Sun W, Sugiyama K (2001) Structural characteristics of Al–Co–Ni decagonal quasicrystals and crystalline approximants. Mater Trans 42:2354–2367
5. Taniguchi S, Abe E (2008) Highly-perfect decagonal quasicrystalline $Al_{64}Cu_{22}Co_{14}$ with noncentrosymmetry. Philos Mag 88:1949–1958
6. Ma XL, Li XZ, Kuo KH (1995) A family of τ-inflated monoclinic $Al_{13}Co_4$ phases. Acta Crystallogr, Sect B 51:36–43
7. Sugiyama K, Nishimura S, Hiraga K (2002) Structure of a W-(AlCoNi) crystalline phase related to Al–Co–Ni decagonal quasicrystals, studied by single crystal X-ray diffraction. J Alloys Compd 342:65–71
8. Yasuhara A, Saito K, Hiraga K (2012) In: Proceedings of aperiodic 2012

Chapter 31
Structure of ϵ_{16} Phase in Al–Pd–Co System Studied by HREM and X-Ray Diffraction

K. Yubuta, S. Suzuki, R. Simura, and K. Sugiyama

Abstract A variety of orthorhombic approximant ϵ_n ($n = 6$, 16, 22, and 28)-phases exist in Al–Pd–(Mn, Fe, Co, Rh) systems. HREM images and the corresponding electron diffraction patterns show that the ϵ_{16} phase in an annealed $Al_{80}Pd_{11}Co_9$ alloy exhibits a locally disordered structure consisting of pentagonal and banana-shaped tiles with an edge length of 0.76 nm. This paper demonstrates a feasible structural model for the ϵ_{16} phase in the Al–Pd–Co system by single-crystal X-ray diffraction coupled with HREM.

31.1 Introduction

A variety of orthorhombic phases related to the Al_3Pd approximant phase with eight planes along the periodic axis were found in Al–Pd–(Mn, Fe, Co, Rh) systems [1]. Their lattice parameters of $a = 2.35$ and b (pseudo-tenfold axis) $= 1.68$ nm are essentially similar to each other; nevertheless, the corresponding c parameters are 1.2, 3.2, 4.5, and 5.7 nm for ϵ_6 (Al_3Pd phase), ϵ_{16}, ϵ_{22}, and ϵ_{28} phases, respectively [2]. Balanetskyy et al. reported compositional ranges of ϵ_n ($n = 6$, 16, 22, and 28)-phases on the Al–Pd–Fe system [3, 4] as shown in Fig. 31.1(a). As one can notice in Fig 31.1(a), the compositional range of the ϵ_n-phases extends to the Pd-poor region with increasing temperature range.

The crystal structures of the ϵ_6 phase were determined by some groups [5, 6]. On the other hand, the atomic positions for a series of ϵ_n ($n = 16$, 22, and 28)-phases were not studied quantitatively, though their structures were investigated by using transmission electron microscopy [7, 8].

In the present paper, the crystal structure of the ϵ_{16} phase (Fig. 31.1(b)) in annealed Al–Pd–Co alloys [7, 9, 10] was investigated by HREM and single-crystal X-ray diffraction.

K. Yubuta (✉) · S. Suzuki · R. Simura · K. Sugiyama
Institute for Materials Research, Tohoku University, Sendai 980-8577, Japan
e-mail: yubuta@imr.tohoku.ac.jp

S. Schmid et al. (eds.), *Aperiodic Crystals*, DOI 10.1007/978-94-007-6431-6_31,

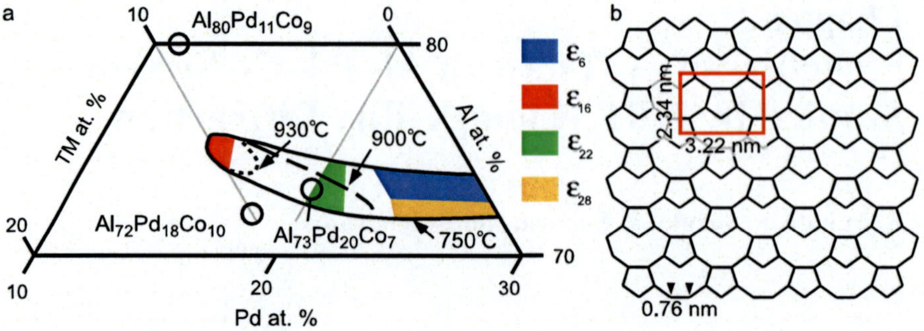

Fig. 31.1 (**a**) Compositional range of the ϵ_n ($n = 6$, 16, 22, and 28)-phases on the Al–Pd–Fe system reported in [3, 4]. Three circles indicate compositions of the Al–Pd–Co system in the present study. (**b**) The tiling image for the ϵ_{16} phase is shown together with a rectangular unit cell for the convenience of discussion

Table 31.1 Annealing conditions of $Al_{80}Pd_{11}Co_9$, $Al_{72}Pd_{18}Co_{10}$, and $Al_{73}Pd_{20}Co_7$ alloys

No.	Phase	Composition	Annealing and cooling conditions
APC1	ϵ_{16}	$Al_{80}Pd_{11}Co_9$	1000 °C for 6.5 h, 790 °C at a rate of 2.3 °C/h
APC2	ϵ_x^*	$Al_{72}Pd_{18}Co_{10}$	1000 °C for 6.5 h, 790 °C at a rate of 2.3 °C/h
APC3	ϵ_6	$Al_{73}Pd_{20}Co_7$	1000 °C for 6.5 h, 700 °C at a rate of 7.5 °C/h

$^*\epsilon_x$—incommensurate phase

31.2 Experimental Procedures

Alloy ingots with compositions $Al_{80}Pd_{11}Co_9$ (APC1), $Al_{72}Pd_{18}Co_{10}$ (APC2), and $Al_{73}Pd_{20}Co_7$ (APC3) were prepared from high-purity Al, Pd, and Co metals by conventional arc-melting in a purified Ar atmosphere. Three circles in Fig. 31.1(a) indicate the compositions of Al–Pd–Co alloy ingots prepared in the present study. Fragments of the as-prepared ingot were charged into a carbon crucible and annealed in a high vacuum furnace. Conditions of the annealing for the alloy samples are listed in Table 31.1.

HREM images and selected area electron diffraction (SAED) patterns were taken by using a 200 kV electron microscope (TOPCON EM-002B). Single crystal diffraction data was measured by Mo $K\alpha$ ($\lambda = 0.071073$ nm) radiation with Rigaku RAPID-AUTO equipped with an imaging plate. The least-squares software of *SHELXL-97* [11] was used for the structural analysis.

31.3 Results and Discussion

Figure 31.2 shows SAED patterns of (a) $Al_{80}Pd_{11}Co_9$, (b) $Al_{72}Pd_{18}Co_{10}$, and (c) $Al_{73}Pd_{20}Co_7$ alloys, taken with the incident beam parallel to pseudo-tenfold

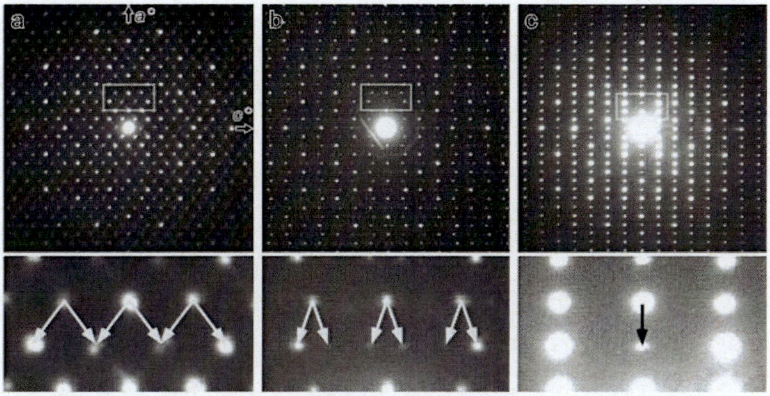

Fig. 31.2 Electron diffraction patterns of (**a**) $Al_{80}Pd_{11}Co_9$, (**b**) $Al_{72}Pd_{18}Co_{10}$, and (**c**) $Al_{73}Pd_{20}Co_7$ alloys, taken with the incident beam parallel to pseudo-tenfold axis. *Arrows* in the enlarged patterns on the *bottom* indicate structural modulations

axis. Patterns on the bottom are enlarged from the areas enclosed by rectangles on the top. Arrows in the enlarged patterns on the bottom indicate structural modulations. Although there is a slight splitting of diffraction spots indicated by arrows in Fig. 31.2(a), new periodicity (2.35 and 3.27 nm with a base-centered symmetry) exists on the long-range. In addition, diffuse streaks are running along the arrows. From SAED patterns, $Al_{80}Pd_{11}Co_9$ (Fig. 31.2(a)) and $Al_{73}Pd_{20}Co_7$ (Fig. 31.2(c)) alloys correspond to the locally disordered ϵ_{16} and ϵ_6 phases, respectively. On the other hand, in $Al_{72}Pd_{18}Co_{10}$ (Fig. 31.2(b)) alloy, zigzag-like shifts of diffraction spots can be readily recognized. This appearance is similar to the Al–Pd–Fe system as reported by Balanetskyy et al. [3, 4]. It is reasonable to suppose that commensurate, incommensurate, and no modulations occur in $Al_{80}Pd_{11}Co_9$ (ϵ_{16} phase), $Al_{72}Pd_{18}Co_{10}$ (intermediate state between ϵ_6 and ϵ_{16} phases), and $Al_{73}Pd_{20}Co_7$ (ϵ_6 phase) alloys, respectively.

Figure 31.3 is an HREM image of the ϵ_{16} phase in the $Al_{80}Pd_{11}Co_9$ alloy taken with the incident beam parallel to the pseudo-tenfold axis. In the image, one can see ring contrasts corresponding to the decagonal atom columns with the 1.68 nm periodicity. As indicated by black lines, the ϵ_{16} phase consists of pentagonal and banana-shaped tiles with an edge length of 0.76 nm, which is the typical distance of edge-sharing of the decagonal atom columns. As indicated by thick lines, one can notice the stacking defects exist on the $(10\bar{1})$ plane. The direction of the stacking defects is consistent with that of the observed diffuse streaks in Fig. 31.2(a).

The observed reflection conditions, hkl: $h + l = 2n$, in the diffraction pattern suggest the B-centered orthorhombic structure with $a = 2.35$ nm, $b = 1.67$ nm, and $c = 3.27$ nm. Although the four space groups of *B*222, *Bm*2*m*, *B*2*mm,* and *Bmmm* fulfill these conditions, the HREM image of Fig. 31.3 leads to the further selection of space group symmetry—the possible space group was concluded as *B*2*mm*. A single crystal sample of size $75 \times 90 \times 150\ \mu m^3$ was cut out from the annealed $Al_{80}Pd_{11}Co_9$ alloy and used for the X-ray examination. Since the dif-

Fig. 31.3 HREM image of the $Al_{80}Pd_{11}Co_9$ alloy taken with the incident beam parallel to the pseudo-tenfold axis. The *bright circles* in the image correspond to the projection of the decagonal atom columns with the 1.68 nm periodicity. *Gray rectangles* indicate the unit cells of the ϵ_{16} phase

fuse scattering feature in the observed diffraction pattern of the ϵ_{16} phase could not allow us to measure an accurate intensity data set, the least-squares refinement was performed only by using the observed reflections with $F > 4.0\sigma(F)$. During the refinement of atomic positions and corresponding isotropic atomic displacement parameters, several chemically disordered sites of Pd/Co, Pd/Al, and Co/Al were introduced so as to produce the reasonable atomic distance together with the isotropic displacement parameters. The refinement converged with $R_1 = 0.175$ for the observed 8325 reflections. The refined cell parameters are $a = 2.321(1)$ nm, $b = 1.6426(1)$ nm, $c = 3.1941(8)$ nm. The chemical composition for $Al_{70}Pd_{17}Co_{13}$ of the obtained model deviates in Al-poor region from an analyzed one, namely that of $Al_{74.4}Pd_{15.9}Co_{9.7}$ determined by EPMA. This suggests that several additional Al sites should be considered in the structural model. Nevertheless, the heavy metal sites of the obtained model agree well with those proposed for the ϵ_{16} structure of the Al–Rh system[8].

The structure model of the ϵ_{16} phase along the a-axis is shown in Fig. 31.4(a). The structure could be described by the stack of eight layers perpendicular to the b-axis. The layers at $y = 0$ (***A***), $\simeq 0.125$ (***B***), $\simeq 0.25$ (***C***), $\simeq 0.375$ (***D***), and $1/2$ (***E***) are independent (Figs. 31.4(b)–(f)) and the other three can be generated by a mirror symmetry located at $y = 1/2$. Each layer structure indicates a pentagonal tiling with heavy metal apexes. The flat layers of ***A***, ***C***, and ***E*** indicate a pentagonal tiling with an edge length of 0.76 (=0.47τ) nm so as to form ship-shaped octagons as indicated by broken lines. On the other hand, the puckered layers of ***B*** and ***D*** show a tiling with an edge length of 0.47 nm and star-shaped pentagons, ship-shaped octagons and hexagons are realized in the tiling as indicated by broken lines. As shown in Fig. 31.4, the structure of the ϵ_{16} phase could be described by two types of columnar structures along the b-axis: a decagonal column and a pentagonal column. The connection of five decagonal columns forms the pentagonal tile with a pentagonal column inside and that of nine decagonal columns forms the banana-shaped tile, which could be observed in the HREM image, as indicated by red lines.

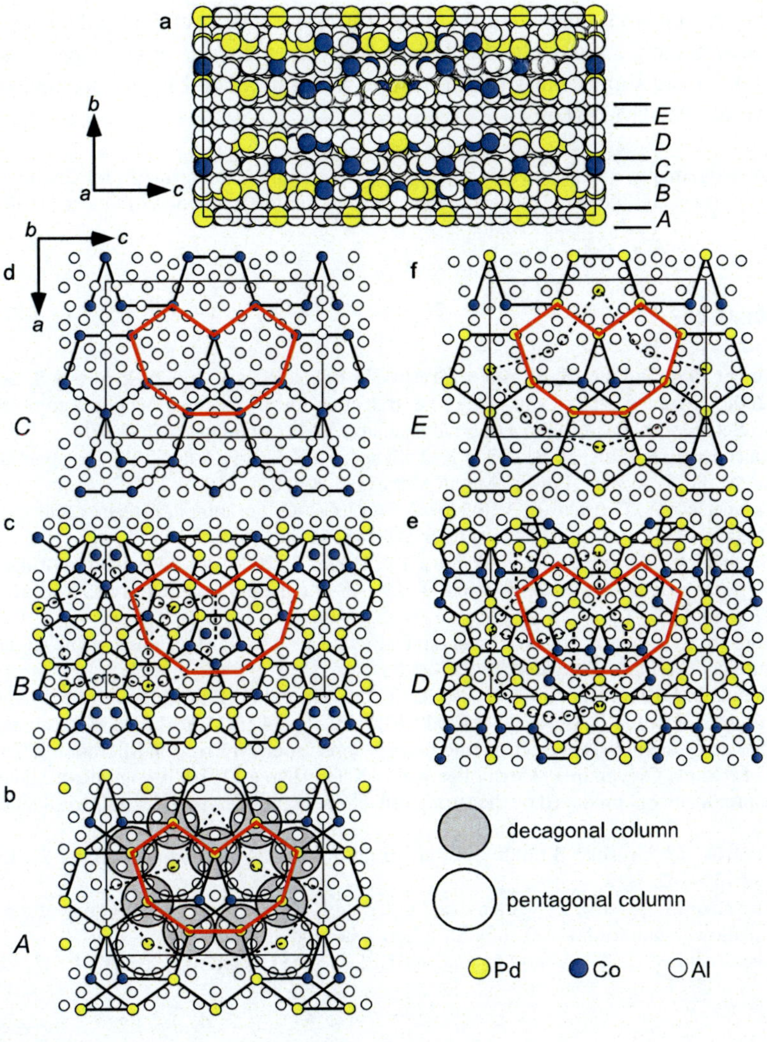

Fig. 31.4 The refined structure model of the ϵ_{16} phase

31.4 Summary

The present paper demonstrates the structural model of the ϵ_{16} phase in the Al–Pd–Co system. The structure could be described by the layered structure stacked along the pseudo-tenfold axis and the transition metal positions represent two different ways of a pentagonal tiling in the flat and puckered layers. The alternate stack of these pentagonal tiling produces the characteristic atomic arrangement of decagonal and pentagonal columns running parallel to the pseudo-tenfold axis whose linkage produces the pentagonal and the banana-shaped tiles with an edge length of 0.76 nm observed frequently in the decagonal quasicrystals of the Al–Pd–TM systems. It should be added that our structural model introduces the chemically disordered sites between Pd and Co. Such chemical disorder by the elements with different size may account for the significant structural imperfection realized in the ϵ_{16} phase of the ternary Al–Pd–TM systems.

Acknowledgements A part of this work was supported by a Grant-in-Aid for Challenging Exploratory Research (22656152) from the Japan Society for the Promotion of Science (JSPS).

References

1. Smontara A, Smiljanić I, Bilušić A, Grushko B, Balanetskyy S, Jagličić Z, Vrtnik S, Dolinšek J (2008) Complex ϵ-phases in the Al–Pd–transition-metal systems: towards a combination of an electrical conductor with a thermal insulator. J Alloys Compd 450:92–102
2. Balanetskyy S, Grushko B, Urban K, Velikanova TY (2005) A new rhombic approximant in the Al–Pd–Fe system. Powder Metall Met Ceram 44:282–286
3. Balanetskyy S, Grushko B, Velikanova TY (2004) Orthorhombic ϵ-phases and transitional structures in Al–Pd–(Fe). Z Kristallogr 219:548–553
4. Balanetskyy S, Grushko B, Velikanova TY, Urban K (2004) An investigation of the Al–Pd–Fe phase diagram between 50 and 100 at.% Al: phase equilibria at 750 °C. J Alloys Compd 376:158–164
5. Matsuo Y, Hiraga K (1994) The crystal structure of Al_3Pd: close relationship to decagonal quasicrystal. Philos Mag Lett 70:155–161
6. Boudard M, Klein H, de Boissieu M, Audier M, Vincent H (1996) Mag A 74:939–956
7. Hiraga K, Ohsuna T, Kawasaki M (2000) The structure of an Al_3Pd-based modulated structure studied by atomic-scale electron microscopy observations. J Electron Microsc 49:729–733
8. Li M, Sun J, Oleynikov P, Hovmöller S, Zou X, Grushko B (2010) A complicated quasicrystal approximant ϵ_{16} predicted by the strong-reflections approach. Acta Crystallogr, Sect B 66:17–26
9. Yurechko M, Grushko B (2000) A study of the Al–Pd–Co alloy system. Mater Sci Eng 294–296:139–142
10. Yurechko M, Grushko B, Velikanova TY, Urban K (2002) Isothermal sections of the Al–Pd–Co alloy system for 50–100 at.% Al. J Alloys Compd 337:172–181
11. Sheldrick GM (2008) A short history of SHELX. Acta Crystallogr, Sect A 64:112–122

Chapter 32
Structure of τ^2-Al_3Co, a Monoclinic Approximant of the Al–Co Decagonal Quasicrystal

K. Sugiyama, A. Yasuhara, and K. Hiraga

Abstract The structure of the approximant τ^2-Al_3Co ($P2/m$: $a = 3.9831(3)$ nm, $b = 0.8127(1)$ nm, $c = 3.2182(3)$ nm, and $\beta = 108.03(1)°$), associated with the decagonal quasicrystals with a period of 0.8 nm, was analyzed using a high-angle annular detector dark-field (HAADF) observation with *C*s-corrected scanning transmission electron microscopy (STEM). The HAADF-STEM image clearly showed the arrangement of individual Co atoms as bright dots. The contrast among the atoms in the lattice led to an image of the fundamental structure of the τ^2-Al_3Co phase, composed of an ordered arrangement of pentagonal columnar units with edge lengths of 0.47 nm. The arrangements of atoms in the columnar units were quantitatively determined by single crystal X-ray diffraction (XRD). The results demonstrate that the pentagonal columnar units form common tiles in the shape of a squashed hexagon, a pentagonal star, and a crown. Among the tiles, the pentagonal star composed of 10 pentagonal units was similar to that found in the W-(AlNiCo) approximant for the Al–Ni–Co decagonal quasicrystal (DQC).

32.1 Introduction

A variety of intermetallic crystalline phases, structurally associated with quasicrystals, have been found while researching Al-based quasicrystals. They are referred to as crystalline approximants for quasicrystals, and their structural features enable the elucidation of atomic arrangements in quasicrystals. In particular, decagonal quasicrystals (DQCs) and their approximants share a common decagonal columnar

K. Sugiyama · K. Hiraga
Institute for Materials Research, Tohoku University, Sendai 980-8577, Japan

K. Sugiyama
e-mail: kazumasa@imr.tohoku.ac.jp

K. Hiraga
e-mail: hiraga@imr.tohoku.ac.jp

A. Yasuhara (✉)
EM Application Group, EM Business Unit, JEOL Ltd., Akishima 196-8558, Japan
e-mail: ayasuhar@jeol.co.jp

S. Schmid et al. (eds.), *Aperiodic Crystals*, DOI 10.1007/978-94-007-6431-6_32,

structure running parallel to the periodic axis [1]. Recent atomic-scale observations made using high-resolution electron microscopy (HREM) coupled with high-angle annular detector dark-field scanning transmission electron microscopy (HAADF-STEM) have revealed the structural features of DQCs in terms of their columnar structures [2].

In the binary Al–Co system, DQCs with periodicities of about 0.4, 0.8, and 1.6 nm have previously been reported for rapidly solidified samples [3–6]. The (100) twins of the τ^2-Al_3Co phase have been achieved by annealing DQCs in particular, and the [010] electron diffraction pattern showed a strong resemblance to the tenfold electron diffraction pattern of DQCs [7]. This led to an interesting discussion about whether DQCs and the τ^2-Al_3Co phase consisted of similar columnar structures, and a structural model for the τ^2-Al_3Co phase was proposed based on HREM images [8–11]. A pentagonal unit, with an edge length of 1.2 nm, composed of six smaller pentagonal subunits with edge lengths of 0.47 nm, was proposed in the structural model. This configuration of the smaller pentagons is also found in the Penrose pattern, which is frequently used as a structural model for quasicrystals.

In this context, single crystal X-ray diffraction (XRD) is necessary in order to understand the atomic-level structures of the pentagonal columnar structures in the τ^2-Al_3Co phase. Since the advantages of using single crystal XRD coupled with HAADEF-STEM was demonstrated in the analysis of the complex approximant W-(AlCoNi) [12], we used it to analyze the structure of the τ^2-Al_3Co phase.

32.2 Experimental Procedure

An alloy ingot of $Al_{79}Co_{21}$ was prepared from pure Al (99.999 %) and Co (99.9 %) in Ar atmosphere in a conventional arc furnace. The solidified ingot was remelted several times to ensure sample homogeneity. Fragments of the prepared sample were charged into an Al_2O_3 crucible and were subsequently sealed in an evacuated silica tube. The sample tube was heated at 1180 °C for 24 h and then cooled to 1150 °C at the rate of 10 °C/h. Electron probe microanalysis (EPMA, JEOL JXA-8621MX) showed that the chemical composition of the present phase was $Al_{74.8}Co_{25.2}$. Thin samples for transmission electron microscopy were prepared by dispersing crushed samples onto holey carbon films. HAADF-STEM images were taken using a spherical aberration (Cs)-corrected electron microscope (JEOL-ARM200F) operated in scanning transmission mode. It should be added that recent Cs-corrected STEM has high-enough resolution to produce images of individual transition metal atoms as bright dots when the images are taken with the incident beam parallel to the periodic axis of the DQCs. A single crystal sample with a volume of $0.03 \times 0.04 \times 0.05$ mm^3 was cut from the annealed sample and was used for the XRD analysis. Intensity data sets for the structural analysis were collected in the ω-scan mode on a Bruker SMART APEX-II system by using Mo $K\alpha$ radiation ($\lambda = 0.71073$ Å). After Lorentz and polarization corrections, an absorption correction was performed using an SADABS algorithm [13]. The refined cell parameters were $a = 3.9831(3)$ nm,

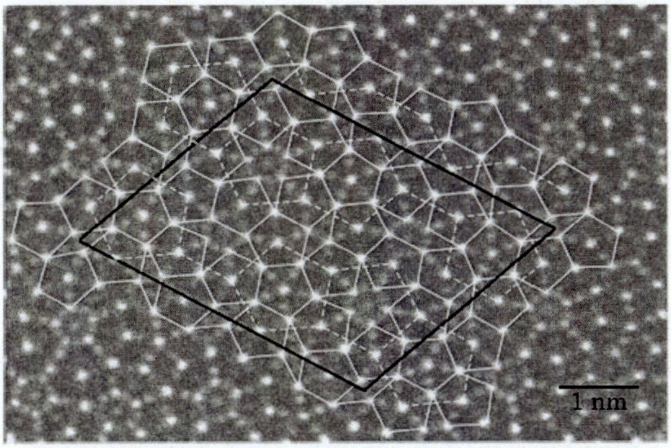

Fig. 32.1 HAADF-STEM image of the τ^2-Al_3Co phase and tilings formed by connecting bright dots corresponding to Co atoms in A (*solid*) and B (*dotted*) layers stacked along pseudo-tenfold axis. *Black lines* indicate unit cell of the τ^2-Al_3Co phase

$b = 0.8127(1)$ nm, $c = 3.2182(3)$ nm, and $\beta = 108.03(1)°$, consistent with those reported in previous research [8–11]. For the present analysis, the least-squares software SHELXL-97 [14] was used.

32.3 Results and Discussion

Figure 32.1 shows an HAADF-STEM image of the τ^2-Al_3Co phase, taken with the incident beam parallel to the pseudo-tenfold axis. Almost all of the bright dots (corresponding to Co atoms) in the image can be connected as pentagonal tiles with edge lengths of 0.47 (solid lines) and 0.76 nm (dashed lines), which were later revealed by single crystal XRD to be those in the A and B layers, respectively, stacked along the b-axis. It should be added that pentagonal tilings with edge lengths of 0.47 and 0.76 nm also exist in the A and B layers of the W-(AlCoNi) approximant phase [12]. Although the structure of the W-(AlCoNi) phase consists of one type of 1.2 nm columns connected at an interval of 2 nm, the structure of the τ^2-Al_3Co phase appears to be formed by two kinds of 1.2 nm decagonal columns, whose detailed arrangements of atoms is well characterized in the following single crystal XRD image.

The XRD pattern indicates no systematic absence of reflections, and the space group $P2/m$ was selected for the τ^2-Al_3Co phase. However, a number of weak intensity peaks for the reflection condition $h + k = 2n$ indicated pseudo-translational symmetry in the structure. Starting from the structural model produced from the HAADF-STEM image, the least-squares refinement revealed the atomic positions of 73 Co, 12 Co/Al, and 168 Al sites, among which 2 Co/Al and 5 Al sites were split into several additional sites. During the refinement, the atomic positions and corresponding isotropic displacement parameters were refined, and several chemically disordered Co/Al sites were introduced so as to obtain the reasonable isotropic displacement parameters. The reliable factor for the present structural model converged to $R_1 = 0.104$ for the observed 27201 reflections, with $F > 4.0\sigma(F)$. The chemical composition of the structural model was $Al_{511.6}Co_{175.6}$, which corresponds well

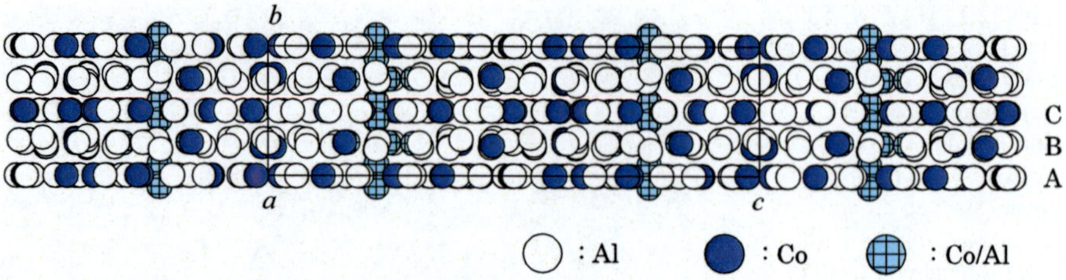

Fig. 32.2 The a-axis projection of τ^2-Al_3Co structure

with Al_3Co. It should be added that a pseudo-merohedral (001) twin was also considered in the present analysis.

The structural model of the τ^2-Al_3Co phase viewed along the a-axis is shown in Fig. 32.2. There are four atomic layers perpendicular to the b-axis. Three of them (at $y = 0$ (A), ≈ 0.25 (B), and $= 1/2$ (C)) are independent. The other one can be generated through a mirror symmetry located at $y = 1/2$. The atoms within the B layer deviate from $y = 1/4$, so it is considered to be a puckered layer. Several noticeably chemically disordered sites are located outside these layers. The sites correspond to the centers of pentagonal units with edge lengths of 1.2 nm, composed of six smaller pentagonal subunits with edge lengths of 0.47 nm.

Figure 32.3 shows the structures of the three layers, (A), (B), and (C). As shown in the HAADF-STEM image, the locations of Co atoms in layers (A) and (C) are well-defined by pentagonal tilings with edge lengths of 0.47 nm, and those in layer B are well-defined by a pentagonal tiling with an edge length of 0.67 nm. The association between the two types of pentagonal tiling is shown in the HAADF-STEM image by solid and dotted lines. It is clear that the six smaller Co pentagons form a τ^2-inflated pentagon with edge lengths of 1.2 nm, as shown in Figs. 32.3(A) and (C) [8–11, 15]. The tessellation of the τ^2-inflated pentagons in the τ^2-phase is exactly the same as that of the smaller pentagons in monoclinic $Al_{13}Co_4$ [16]. This is attributed to the fact that the a and c parameters of the τ^2-phase are τ^2 times ($\tau = (1 + 5^{1/2})/2$) greater than those of monoclinic $Al_{13}Co_4$. Further, almost all of the Co pentagons with edge lengths of 0.47 nm in τ^2-phase are similar to those found in monoclinic and orthorhombic $Al_{13}Co_4$ [16–18]. This suggests that the structure of the τ^2-phase is achieved by linking the pentagonal columnar units found in monoclinic and orthorhombic $Al_{13}Co_4$. The pentagonal columnar units produce three common basic tilings: a squashed hexagon (H), a pentagonal star (S), and a crown (C) (marked in Fig. 32.3(A)). Two H tiles linked together with one C tile form a decagonal ring, and each S tile is surrounded by three decagonal rings. It is interesting to note that the chemically disordered sites are located at the junctions of the H, S, and C tiles.

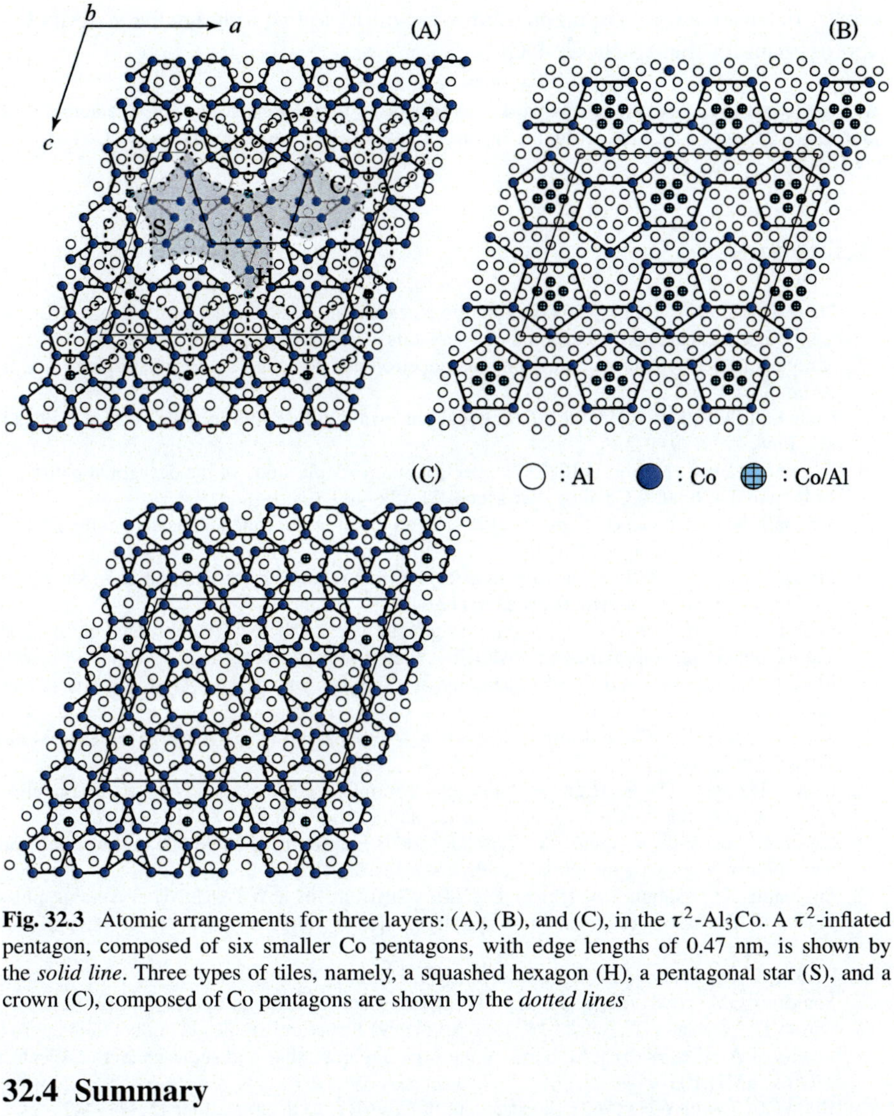

Fig. 32.3 Atomic arrangements for three layers: (A), (B), and (C), in the τ^2-Al_3Co. A τ^2-inflated pentagon, composed of six smaller Co pentagons, with edge lengths of 0.47 nm, is shown by the *solid line*. Three types of tiles, namely, a squashed hexagon (H), a pentagonal star (S), and a crown (C), composed of Co pentagons are shown by the *dotted lines*

32.4 Summary

The crystalline approximant τ^2-Al_3Co was analyzed using Cs-corrected HAADF-STEM coupled with single crystal XRD. Its structure was found to be a monoclinic unit cell ($P2/m$: $a = 3.9831(3)$ nm, $b = 0.8127(1)$ nm, $c = 3.2182(3)$ nm, and $\beta = 108.03(1)^\circ$), associated with decagonal quasicrystals with a period of 0.8 nm. The structure consists of alternating flat and puckered layers, which are well-defined by pentagonal tilings with edge lengths of 0.47 and 0.76 nm, respectively. Single crystal XRD quantitatively revealed fundamental columnar units, similar to those found in monoclinic and orthorhombic $Al_{13}Co_4$. These pentagonal columnar units form three common basic tilings: a squashed hexagon, a pentagonal star, and a crown. In particular, the pentagonal star was similar to that found in the W-(AlNiCo) approximant for the Al–Ni–Co DQC.

Acknowledgement A part of this work was supported by a Grant-in-Aid for Challenging Exploratory Research (22656152) from the Japan Society for the Promotion of Science (JSPS).

References

1. Steurer W (2004) Twenty years of structure research on quasicrystals, part I: pentagonal, octagonal, decagonal and dodecagonal quasicrystals. Z Kristallogr 219:391–446
2. Hiraga K (2002) Advances in imaging and electron physics, vol 122. Elsevier Science, Amsterdam
3. Dong C, Li GB, Kuo KH (1987) Decagonal phase in rapidly solidifies Al_6Co alloy. J Phys F, Met Phys 17:L189–192
4. Suryanarayana C, Menon J (1987) An electron microscopic study of the decagonal phase in a melt-spun Al-26 wt.% Co alloy. Scr Metall 21:459–460
5. Menon J, Suryanarayana C, Singh G (1989) Polytypism in a decagonal quasicrystalline Al–Co phase. J Appl Crystallogr 22:96–99
6. Ma XL, Kuo KH (1992) Decagonal quasicrystal and related crystalline phases in slowly solidified Al–Co alloys. Metall Trans 23A:1121–1128
7. Ma XL, Kuo KH (1994) Crystallographic characteristics of the Al–Co decagonal quasicrystal and its monoclinic approximant τ^2-$Al_{13}Co_4$. Metall Trans 25A:47–56
8. Ma XL, Li XZ, Kuo KH (1995) A family of τ-inflated monoclinic $Al_{13}Co_4$ phases. Acta Crystallogr, Sect B 51:36–43
9. Ma XL, Kuo KH (1995) Multiple twins of τ^2-$Al_{13}Co_4$ showing fivefold symmetry. Metall Trans 26A:757–763
10. Li XZ, Hiraga K (1998) High-resolution electron microscopy of the ε-Al_3Co, a monoclinic approximant of the Al–Co decagonal quasicrystal. J Alloys Compd 269:L13–L16
11. Saitoh K, Yokosawa T, Tanaka M, Tsai AP (2004) Formation of an Al–Co quasicrystal from a τ^2-inflated $Al_{13}Co_4$ approximant. J Phys Soc Jpn 68(9):2886–2889
12. Sugiyama K, Nishimura S, Hiraga K (2002) Structure of a W-(AlCoNi) crystalline phase related to Al–Co–Ni decagonal quasicrystals, studied by single crystal X-ray diffraction. J Alloys Compd 342:65–71
13. Bruker (2001) Bruker AXS Inc., Madison, WI, USA
14. Sheldrick GM (2008) A short history of SHELX. Acta Crystallogr, Sect A 64:112–122
15. Saitoh K, Yokosawa T, Tanaka M, Tsai AP (1999) Structural studies of monoclinic approximants of $Al_{13}Fe_4$ and τ^2-$Al_{13}Co_4$ by the high-angle annular dark-field method. J Electron Microsc 48(2):105–114
16. Hudd RC, Taylor WH (1962) The structure of Co_4Al_{13}. Acta Crystallogr 15:441–442
17. Widom M, Phillips R, Zou J, Carlsson AE (1995) Structural model of orthorhombic Al_3Co. Philos Mag B 71:397–406
18. Grin J, Burkhardt U, Ellner M, Peters K (1994) Crystal structure of orthorhombic Co_4Al_{13}. J Alloys Compd 206:243–247

Chapter 33
Reverse Monte Carlo Study of Diffuse Scattering from a Frustrated Protein System

T.R. Welberry, A.P. Heerdegen, and P.D. Carr

Abstract Distinctive diffuse rings around Bragg positions have been observed in the diffraction patterns of a crystal of the N-terminal fragment of the Gag protein from Feline Foamy Virus. It is shown that these are caused by geometric frustration as molecules try to pack on the triangular a–b mesh of the space group $P6_122$. The disorder prohibits conventional structure solution. The possibility of using the diffuse scattering to aid solution is explored using Reverse Monte Carlo modelling.

33.1 Introduction

Strong diffuse scattering has been observed in Bragg data collected from a crystal of the N-terminal fragment of the Gag protein from Feline Foamy Virus. Reciprocal space sections reconstructed from these data show diffuse scattering, largely confined to thin layers normal to **c** corresponding to the sections $h\ k\ nl$, with n integral. Within these layers the scattering lies in diffuse rings around each Bragg position. For layers with $n = 5m$ (m integral) the diffuse intensity is virtually absent and the sections contain only Bragg peaks; for $n = 5m \pm 1$ the layers contain moderately strong Bragg peaks and weak diffuse scattering; while for $n = 5m \pm 2$ the sections contain very weak Bragg peaks but strong diffuse scattering.

These patterns have been shown [1] to be caused by frustration on a triangular lattice (see, e.g. [2]). The protein is occupationally disordered and the crystal contains two different molecular entities, **A** and **B**. Neighbouring pairs on the triangular mesh of the a–b plane prefer to be of type **AB** or **BA** rather than **AA** or **BB**. For the diffuse scattering to be confined to layers of integral l, neighbours along **c** of type **AA** and **BB** are preferred. This is shown schematically in Fig. 33.1(a) where the motifs shown in two different colours represent (in projection) columns of **AAA**... or **BBB**... molecules. Figure 33.1(c) shows a representation of a region of crystal in which the frustration has been built-in using Monte Carlo simulation (see [1]).

To explain the distributions of Bragg reflections and diffuse features described above, **A** and **B** must have the same molecular shape but differ in their position along

T.R. Welberry (✉) · A.P. Heerdegen · P.D. Carr
Research School of Chemistry, Australian National University, Canberra, ACT 0200, Australia
e-mail: welberry@rsc.anu.edu.au

S. Schmid et al. (eds.), *Aperiodic Crystals*, DOI 10.1007/978-94-007-6431-6_33,

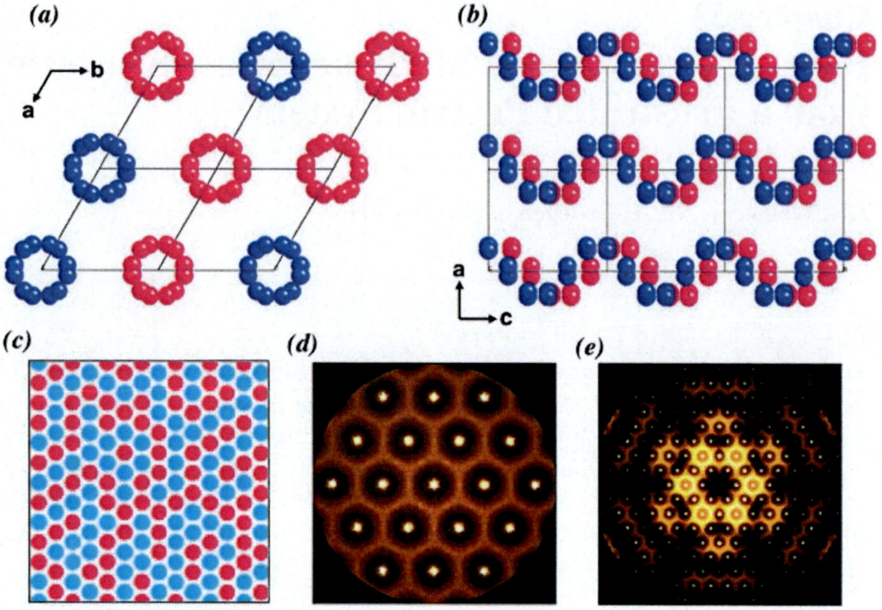

Fig. 33.1 (**a**) and (**b**) show projections of a disordered structure containing molecules **A** and **B** with two different shifts along **c**. (**c**) shows a region of the lattice illustrating the frustrated distribution. (**d**) shows the diffraction pattern of this distribution when **A** and **B** are single atoms. (**e**) shows the same result when **A** and **B** correspond to the 12 atom helical molecules shown in (**a**) and (**b**)

c by 0.2**c** (Fig. 33.1(b)). If F_A and F_B are the two molecular scattering factors,

$$\begin{aligned} F_B &= F_A \exp(2\pi i \mathbf{S} \cdot \mathbf{c}/5), \\ F_A - F_B &= F_A\big(1 - \exp(2\pi i \mathbf{S} \cdot \mathbf{c}/5)\big), \\ F_A + F_B &= F_A\big(1 + \exp(2\pi i \mathbf{S} \cdot \mathbf{c}/5)\big). \end{aligned} \tag{33.1}$$

Since the diffuse scattering is proportional to the *difference* of the scattering factors and the Bragg scattering to the *average* [3], Eq. (33.1) explains the observations.

Figures 33.1(d) and (e) show diffraction patterns computed from the distribution shown in Fig. 33.1(c) when the two different scatterers are: (d) single atoms and (e) helical 12 atom molecules (as shown in Figs. 33.1(a) and (b)) [1]. This model explains the basic form of the observed diffraction patterns. We suggest that in the real protein crystal the same basic physics is occurring but the scatterer is the protein molecule, with its highly complicated molecular scattering factor giving rise to the far richer detail in the diffuse scattering.

Since only one reciprocal section in five contains normal Bragg peak intensities, two contain reduced intensity and two contain little or no Bragg intensity, the number of reliable observations available for structure determination is much reduced compared to a normal data set and this has so far prevented structure solution. In the work described herein, we aim to use the rich detail in the diffuse scattering to help solve the protein structure. We use Reverse Monte Carlo (RMC) simulation to simultaneously fit a generic scattering model to both the observed diffuse scattering *and* to the Bragg reflection data.

33.2 X-Ray Data

Data were collected and processed as noted in [1]. For modelling, diffuse scattering from eight reciprocal sections was used ($h\ k\ nl$, with $n = 1, 2, 3, 4, 6, 7, 8, 9$), totalling ~112,000 unique pixels with $(\sin\theta/\lambda) < 0.111$ Å^{-1} (~4.5 Å resolution). Bragg data from 3941 reflections ($(\sin\theta/\lambda) < 0.17$ Å^{-1}, ~3.0 Å resolution) were used.

33.3 Reverse Monte Carlo Simulation

Reverse Monte Carlo (RMC) [4] is a variation of the Monte Carlo (MC) algorithm [5] where the difference between observed and calculated intensity is minimised, rather than the system energy. A model is set up in terms of variables such as atomic occupancies or displacements and these are adjusted as the RMC proceeds. In each step, a site in the model is selected at random and its variables are changed by a random amount. The scattering is calculated before and after the shift and the change in the goodness-of-fit parameter (Eq. (33.2)) is computed:

$$\chi^2 = \sum_{i=1}^{N} \frac{[I_e(\mathbf{h}_i) - I_c(\mathbf{h}_i)]^2}{\sigma^2}. \tag{33.2}$$

The sum is over all measured data points $\mathbf{h}_i$. I_e is experimental and I_c calculated intensity. The change in χ^2 is $\Delta\chi^2 = \chi^2_{\text{old}} - \chi^2_{\text{new}}$. If $\Delta\chi^2 < 0$ the move is accepted. If $\Delta\chi^2 > 0$ the move is accepted with a probability of $P = \exp(-\Delta\chi^2/2)$. σ (independent of $\mathbf{h}$) is analogous to the temperature $\mathbf{T}$ in the normal Monte Carlo method.

33.3.1 Development of RMC Model

The choice of the generic scattering model used in the study was influenced by several considerations. In the first stage of modelling, static sites were created on a grid within the asymmetric unit of the cell and the RMC process adjusted only the occupancies of these sites.

Parasitic scattering precluded reliable measurements of diffuse intensities at high $\mathbf{Q}$ and usable diffuse scattering data were confined to moderately low values of $\mathbf{Q}$

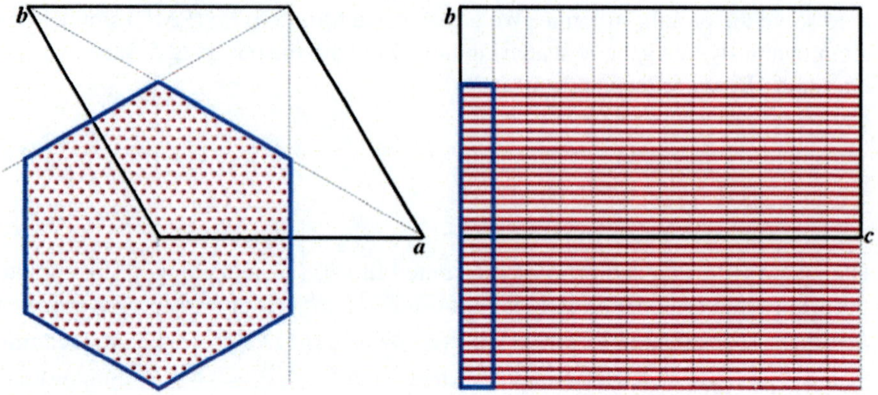

Fig. 33.2 The generic occupancy scattering model. *Small discs* represent possible scattering sites. The asymmetric unit in $P6_122$ is outlined by the *heavy* (*blue*) *lines*. Each point in the asymmetric unit is duplicated in each of the 11 other symmetry related units

within the protein *solvent ring* (minimum *d*-spacing of ∼5 Å). Within a given *a*–*b* layer of the 73 × 73 Å cell 25 × 25 grid points were used, giving a real-space resolution of ∼3 Å—somewhat better that of the observed diffuse scattering data. Correspondingly, of the full set of Bragg reflection data available, only those with a resolution of >∼3 Å were used (3941 reflections).

In the **c** direction, the asymmetric unit occupies only $\frac{1}{12}$ of the 109 Å cell. Ten grid points per asymmetric unit were used along **c**—a resolution of better than 1 Å. This was considered necessary to allow the fitting of different sections of data normal to **c** to develop independently. The total number of sites within the asymmetric unit (see the heavy (blue) lines in Fig. 33.2) was 6760.

The real molecule possesses a large number of non-H atoms (>1000), and so as a starting model we allowed the 6760 possible scattering sites to be occupied at random with ∼1000 scatterers. For this preliminary study, since almost all the non-H atoms in the protein are C, N or O (i.e. having a similar size), the same atomic scattering factor was used for all scatterers. At each RMC step, an occupied site was swapped with an unoccupied one, preserving the total number throughout.

Once the first stage of refinement had reached a χ^2 minimum, the scatterers were then allowed to shift away from the fixed grid points in stage two.

33.3.2 Results of RMC for Model 1

The RMC simulation was carried out for ∼350,000 steps to minimise χ^2. Table 33.1 lists agreement factors for 8 diffuse sections and 3941 Bragg reflections.

The final configuration of scatterers in molecule **A** is shown in Fig. 33.3(a). Inter-scatterer vectors <4 Å have been plotted to give a qualitative rendering of the electron density. Figure 33.4 shows observed and calculated patterns for sections *h k* 1–*h k* 4. Note that sections *h k* 6–*h k* 9 (not shown) were also used in the fitting.

Table 33.1 Agreement factors for observed and calculated diffuse intensities in the 8 reciprocal sections used in the RMC fit as well as the value of R_{Bragg}. Note the sections $hk1$–$hk4$ are shown in Fig. 33.4. $R_{\text{Diffuse}} = \sqrt{\sum(\Delta I)^2/\sum(I_{\text{obs}})^2}$, $R_{\text{Bragg}} = \sqrt{\sum(\Delta F)^2/\sum(F_{\text{obs}})^2}$

Section	Over all	$hk1$	$hk2$	$hk3$	$hk4$	$hk6$	$hk7$	$hk8$	$hk9$	R_{Bragg}
R (Model 1)	0.357	0.358	0.321	0.298	0.378	0.364	0.384	0.347	0.404	0.246
R (Model 2)	0.351	0.374	0.333	0.294	0.383	0.366	0.343	0.329	0.386	0.324[a]

[a]This value does not correspond to an agreement between observed and calculated data. See text

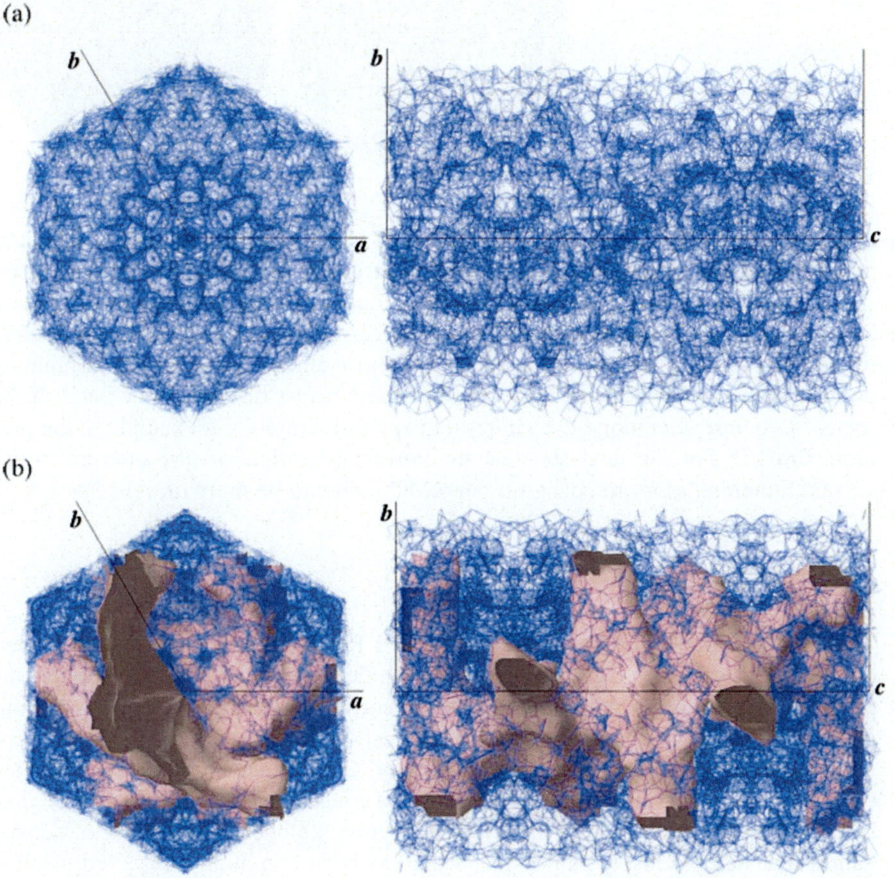

Fig. 33.3 Representations of the two different RMC-fitted scattering models (see text): (**a**) Model 1; (**b**) Model 2. All vectors between pairs of scatterers (atoms) <4 Å are plotted. The *left panels* correspond to projection down c and the *right panels* to projection down a. In (**a**), the scatterers are distributed over the whole cell. In (**b**), they are confined to the space outside of the *pink* isosurface that bounds the 'solvent' region

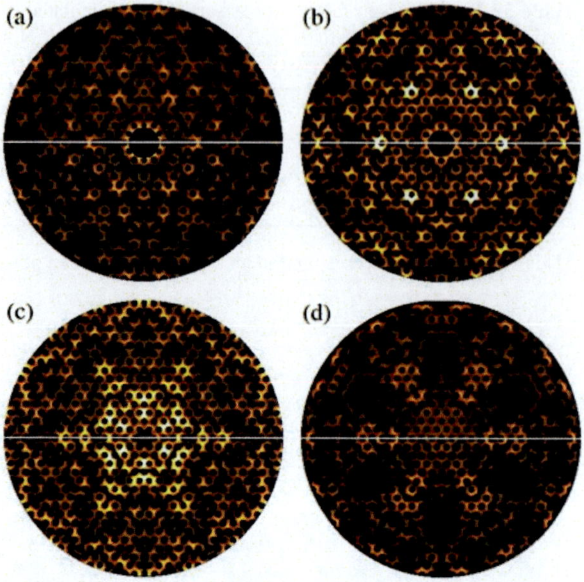

Fig. 33.4 Observed and calculated diffraction patterns for the four reciprocal sections: (**a**) $h\,k\,1$; (**b**) $h\,k\,2$; (**c**) $h\,k\,3$; (**d**) $h\,k\,4$. The *upper half* of each figure is observed data and the *lower half* is calculated from the RMC Model 1. All four sections are on the same intensity scale

In protein crystals, the polypeptide chain usually forms into compact structural domains that can be delimited by a molecular envelope. These regions are interspersed with large solvent channels of a comparable total volume, within these channels the solvent is considered to be totally disordered leading to a uniform electron density outside the molecular envelope. Improvement of initial error-containing phases can be achieved using knowledge of these properties and is the basis of the *solvent flattening* technique used in protein crystallography, for example in the program **dm** [6]. For the next stage of the current investigation, we attempt to use solvent flattening to try to constrain our RMC model to be more *protein-like*.

33.4 Solvent Flattening

The molecular coordinates **A** from the Model 1 RMC refinement were used to generate a list of structure factors and an electron density map which was used as the preliminary electron density model. The **dm** software essentially identifies regions of the map that contain errors (i.e. below threshold within the molecular envelope or significantly above the mean solvent density outside the envelope). The software replaces the values in the solvent region with a constant value and at the same time manipulates the values within the envelope so that the overall histogram of density values matches as closely as possible what has been commonly observed in other proteins. The chosen threshold dictates what percentage of the cell volume is occupied by solvent. For the present Gag protein, the solvent region is expected to be ~44 % [7]. However, using this percentage the solvent flattened map appeared quite disjointed with no clearly defined single molecular envelope. Consequently, for a first pass through the process a lesser percentage (30 %) was used.

33.5 Reverse Monte Carlo—Model 2

RMC simulation was repeated using the molecular envelope derived from the solvent flattening (Sect. 33.4). The ∼1000 scatterers were now constrained to lie initially within the molecular envelope although subsequently the displacements in stage 2 of the refinement allowed them to encroach into the solvent region, thus enabling the envelope to evolve with time. It might be anticipated that by repeating the cycle of solvent flattening followed by RMC simulation a self-consistent molecular envelope will be established.

This procedure assumes that the observed diffuse patterns result from the molecular region alone. Any diffuse scattering from the solvent would be a relatively structureless background and this would most likely have been removed by the data reduction algorithms. Thus it is appropriate to ignore the solvent regions in modelling the diffuse scattering. However, the Bragg scattering still contains a contribution from the solvent. In carrying out the new RMC simulation, Bragg scattering was calculated for a crystal of only **A** molecules and this was compared to the structure factors calculated from the solvent flattened electron density map. Consequently, the value of R_{Bragg} given in Table 33.1 is not a true reflection of the Model 2 fit and cannot be related to the value given for Model 1.

33.5.1 Results of RMC for Model 2

RMC simulation 2 proceeded much as the Model 1 simulation. For the diffuse scattering, the agreement factors (Table 33.1) and the observed and calculated intensities (not shown) are very similar to those for Model 1. The agreement factors for the higher layers, $h\,k\,5$–$h\,k\,9$, are somewhat better than for Model 1. Figure 33.3(b) shows a plot of the resulting structure, much like that of Model 1 (Fig. 33.3(a)) but including the isosurface marking the partition between the solvent (inside) and molecule regions.

The Fourier transformation of the Model 2 molecule **A** coordinates gives a revised electron density map which has an appearance that is much more *protein-like* than Model 1, with the solvent and protein regions clearly delineated.

As an initial step towards interpretation, we ran software [9, 10] that uses algorithms to recognize peptide chains within electron density maps, and where possible to include side chains in agreement with the known protein sequence. This produced a partial model containing 106 residues in 8 chains (longest containing 21 residues, 17 residues with side chains from the sequence). Our GAG protein fragment sequence is known to contain 154 residues.

As an additional piece of evidence we used the molecular envelope produced from small angle scattering data [8]. This was manually manipulated onto our electron density map. It was possible to find a region coinciding approximately in shape and overall dimensions with the strong electron density without large excursions into the solvent channels. This is encouraging as these are independent data and no knowledge of this small angle scattering model was used in producing our maps.

Although the resulting model coincided well with the molecular shape derived from small angle scattering, it did not match very well any known protein structure available in the protein data bank (PDB) [11]. This may mean we have an unusual and therefore interesting structure.

33.6 Conclusion

An initial model, consisting of a set of ∼1000 identical scatterers (atoms), has been derived that gives good agreement between observed and calculated data for both the diffuse scattering and the Bragg reflections for the N-terminal fragment of the Gag protein from Feline Foamy Virus. However, the derived model did not comprise a clearly discernible molecular envelope with interspersed solvent regions which is typical of a protein structure. While in our view the good agreement indicates that the derived structure must contain a great deal that is right, it must still contain extraneous features or noise that preclude its further interpretation as a protein molecule. In order to move forward, we have applied solvent flattening and density modification techniques to the electron density maps given by this model to derive a first estimate of a molecular envelope. Subsequently, a second RMC simulation was carried out in which the scatterers were constrained to lie within this envelope. This second model gave equally good agreement for the diffuse scattering patterns. Using algorithms to recognize peptide chains within the newly generated electron density map produced a partial model containing 106 residues (of 154 residues) in 8 chains (longest containing 21 residues). Moreover, this partial model fitted well within a molecular envelope derived independently from small angle scattering.

In a situation where the presence of disorder has drastically reduced the reliability of Bragg peak observations and solving the structure from the Bragg peaks alone has failed, using the rich detail in the diffuse scattering intensity distributions has been shown to have good potential to help solve the structure. Although the present model is clearly only the first step in obtaining such a solution, the indications are sufficiently encouraging to suggest that further application of the iterative strategy outlined should eventually lead to structure solution. Further work is in progress.

Acknowledgements The authors wish to thank David Goldstone for allowing us to use his X-ray data and Dr. Darren Goossens for assistance with the manuscript preparation.

References

1. Welberry TR, Heerdegen AP, Goldstone DC, Taylor IA (2011) Acta Crystallogr, Sect B 67(6):516
2. Wannier GH (1950) Phys Rev 79(2):357
3. Welberry TR (1985) Rep Prog Phys 48(11):1543
4. McGreevy RL, Pusztai L (1988) Mol Simul 1(6):359
5. Metropolis N, Rosenbluth AW, Rosenbluth MN, Teller AH, Teller E (1953) J Chem Phys 21:1087

6. Cowtan K (1994) Joint CCP4 and ESF-EACBM. Newsl Protein Crystallogr 31:34
7. Mathews BW (1968) J Mol Biol 3:491
8. Goldstone DC (2011) Personal communication
9. Cowtan K (2006) Acta Crystallogr, Sect B 62:1002
10. Cowtan K (2008) Acta Crystallogr, Sect B 64:83
11. Bernstein FC, Koetzle TF, Williams GJ, Meyer EEJ, Brice MD, Rodgers JR, Kennard O, Shimanouchi T, Tasumi M (1977) J Mol Biol 112:535

Chapter 34
Dynamical Flexibility in the Periodic Zn_6Sc 1/1-Approximant

Holger Euchner, Tsunetomo Yamada, Helmut Schober, Stephane Rols, Marek Mihalkovič, Ryuji Tamura, Tsutomu Ishimasa, and Marc de Boissieu

Abstract Using quasielastic neutron scattering (QENS) and molecular dynamics (MD) simulations, dynamical disorder was shown to be present in the Zn_6Sc cubic 1/1-approximant to Tsai type quasicrystals. This dynamical disorder originates from reorientations of the innermost tetrahedron shell inside the Tsai type clusters' building blocks. To enable such a rotational motion inside a close-packed alloy, a unique dynamical flexibility is necessary. We present a study of the tetrahedron dynamics with respect to this structural flexibility.

H. Euchner (✉)
Institut für Theoretische und Angewandte Physik, Universität Stuttgart, Stuttgart, Germany
e-mail: euchner@itap.physik.uni-stuttgart.de

H. Euchner · T. Yamada · M. de Boissieu (✉)
Sciences et Ingénierie des Matériaux et Procédés, Grenoble-INP, CNRS, UJF, St Martin d'Hères Cedex, France
e-mail: Marc.de-boissieu@simap.grenoble-inp.fr

T. Yamada · R. Tamura
Department of Materials Science and Technology, Tokyo University of Science, Tokyo, Japan

H. Schober · S. Rols
Institut Laue Langevin, Grenoble, France

H. Schober
UFR de Physique, Université Joseph Fourier, Grenoble, France

M. Mihalkovič
Institute of Physics, Slovak Academy of Sciences, Bratislava, Slovakia

T. Ishimasa
Division of Applied Physics, Graduate School of Engineering, Hokkaido University, Sapporo, Japan

S. Schmid et al. (eds.), *Aperiodic Crystals*, DOI 10.1007/978-94-007-6431-6_34,

34.1 Introduction

Binary Tsai type quasicrystals (QC) are to date the structurally best understood icosahedral QC phases [3]. Due to the existence of periodic approximants featuring the same structural building blocks, important information on the structure and physical properties [4] could be gained and in parts transferred to the QCs. For many Tsai type QCs, a periodic approximant can be synthesized with almost the same chemical composition [5, 6], e.g. $Cd_{5.7}Yb$ and $Mg_{4.2}Zn_{80.5}Sc_{15}$ in case of the icosahedral QCs with their Cd_6Yb and Zn_6Sc cubic 1/1-approximants. Recently, a description of the structure of a QC and its approximant was achieved using a packing of the same large rhombic triacontahedral (RTH) unit [3]. An RTH cluster consists of 158 atoms arranged in a close packed manner on successive shells with icosahedral symmetry, surrounding the central, symmetry-breaking tetrahedron (Fig. 34.1). In QCs and 1/1-approximants, these RTH units are connected along their twofold axes by shared faces and overlap along their threefold axes. While in the 1/1-approximant they are located on the vertices of a bcc lattice with space group $Im\bar{3}$ (lattice parameter $a = 1.57$ and 1.38 nm for Cd_6Yb and Zn_6Sc, respectively), in the QC the RTH units are arranged on the vertices of a quasiperiodic network.

In this study, we investigate the dynamical flexibility in the Zn_6Sc 1/1-approximant which is isostructural to Cd_6Yb and also evidence the, for this structure type well-known, low-temperature order disorder phase transition [7, 8]. This phase transition was attributed to the central tetrahedron ordering, and in the case of Zn_6Sc occurs at a transition temperature (T_c) of about 160 K [9–11]. Above T_c, the tetrahedral shell in Zn_6Sc is randomly disordered and occupies six orientations which are equivalent under the cubic symmetry group [5, 6]. In Fig. 34.1, the different positions of the atoms of the tetrahedral shell are shown together with their thermal ellipsoids and one possible tetrahedron orientation. At 160 K, the phase transition from the cubic high temperature (HT) phase to a monoclinic low-temperature (LT) phase with space group C2/c takes place. In this LT phase, the tetrahedra are ordered anti-parallel along the [110] direction of the HT phase [11] (see also Fig. 34.1). The structure analysis of the LT phase [11] reveals strong distortions of the shells that surround the innermost tetrahedron and therefore depart significantly from perfect icosahedral symmetry. Since its discovery, this phase transition has been suggested to be driven by a new kind of mechanism [7]. The low T_c at which the transitions occur in Zn_6Sc and related intermetallic compounds (between 100 and 160 K) excludes atomic diffusion as its origin. However, the ordering of the tetrahedra below T_c goes along with large atomic displacements not compatible with a simple displacive transition. Therefore, it was speculated that these transitions are indeed induced by the dynamical reorientation of the tetrahedra above T_c. This hypothesis was supported by total energy ab-initio calculations and long term MD simulations [12–15].

Recent measurements of diffuse scattering [16] in combination with our quasielastic neutron scattering (QENS) results and atomic scale simulations are clear evidence of tetrahedron short range order above T_c with almost diverging correlation length when T_c is approached. With this knowledge it could be demonstrated that the

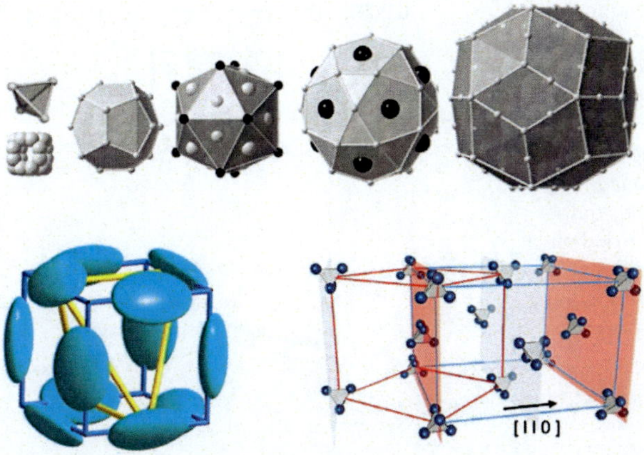

Fig. 34.1 (*Top panel*) Successive shells of the large RTH found in 1/1-approximant and quasicrystal in the ZnSc (or CdYb) system. *Light grey* and *dark colour* stand for Zn and Sc atoms, respectively; from left to right: Zn thetrahedron, Zn dodecahedron, Sc icosahedron, Zn icosidodecahedron, Zn triacontahedron. (*Bottom panel*) (i) Positions and atomic displacement parameters (ADP) of the central Zn atoms, constituting the disordered tetrahedron as determined in [8]. A *cube* is given as a guide for the eye. One tetrahedron orientation is given by the *yellow lines*. (ii) Schematic representation of the tetrahedron ordering along the 110 direction. The two different [110] planes are highlighted with different colours

disorder in the HT phase is indeed dynamical in nature with tetrahedra 'jumping' as rigid objects between energetically equivalent orientations. Such tetrahedron reorientations necessitate an exceptional dynamical flexibility of the structure, a feature unique to this system.

34.2 Quasielastic Neutron Scattering (QENS)

Diffusive or relaxational processes in materials can be investigated by use of QENS (see [17] for an introduction). Close to the elastic line, the dynamic response, $S(Q, E)$, contains information on such processes, the so-called quasielastic contribution. Typically, the QENS signal stems from processes that decay with time and thus is of a Lorentzian lineshape, which in turn is characterized by temperature and Q-dependence of its widths and intensity.

In Fig. 34.2, the Q-integrated $S(Q, E)$ response function is depicted for four different temperatures, as it was obtained from the IN5 time-of-flight spectrometer at the ILL in Grenoble, using high resolution geometry. The grey shaded curve in Fig. 34.2 is obtained for a measurement at 150 K, i.e. below T_c, thus in a state where the tetrahedra are ordered. For this temperature no quasielastic signal is visible and we only find elastic (Bragg scattering) and inelastic (phonons) contributions. However, if the temperature is increased above T_c, a clear quasielastic contribution becomes visible. Furthermore, we evidence that the QENS signal broadens from 175

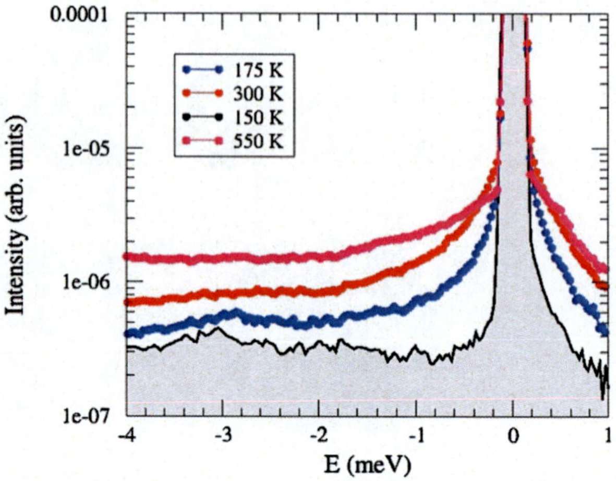

Fig. 34.2 Evolution of the quasielastic signal measured at $T = 175$, 300 and 550 K. *Filled circles* stand for the measured data; *error bars* are smaller than the size of the circles. The *shaded, grey curve* corresponds to the measurement carried out at 150 K. The additional quasielastic contribution at 175 K is clearly visible

to 550 K, indicating a classical over barrier jump. Indeed, the Lorentzian width can be shown to follow an Arrhenius law (activation energy of 60 meV) with temperature, while being independent of Q [1]. Altogether these are typical characteristics of a spatially limited jump process. Since moreover the QENS signal is only present above T_c, these is a strong proof that we indeed deal with tetrahedra that are constantly in motion above T_c. The characteristic jump distance is of the order of 1.6 Å, while the tetrahedron jumps occur on a ps timescale.

34.3 Simulation

Due to the size limit imposed by DFT, a study of the dynamics in this system is only possible using MD simulations with effective interaction potentials. Oscillating pair potentials, which have proven to be well suited to access the dynamics of the ZnSc system [4], are used to conduct simulations with the IMD code [18].

The simulation reproduces perfectly the average structure, as obtained from X-ray diffraction. Figure 34.3 shows the time averaged probability to find a tetrahedron atom at a certain position extracted from an MD simulation. This distribution is almost identical to what has been observed experimentally [6] (see Fig. 34.1). Not only the position of the randomly occupied sites are well reproduced, but also their distribution, which in the Lin and Corbett model is given by the anisotropic thermal displacement (ADP) parameters. Notice in particular that the ellipsoids are flat and with different orientation along a cube, which serves as a guide for the eye. Both simulation and experiment fit nicely. Moreover, it could be shown recently [1] that MD simulations are indeed able to capture the dynamics of the system. For example, the activation energy extracted from the QENS measurement is perfectly reproduced by simulations such that it is reasonable to assume that we capture the main dynamical features. Therefore, with our knowledge about the dynamical dis-

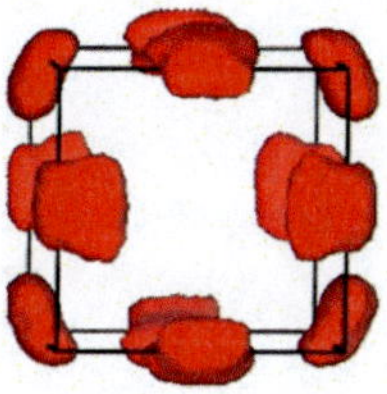

Fig. 34.3 Time averaged distribution of the central Zn tetrahedron from MD

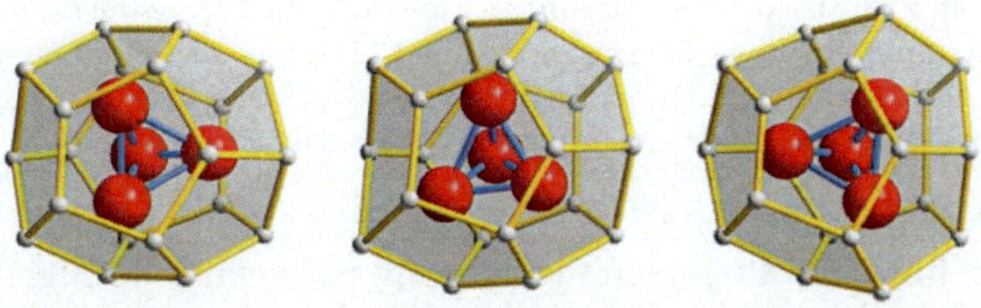

Fig. 34.4 Tetrahedral and dodecahedral shell at $t = 0$, 1.0 and 2.6 ps during a simulation. The displacement of the tetrahedron is almost a clock-wise rotation around a threefold axis passing through the atom in the back. Notice the strong distortion of the dodecahedron as the rotation proceeds. Qualitative investigations of the tetrahedron motion indicate a shorter lifetime of the central configuration

order with corresponding tetrahedron jumps, a next step is the investigation of the dynamical flexibility of the Zn_6Sc approximant on an atomistic level.

To do so, the trajectories of the atoms belonging to tetrahedron, as well as the surrounding dodecahedral, icosahedral and icosidodecahedral shells have been evaluated throughout long MD simulations. It was already discussed that the tetrahedron motion is accomplished on a very short time scale, with all four atoms moving together and 'jumping' on a distance of the order 1.6 Å. For a given orientation, there are two positions for which the tetrahedron is in a stable configuration, of which one has a shorter lifetime than the other (see also [13, 15]). This is exemplified in Fig. 34.4 which depicts three snapshots of the tetrahedron orientations taken during a MD run at $t = 0$, 1.0 and 2.6 ps. The positions at $t = 0$ and 2.6 ps have a lifetime which is roughly of the order of 3 ps, whereas the other one at 1.0 ps is of the order of about 0.5 ps. During its motion, the tetrahedron 'pushes' the atoms in its surrounding away to avoid too short Zn–Zn distances and thus induces large displacement to the surrounding shells which can be nicely seen from our MD simulation. The tetrahedron rotation thus distorts the surrounding shells, meaning that the corresponding atoms are displaced strongly out of their equilibrium positions. This in turn is only possible due to a 'dynamical flexibility' of the whole structure, which is a unique feature of this system. In Fig. 34.5, the probability distribution of the cluster shells surrounding the tetrahedron are depicted. As expected the dodecahedral shell evidences elongated displacements which are a consequence of atoms being pushed away during at tetrahedron rotation. The Sc atoms of the icosahedral shell are less prone to displacements, but they as well are affected when the tetrahedron is moving. Interestingly the next shell, the Zn icosidodecahedron, shows again sites which are strongly displaced throughout a MD run. This indeed shows that

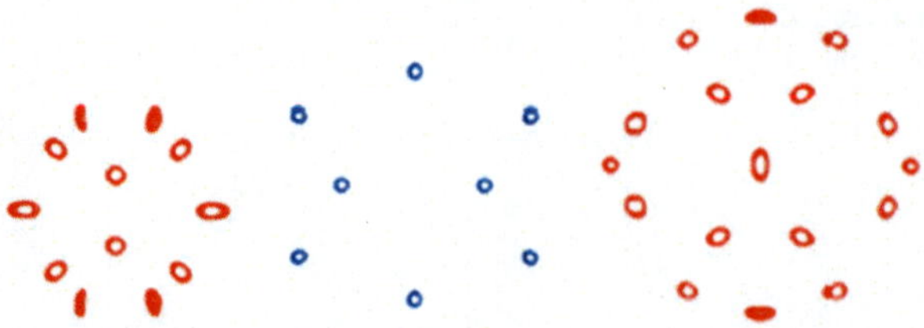

Fig. 34.5 Probability distribution of the atoms surrounding the tetrahedral shell around there equilibrium position, projected along the z-axis; from left to right: dodecahedron, icosahedron, icosidodecahedron. A probability up to 5 % is depicted in *red* (Zn) and *blue* (Sc). Higher probabilities, i.e. close to the equilibrium, are inside the coloured areas and depicted in *white*

the distortion, introduced by a tetrahedron flip, propagates throughout the different cluster shells and thus is able to mediate a tetrahedron–tetrahedron interaction. In the following, the different cluster shells and their displacements are investigated in more detail. The dodecahedral shell (Fig. 34.5, left) shows two different kinds of displacements. While atoms on the 8 cubic sites are uniformly displaced, we find elongated displacements for the 12 remaining sites, which point away from the cluster centre in radial direction (RMSD ~0.22 Å). For the Sc icosahedron, the displacements are almost spherical (Fig. 34.5, middle) and also less pronounced than for the other shells (RMSD of ~0.1 Å). Finally, the Zn icosidodecahedron shows again anisotropic displacements, yet elongated along the cluster surface, thus orthogonal to the radial direction. The RMSD of these displacement amounts up to ~0.25 Å, however, depends on the symmetry of the site.

34.4 Conclusion

The above findings thus indeed support the picture of a structure with a high dynamical flexibility, with distortions that are introduced by the tetrahedron motion. The introduced distortions then propagate throughout the different cluster shells and thus are involved in the long range ordering of the structure. These propagative distortions most likely are strongly involved in the ordering mechanism at T_c. The large distortions of the successive shells are certainly what drives the 'interaction' between the neighbouring tetrahedra despite the fact that they are located at a rather large distance of 1.2 nm. Since especially the Zn atoms show strong displacements out of their equilibria, it is likely that the evidenced flexibility originates in peculiarities in the Zn–Zn interaction as also observed, e.g. in $Zn_{11}Mg_2$ [19]. Due to the same cluster building blocks the here evidenced dynamical flexibility is also expected to be in play for QCs in the high temperature range. However, in the case of QCs the tetrahedron–tetrahedron linkages are different and this may be expected to introduce a 'frustration' term.

Moreover, it is worthwhile to realize that such tetrahedron flips may correspond to phason 'flips' in the QC [2, 20–22] since the occupation domain which describes the central tetrahedron in the 6D model of the i-CdYb QC is located on the external part of the atomic surface [3]. It has to be pointed out that a single 'phason flip' can exist in QC and 1/1-approximant, while long-wavelength phason fluctuations [23] are unique to QCs [24]. Yet, phason hopping has only been observed for temperatures of the order 0.8 T_m (with melting temperature T_m), whereas here we evidence tetrahedron jumps even below room temperature, and as low as 0.2 T_m (T_m is of the order 1000 K for Zn_6Sc). Furthermore, phason hopping has been found to be phonon assisted [20], which in our case could not be evidenced.

Finally, our results point to the importance of the cluster description in Tsai type phases and their relevance for the dynamics of these systems.

Acknowledgement This work is part of a common project within the European C-MAC network.

References

1. Euchner H, de Boissieu M, Schober H, Rols S, Ishimasa T, Tamura R, Yamada T, Mihalkovic M (2012) J Phys Condens Matter
2. Janssen T, Chapuis G, de Boissieu M (2007) Aperiodic crystals. From modulated phases to quasicrystals. Oxford University Press, Oxford
3. Takakura H, Gomez CP, Yamamoto A, de Boissieu M, Tsai AP (2007) Nat Mater 6:58
4. de Boissieu M, Francoual S, Mihalkovic M, Shibata K, Baron AQR, Sidis Y, Ishimasa T, Wu D, Lograsso T, Regnault LP, Gähler F, Tsutsui S, Hennion B, Bastie P, Sato TJ, Takakura H, Currat R, Tsai AP (2007) Nat Mater 6:977
5. Gomez CP, Lidin S (2003) Phys Rev B 68:024203
6. Lin QS, Corbett JD (2004) Inorg Chem 43:1912
7. Tamura R, Murao Y, Takeuchi S, Ichihara M, Isobe M, Ueda Y (2002) Jpn J Appl Phys 41:L524
8. Tamura R, Edagawa K, Aoki C, Takeuchi S, Suzuki K (2003) Phys Rev B 68:174105
9. Tamura R, Nishimoto K, Takeuchi S, Edagawa K, Isobe M, Ueda Y (2005) Phys Rev B 71:092203
10. Yamada T, Tamura R, Muro Y, Motoya K, Isobe M, Ueda Y (2010) Phys Rev B 82:134121
11. Ishimasa T, Kasano Y, Tachibana A, Kashimoto S, Osaka K (2007) Philos Mag 87:2887
12. Ishii Y, Fujiwara T (2001) Phys Rev Lett 87(20):206408
13. Nozawa K, Ishii Y (2008) J Phys Condens Matter 20:315206
14. Hatakeyama T, Nozawa K, Ishii Y (2008) Z Kristallogr 223:830
15. Mihalkovic M, Henley CL (2011) Philos Mag 91:2548
16. Yamada T, Euchner H, Tamura R, de Boissieu M (2012) In preparation
17. Bee M (1988) Quasielastic neutron scattering: principles and applications in solid state chemistry, biology and materials science. Hilger, Bristol
18. Stadler J, Mikulla K, Trebin HR (1997) Int J Mod Phys C 8:1131
19. Euchner H, Mihalkovic M, Gahler F, Johnson MR, Schober H, Rols S, Suard E, Bosak A, Ohhashi S, Tsai AP, Lidin S, Gomez CP, Custers J, Paschen S, de Boissieu M (2011) Phys Rev B 83:144202
20. Lyonnard S, Coddens G, Calveyrac D, Gratias D (1996) Phys Rev B 53:31050
21. Edagawa K, Suzuki K, Takeuchi S (2000) Phys Rev Lett 85(8):1674
22. Abe E, Pennycook SJ, Tsai AP (2003) Nature 421:347
23. de Boissieu M, Currat R, Francoual S (2008) In: Fujiwara T, Ishii Y (eds) Handbook of metal physics: quasicrystals, vol 107. Elsevier Science, Amsterdam
24. de Boissieu M, Francoual S, Kaneko Y, Ishimasa T (2005) Phys Rev Lett 95:105503

Chapter 35
Trajectories of Colloidal Particles in Laser Fields with Eight-, Ten-, or Twelve-Fold Symmetry and Phasonic Drift

Matthias Sandbrink and Michael Schmiedeberg

Abstract Quasicrystals are structures with long range order but no translational symmetry. Besides phonons, quasicrystals posses additional hydrodynamic modes called phasons. In a recent article (Kromer et al., Phys. Rev. Lett. 108:218301, 2012), the trajectories of colloidal particles in a laser field with decagonal symmetry were studied when the phasonic displacement was changed. Here we generalize the results to laser fields with eight- and twelve-fold symmetry. In principle, the method can also be used to predict collective rearrangements of atoms due to phasons in intrinsic quasicrystalline systems.

35.1 Introduction

Quasicrystals have long range order but are not periodic [9, 18]. Therefore, they might possess rotational point symmetries that are forbidden in periodic structures. Furthermore, there are additional hydrodynamic modes called phasons that correspond to correlated rearrangements of the atoms of a quasicrystal. Many properties of phasons are still a main topic of research [6].

Colloidal particles are widely employed to study ordering, crystallization and dynamics in external potentials [10]. In a laser field, a force in the direction towards the highest light intensity acts on the colloids [1, 2]. Therefore, colloidal particles can be forced to form complex structures [4]. In particular, quasicrystalline ordering of the colloids can be obtained and studied by employing an interference pattern with quasicrystalline symmetry [3, 7, 11–17]. By tuning the phases of the laser beams appropriately it is possible to change the phasonic displacements of the laser field in a controlled way [7, 14, 16].

In a recent article [7], Kromer et al. presented a method to predict the trajectories of colloidal particles in a laser field with phasonic drift. The laser field is created by

M. Sandbrink (✉) · M. Schmiedeberg
Institut für Theoretische Physik 2: Weiche Materie, Heinrich-Heine-Universität Düsseldorf, 40204 Düsseldorf, Germany
e-mail: sandbrink@thphy.uni-duesseldorf.de

M. Schmiedeberg
e-mail: schmiedeberg@thphy.uni-duesseldorf.de

S. Schmid et al. (eds.), *Aperiodic Crystals*, DOI 10.1007/978-94-007-6431-6_35,

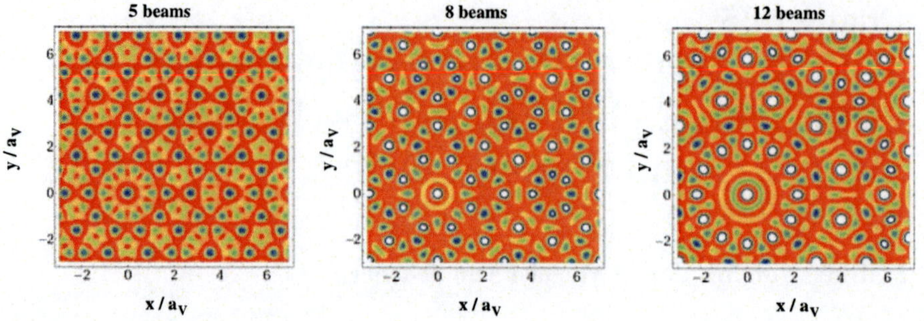

Fig. 35.1 Interference patterns calculated from Eq. (35.1) for different numbers of laser beams. The length unit is given by $a_V = 2\pi/|\mathbf{G}_j|$

five laser beams, which corresponds to an external potential with decagonal symmetry. Kromer et al. showed that every colloid can be mapped into characteristic areas of reduced phononic and phasonic displacement. Therefore, only the paths of colloids inside the characteristic areas have to be studied. Here we generalize this method to laser fields with eight- and twelve-fold symmetry. The corresponding interference patterns are depicted in Fig. 35.1.

In Sect. 35.2, we introduce the model system. In Sect. 35.3, we determine the characteristic areas of reduced phononic and phasonic displacement before we study the behavior of a colloid inside these areas in order to predict colloidal paths in general in Sect. 35.4. Finally, we conclude in Sect. 35.5.

35.2 Laser Fields and Phasonic Drifts

A laser field acts on colloidal particles like an external potential. The interference pattern created by N laser beams corresponds to an external potential [5, 15, 17]

$$V(\mathbf{r}) = -\frac{V_0}{N^2}\sum_{i=0}^{N-1}\sum_{j=0}^{N-1}\cos\big((\mathbf{G}_i - \mathbf{G}_j)\cdot\mathbf{r} + \phi_i - \phi_j\big) \tag{35.1}$$

where $\mathbf{G}_j = (2\pi/a_V \cos[2\pi j/N], 2\pi/a_V \sin[2\pi j/N])$ are the wave vectors of the laser beams projected onto the sample plane, a_V denotes the length scale of the potential and ϕ_j the phases of the laser beams. For N beams the resulting interference pattern has $2N$-fold rotational symmetry if N is odd and N-fold if N is even. The laser fields that we consider in this article are shown in Fig. 35.1.

The phases are parametrized in accordance with [8]

$$\phi_j = \mathbf{u}\cdot\mathbf{G}_j + \mathbf{w}\cdot\mathbf{G}_{kj \bmod N} + \gamma/N \tag{35.2}$$

where we use $k = 3$ for $N = 5$ or $N = 8$ and $k = 5$ for $N = 12$. The vector $\mathbf{u} = (u_x, u_y)$ describes the phononic and $\mathbf{w} = (w_x, w_y)$ the phasonic displacement. In contrast to **u** and **w**, the global phase γ is not a hydrodynamic variable and vanishes in the interference patterns. We denote a potential with a phasonic displacement **w** by $V_{\mathbf{w}}(r)$.

Note that the laser fields are similar to the Fourier description that is often used to describe a density field $\rho(\mathbf{r})$ of a typical quasicrystal [8]

$$\rho(\mathbf{r}) = \sum_{j=0}^{N-1} \cos(\mathbf{G}_j \cdot \mathbf{r} + \phi_j). \tag{35.3}$$

The reciprocal lattice vectors $\mathbf{G}_j$ and the phases ϕ_j are the same as in (35.1) and (35.2). While in the laser fields the colloids usually occupy the minima, atoms are most-likely located close to maxima in the Fourier description.

In the following, we consider the limit of high potential strengths or low temperatures, e.g., $V_0 \gg k_B T$, such that all colloids are always located in local minima of the external potential. We apply a phasonic drift to the potential, i.e., the phasonic displacement is changed at a constant rate in time.

35.3 Characteristic Areas

All colloidal positions can be mapped onto particles inside characteristic areas of small phononic and phasonic displacements. The reason is that there are combinations of phononic and phasonic displacements $\Delta\mathbf{u}$ and $\Delta\mathbf{w}$ that do not change the potential, i.e., $V_{\mathbf{w}}(\mathbf{r}) = V_{\mathbf{w}+\Delta\mathbf{w}}(\mathbf{r}+\Delta\mathbf{u})$. For example, for $j = 0, 1, \ldots, M$ displacements with

$$\begin{aligned} \Delta\mathbf{u} &= \big(u_r \cos[2\pi j/M], u_r \sin[2\pi j/M]\big) \quad \text{and} \\ \Delta\mathbf{w} &= \big(w_r \cos[2\pi kj/M], w_r \sin[2\pi kj/M]\big) \end{aligned} \tag{35.4}$$

change the differences between phases $\phi_j - \phi_k$ only by integer multiples of 2π if the step lengths u_r and w_r are chosen appropriately (see Table 35.1 for possible values). The number of symmetry axes is $M = 2N$ for an odd number of beams and $M = N$ for an even one.

By using suitable combinations obeying Eq. (35.4), we can map a colloid at position **r** in a potential $V_{\mathbf{w}}(\mathbf{r})$ onto a particle at reduced position $\mathbf{r}^{(\mathrm{red})} = \mathbf{r} - \Delta\mathbf{u}$ in a potential $V_{\mathbf{w}^{(\mathrm{red})}}(\mathbf{r}^{(\mathrm{red})})$ with reduced phasonic displacement $\mathbf{w}^{(\mathrm{red})} = \mathbf{w} + \Delta\mathbf{w}$ within characteristic areas whose size is determined as follows: We first consider the mapping along a phononic direction given by $\mathbf{e}_j = (\cos[2\pi j/M], \sin[2\pi j/M])$ and its phasonic counterpart. The sizes of the characteristic intervals δu and δw have to be chosen at least such that the two-dimensional subspace spanned by the considered phononic and phasonic directions can be completely covered with rectangles of dimensions δu and δw that are displaced by Δu in phononic and Δw in phasonic direction. We list the appropriate interval lengths δu and δw in Table 35.1. Considering all symmetry directions, we end up on reduced values $\mathbf{r}^{(\mathrm{red})}$ and $\mathbf{w}^{(\mathrm{red})}$ within

Table 35.1 The step lengths u_r and w_r of phononic and phasonic displacement vectors used in Eq. (35.4) such that there is no change to the laser field for all integer numbers n and m. In the last two columns, the size of the characteristic areas is given. Note, the same step lengths and area sizes occur for a density field $\rho(\mathbf{r})$ given by (35.3). We denote the corresponding symmetry in the second column

Laser	Fourier	u_r/a_V	w_r/a_V	$\delta u/2a_V$	$\delta w/2a_V$
5		$\frac{2}{5}n + (\frac{1}{5} + \frac{1}{\sqrt{5}})m$	$\frac{2}{5}n + (\frac{1}{5} - \frac{1}{\sqrt{5}})m$	$(1+\sqrt{5})/10$	$(1+\sqrt{5})/10$
10	5; 10	$\frac{2}{\sqrt{5}}n + (1 + \frac{1}{\sqrt{5}})m$	$-\frac{2}{\sqrt{5}}n + (1 - \frac{1}{\sqrt{5}})m$	$\frac{1}{2}(1 + \frac{1}{\sqrt{5}})$	$\frac{1}{2}(1 + \frac{1}{\sqrt{5}})$
8	8	$\frac{1}{2}n + \frac{1}{\sqrt{2}}m$	$\frac{1}{2}n - \frac{1}{\sqrt{2}}m$	$\frac{1}{2\sqrt{2}}$	$\frac{1}{2}(\frac{1}{2} + \frac{1}{\sqrt{2}})$
12	12	$n + \frac{1}{\sqrt{3}}m$	$n - \frac{1}{\sqrt{3}}m$	$\frac{1}{2}$	$\frac{1}{2}(1 + \frac{1}{\sqrt{3}})$

areas of polygonal shape defined by

$$\left|\mathbf{r}^{(\mathrm{red})} \cdot \mathbf{e}_j\right| \le \frac{\delta u}{2a_V} \quad \text{and} \quad \left|\mathbf{w}^{(\mathrm{red})} \cdot \mathbf{e}_j\right| \le \frac{\delta w}{2a_V} \tag{35.5}$$

for all $j = 1, \dots, M$ (see also black polygons in Fig. 35.2).

To gain a complete understanding of the colloidal dynamics, it is sufficient to study the colloidal behavior within the characteristic areas. We follow the paths of colloids that are started at the origin for phasonic drifts in all directions. In Fig. 35.2, the important points of these trajectories are shown as described in more detail in the caption of the figure. Note the color is always the same for the reduced phononic displacement and its corresponding reduced position. Therefore, the color indicates in what direction a colloid is moving depending on the direction of the phasonic drift. A detailed description of the analysis for a laser field created by five beams is given in [7]. In the next section, we will show that the path of a colloid started at an arbitrary position can be predicted by using the diagrams shown in Fig. 35.2.

35.4 Colloidal Trajectories

With the diagrams deduced in the previous section one can predict colloidal trajectories due to phasonic drifts. Since the ten-fold case has been studied before [7], we focus here on the potentials with eight- and twelve-fold rotational symmetry. Each column in Fig. 35.3 contains the description of a single trajectory.

A trajectory depends on the starting value of $\mathbf{w}^{(\mathrm{red})}$. All initial conditions that lead to similar trajectories are marked by the crosshatched areas in Fig. 35.3. The procedure of vanishing minima, sliding colloids and mapping is explained in detail in the caption of Fig. 35.3. We obtain a straight path if this procedure repeats after one slide and a zigzag path in case it repeats after two slides.

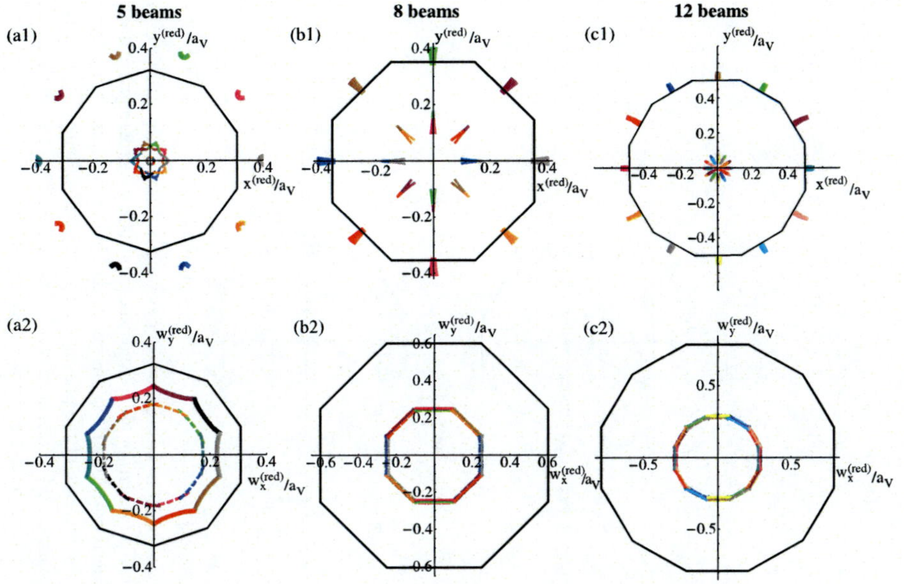

Fig. 35.2 The areas of reduced positions $\mathbf{r}^{(\text{red})}$ (**a1**, **b1**, **c1**) and phasonic displacements $\mathbf{w}^{(\text{red})}$ (**a2**, **b2**, **c2**) are depicted for different laser potentials. The characteristic areas are marked by black polygons. The threshold value of $\mathbf{w}^{(\text{red})}$ where a minimum disappears is shown by solid colored lines inside the polygons [in (**b2**) and (**c2**) drawn on *top*]. If this threshold value is surpassed, the colloid slides into a new minimum whose position is marked by the outermost colored spots in the diagrams for $\mathbf{r}^{(\text{red})}$. Note, the reduced positions are drawn with the same color as the associated reduced phasonic displacements. Afterwards $\mathbf{r}^{(\text{red})}$ and $\mathbf{w}^{(\text{red})}$ are mapped back into the characteristic areas as illustrated by the innermost colored dashed lines or spots that lie on the opposite site of the lines where the sliding started. In (**b2**) and (**c2**), the remapping ends exactly underneath the lines where a slide for opposite phasonic drift starts. Note, the mapping does not change the real position or the real phasonic displacement. The diagrams (**a1**, **a2**) correspond to the ones already shown in [7]. Examples of colloidal paths within the diagrams are given in Fig. 35.3

35.5 Conclusions

We have extended previous studies [7] to quasicrystals with eight- and twelve-fold rotational symmetry and showed how one can predict colloidal trajectories that are caused by phasonic drifts. In the future, we want to investigate the particle motion that originates from phasonic modes with finite wavelength. In principle, all rearrangements of colloidal particles due to any given phasonic mode can be predicted with the diagrams shown in Fig. 35.2.

As indicated, our method is also applicable to intrinsic quasicrystals, e.g., atomic systems characterized by a density distribution, where atoms reside at the most pronounced maxima. By analyzing the properties of these maxima within the reduced phononic and phasonic areas, in principle, it is possible to predict the correlated rearrangements of atoms due to phasonic fluctuations.

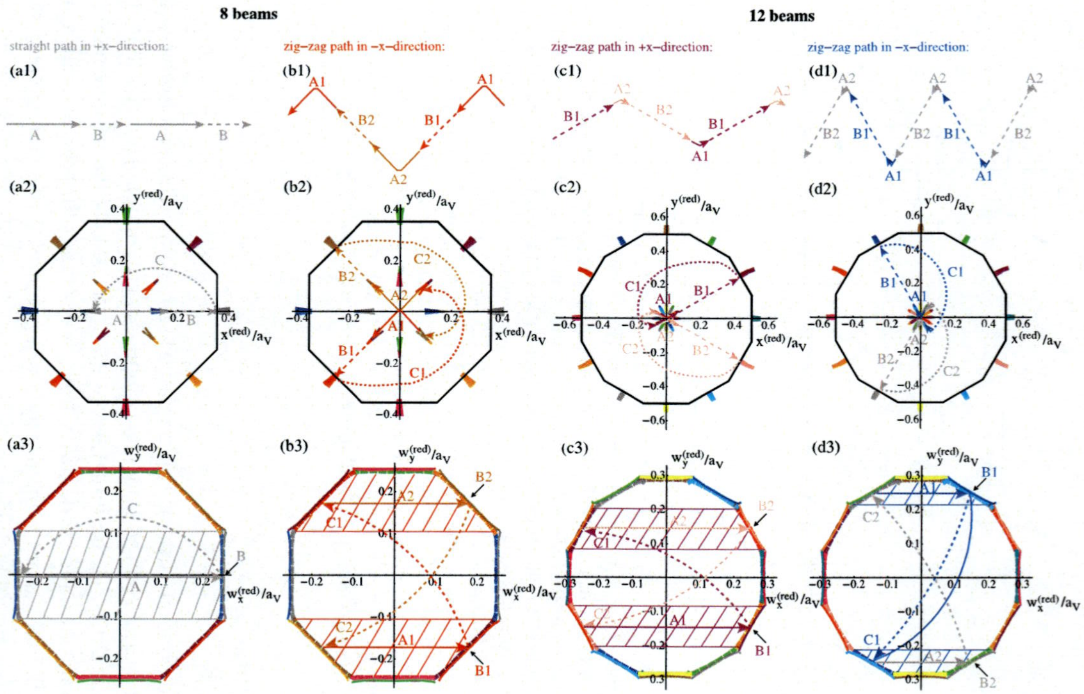

Fig. 35.3 A selection of colloidal trajectories in laser fields created by 8 or 12 beams for a phasonic drift in x-direction. For symmetry reasons, a phasonic drift in the y-direction causes analogue particle movements. In each *column*, a trajectory is shown that is typical for all colloids started inside the crosshatched area. In the *first line*, the paths of the colloids are shown; in the *second line*, the corresponding reduced positions $\mathbf{r}^{(\mathrm{red})}$ are given; and in the *last line*, the reduced phasonic displacements $\mathbf{w}^{(\mathrm{red})}$ are presented. First, $\mathbf{r}^{(\mathrm{red})}$ changes only gradually in response to an increase of the phasonic displacement. When $\mathbf{w}^{(\mathrm{red})}$ reaches the colored solid border, the local minimum occupied by the colloid vanishes. As a result, the colloid slides into another local minimum depicted by the same color. After that $\mathbf{r}^{(\mathrm{red})}$ and $\mathbf{w}^{(\mathrm{red})}$ are mapped such that they are close to the origin and all steps are repeated for the new reduced quantities

Acknowledgements We thank J. Kromer, J. Roth, and H. Stark for helpful discussions and the Deutsche Forschungsgemeinschaft for financial support (Schm 2657/2 and Ro 924/5).

References

1. Ashkin A (1970) Phys Rev Lett 24:156
2. Ashkin A (1980) Science 210:1081
3. Bohlein T, Bechinger C (2012) Phys Rev Lett 109:058301
4. Burns MM, Fournier JM, Golovchenko JA (1990) Science 249:749
5. Gorkhali S, Qi J, Crawford G (2006) J Opt Soc Am B 23:149
6. Henley CL, de Boissieu M, Steurer W (2006) Philos Mag 86:1131
7. Kromer JA, Schmiedeberg M, Roth J, Stark H (2012) Phys Rev Lett 108:218301
8. Levine D, Lubensky TC, Ostlund S, Ramaswamy S, Steinhardt PJ, Toner J (1985) Phys Rev Lett 54:1520
9. Levine D, Steinhardt PJ (1984) Phys Rev Lett 53:2477
10. Löwen H (2001) J Phys Condens Matter 13:R415
11. Mikhael J, Gera G, Bohlein T, Bechinger C (2011) Soft Matter 7:1352
12. Mikhael J, Schmiedeberg M, Rausch S, Roth J, Stark H, Bechinger C (2010) Proc Natl Acad Sci USA 107:7214
13. Reichhardt C, Olson Reichhardt CJ (2011) Phys Rev Lett 106:060603
14. Schmiedeberg M, Mikhael J, Rausch S, Roth J, Helden L, Bechinger C, Stark H (2010) Eur Phys J E 32:25
15. Schmiedeberg M, Roth J, Stark H (2007) Eur Phys J E 24:367
16. Schmiedeberg M, Stark H (2008) Phys Rev Lett 101:218302
17. Schmiedeberg M, Stark H (2012) J Phys Condens Matter 24:284101
18. Shechtman D, Blech I, Gratias D, Cahn JW (1984) Phys Rev Lett 53:1951

Chapter 36
Catalytic Properties of Five-Fold Surfaces of Quasicrystal Approximants

M. Krajčí and J. Hafner

Abstract Recently it has been shown that some low order approximants to decagonal or icosahedral quasicrystals provide excellent activity and selectivity for hydrogenation of alkynes. Our recent works on $Al_{13}Co_4$ and AlPd compounds demonstrated that the catalytically active surfaces in both cases are surfaces with (pseudo-) five-fold symmetry. Ab-initio DFT calculations have been used to identify the reaction centers and to construct a detailed atomistic scenario for the acetylene to ethylene hydrogenation. It was found that the activity of the catalysts is not promoted by the transition metal (TM) atoms alone but by a cluster of Al atoms centered at a slightly protruding TM atom. In the present contribution, we demonstrate that local configurations of Al and TM atoms favorable for selective catalysis of the hydrogenation reactions naturally appear at Al–TM surfaces with pentagonal symmetry. We discuss the possibility to use surfaces of the Al–TM quasicrystals and their approximants as catalysts for hydrogenation reactions.

36.1 Introduction

Surfaces of transition metals are often used to catalyze oxidation and hydrogenation reactions. The possibility to use for chemical catalysis instead of the surfaces of close-packed transition metals the surfaces of complex metallic compounds like quasicrystals and their approximants is a new, promising and so far largely unexplored research area. While surfaces of the close-packed metals have only a few inequivalent adsorption sites, the surfaces of complex intermetallics provide a rich variety of different adsorption sites, leading to a multitude of possible reaction channels for catalytic reactions. Al–TM compounds are a class of intermetallics that can

M. Krajčí (✉)
Institute of Physics, Slovak Academy of Sciences, Bratislava, Slovakia
e-mail: fyzikraj@savba.sk

J. Hafner
Faculty for Physics and Center for Computational Materials Science, Vienna University, Vienna, Austria
e-mail: juergen.hafner@univie.ac.at

S. Schmid et al. (eds.), *Aperiodic Crystals*, DOI 10.1007/978-94-007-6431-6_36,

exhibit very high complexity of their surfaces. A particularly large number of inequivalent sites can be expected on the surfaces of quasiperiodic systems and their approximants. At some surface terminations, the local atomic environments of the surface TM atoms can modify their catalytic properties in a desired way. An isolation of the transition metal sites on the surface of a complex intermetallic compound [1] provides thus a possibility to optimize their catalytic properties.

The potential of using the complex intermetallic compounds for chemical catalysis is demonstrated on hydrogenation of acetylene, the simplest alkyne. We shall consider a chemical reaction where acetylene C_2H_2 is hydrogenated to ethylene C_2H_4. The hydrogenation reaction $C_2H_2 + H_2 \rightarrow C_2H_4$ is a subprocess in the industrial production of the polyethylene. Usually, Pd is used to catalyze the ethylene hydrogenation subprocess. The role of the catalyst in ethylene hydrogenation, as in any other reaction, is to provide an alternative reaction pathway to the reaction product. Without a catalyst, the reaction would proceed only at very high temperatures. The reaction temperature can be substantially reduced and the rate of the reaction can dramatically increase if the alternative path has a lower activation energy than the reaction path not mediated by the catalyst. The activity of the catalyst is its most important property. Pd catalysts exhibit sufficient activity but low selectivity. In the polyethylene production, it is undesired if the hydrogenation reaction continues and ethylene is further hydrogenated to ethane C_2H_6. In addition to the activity, the selectivity of the catalyst thus also plays a very significant role.

Recently, it has been shown that some low order approximants to decagonal or icosahedral quasicrystals provide excellent activity and selectivity for hydrogenation of alkynes [1, 2]. Our recent works on $Al_{13}Co_4$ and AlPd compounds [3, 4] demonstrated that the catalytically active surfaces in both cases are surfaces with the (pseudo-)five-fold symmetry. Ab-initio DFT calculations have been used to identify the reaction centers and to construct a detailed atomistic scenario for the acetylene-to-ethylene hydrogenation. The local configurations of Al and TM atoms favorable for selective catalysis of the hydrogenation reactions naturally appear at the Al–TM surfaces with the pentagonal symmetry. We discuss the possibility to use surfaces of the Al–TM quasicrystals and their approximants as catalysts for hydrogenation reactions.

36.2 Hydrogenation on Surfaces of Quasicrystalline Approximants

The $Al_{13}Co_4$ and AlPd (and isostructural GaPd) compounds have been identified as active and selective catalysts for acetylene hydrogenation [1, 2]. $Al_{13}Co_4$ is an approximant to the decagonal Al–Co–Ni quasicrystal. AlPd crystallizing in the B20 structure can be considered as the smallest approximant to the i-Al–Pd–Mn quasicrystal. We have analyzed the hydrogenation process and found that in both compounds the catalytically active surfaces are the surfaces with the (pseudo-)five-fold symmetry. In $Al_{13}Co_4$, the active surface is the (100) surface. In AlPd, the active surface seems to be the (210) surface with the pseudo-five-fold symmetry.

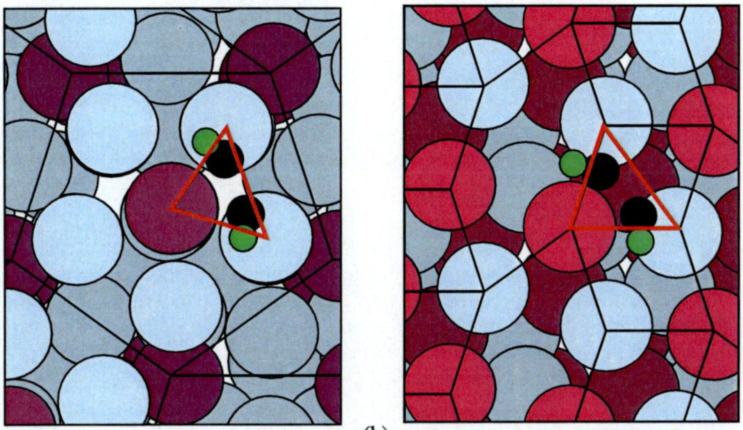

(a) (b)

Fig. 36.1 The atomic structure of the catalytically active surfaces of $Al_{13}Co_4$ (**a**) and AlPd (**b**). The *light blue* and *gray circles*: Al, the *violet circles*: TM atoms. The *red triangles* mark the most active reaction centers

The hydrogenation reaction on the surface of an intermetallic compound proceeds via the Langmuir–Hinshelwood mechanism. Both reactants H_2 and C_2H_2 must to be coadsorbed on the surface. First, the H_2 molecule dissociates. Then the reaction proceeds via thermally activated diffusion of H atoms and two reaction steps. The complete reaction path is a multistep process involving in addition to the chemical reactions $C_2H_2 + H \rightarrow C_2H_3$ and $C_2H_3 + H \rightarrow C_2H_4$, the diffusion of coadsorbed atomic hydrogen towards the hydrocarbon species.

On the studied Al–TM surfaces, the H_2 molecule binds exclusively only on top of a TM atom. Dissociation of H_2 is an activated process with the activation energy of $E_a \leq 60$ kJ/mol. Diffusion of dissociated atomic hydrogen over the surface is also a thermally activated process. We have observed that the diffusion jumps over Al atoms ($E_a \approx$ 40–60 kJ/mol) are more difficult than over TM atoms ($E_a \approx$ 5–20 kJ/mol). Although the C_2H_2 molecule also binds on top of the TM atoms, it was found that the most preferable adsorption sites for this molecule are not the TM atoms, as expected, but certain bridge positions between two neighboring Al atoms.

The most remarkable observation was that the adsorption configurations of the reactants and hydrogenation mechanisms on (100)-$Al_{13}Co_4$ and on the (210) surface of AlPd are essentially the same. In both cases, the catalytically active sites are formed by a pentagonal arrangement of Al atoms around a slightly protruding transition metal atom. The most active reaction centers in both cases are triangular configurations of two Al atoms and one TM atom, see Fig. 36.1. The preferable adsorption geometry for acetylene is the Al–Al bridge position, forming a di-σ-bond. Vinyl C_2H_3 is adsorbed with the bond between the C–H group and an Al atom. The product of the reaction, the ethylene molecule, is adsorbed via a π-bond on top of the TM atom. During the hydrogenation reaction, the molecular complex shifts from the bridge position between the Al sites to the top position on top of the TM site

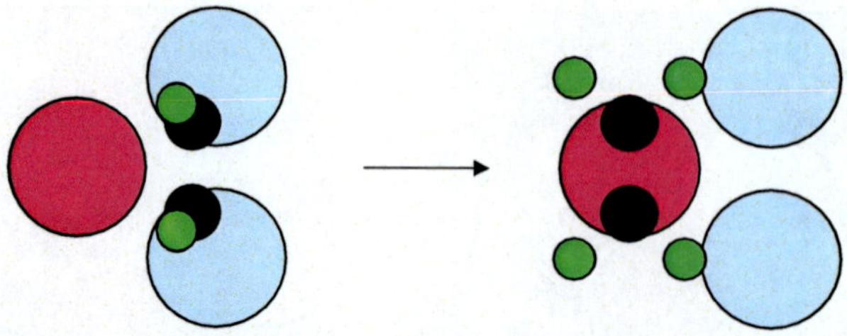

Fig. 36.2 The most active reaction center. The preferable adsorption geometry for C_2H_2 is the Al–Al bridge position. The product of the reaction, the C_2H_4 molecule, is adsorbed on top of the TM atom. During the hydrogenation reaction, the molecular complex shifts from the bridge position between the Al sites to the top position on the TM site

where it is only weakly bound, favoring desorption. The highest activation energy barriers of the whole reaction path for the (100)-$Al_{13}Co_4$ and (210) AlPd surfaces are 63 and 71 kJ/mol, respectively. These values are lower than that of 74 kJ/mol calculated for the reference (111) Pd surface. The selectivity of both investigated intermetallic compounds results from a low desorption energy for C_2H_4 that is by $\approx$10 kJ/mol lower than the activation energy for further hydrogenation to C_2H_5 and ethane.

The vertical position of the TM atom above the surrounding atoms plays an essential role. There is a clear correlation between the height of the TM atom above the surface plane and the adsorption energy of C_2H_4. The binding energy of C_2H_4 increases with the increasing height of the TM atom above the surface plane which leads to a lower selectivity. On the other hand, if the vertical position of the TM atom is below the neighboring atoms, the TM atom does not bind the hydrocarbons and the TM site becomes catalytically inactive. In an optimal configuration, the TM atom is slightly protruding above the surrounding surface atoms.

A catalytic reaction is generally a complex nonlocal process. We have seen that the adsorption site of the reactant and the product need not be the same site. One has to consider also the length of diffusion paths determined by distant positions of the adsorption sites of both reactants. In this short contribution, we have restricted our analysis only on the dependence of the catalytic properties of an active site on structural details of its local environment. The Sabatier principle is known as a qualitative concept in the chemical catalysis. According to this principle, the interactions between the catalyst and the reacting adsorbates should be "just right", that is, neither too strong nor too weak. If the interaction is too weak, the adsorbates fail to bind to the catalyst and no reaction takes place. On the other hand, if the adsorption is too strong, the activation energies of the reaction steps become too large.

A pure Al(111) surface cannot be used as a hydrogenation catalyst. Although C_2H_2 is adsorbed with a sufficient binding energy of $E_b \approx -119$ kJ/mol, the surface does not bind H_2 at all. The presence of the TM atoms at the surface is thus essential for the hydrogenation. At the B20 compounds the studied pseudo-fivefold (210) surface plays the same role as the (110) surface at the B2 compounds. The triangular configuration of the atoms similar to that presented in Fig. 36.2

can be found also on the (110) surfaces of B2 compound AlCo. However, instead of the pentagonal arrangement here each surface atom has 4 nearest neighbors of the opposite chemical type. While on the (100)-$Al_{13}Co_4$ surface the C_2H_2 molecule is adsorbed with a moderate binding energy of $E_b = -184$ kJ/mol, at the AlCo(110) surface the acetylene molecule is adsorbed with much higher binding energy of $E_b \approx -276$ kJ/mol. On this surface, the first hydrogenation step is strongly endothermic ($\Delta E_b \approx +75$ kJ/mol) and the activation energy of this step of ≈125 kJ/mol is very high. The activity of such a catalyst would be very poor. Because of the too strong C_2H_2 adsorption, also the (0001) surface of hcp Co is not a good hydrogenation catalyst. The adsorption energy of acetylene is here even stronger ($E_b \approx -320$ kJ/mol). These results demonstrate how the local environments of the active sites modify their catalytic properties. An increased lateral interaction of atoms in the denser packed pentagonal cluster on (100)-$Al_{13}Co_4$ reduces the too strong adsorption of the C_2H_2 molecule observed at other surfaces. The pentagonal cluster centered at a TM atom provides high activity and good selectivity for acetylene hydrogenation. As the pentagonal environments of TM atoms can be naturally found on surfaces of quasicrystals, we have performed an inspection of the surfaces of some quasicrystals to estimate possible catalytic activity of these intermetallic compounds.

36.3 Discussion

Perhaps best understood is the structure of the five-fold surface of i-Al–Pd–Mn. Theoretical studies based on the KGB model [5–7] of the structure of the i-Al–Pd–Mn quasicrystal confirmed by experimental analysis [8] show that of the five-fold surface is highly Al-rich. The five-fold surface of this quasicrystal is formed by 2 atomic planes separated (in the ideal structure) by 0.48 Å. The chemical composition of the top plane is ≈95 % Al and ≈5 % Mn. Pd atoms thus occur only in the second plane below the top plane. This information about the chemical composition of the top surface planes is sufficient for the conclusion that the clean five-fold surface of i-Al–Pd–Mn is a poor hydrogenation catalyst. The Pd atoms buried in the surface are catalytically inactive. A closer inspection shows that the only catalytically active sites are in the top atomic plane exposed centers of the pseudo-Mackay clusters occupied by Mn atoms. In pentagonal Al environments of these Mn atoms, one can also recognize (see Fig. 1 in [7]) the triangular reaction centers as shown in Fig. 36.2. However, because of the small number of such sites, the surface will have low catalytic activity.

As Al(111) surface has lower surface energy than TM close-packed surfaces, one can generally expect that the surfaces of Al–TM quasicrystals will be Al-rich. As H_2 binds exclusively on the top of TM atoms, the small number of these sites at the surface reduces the hydrogenation activity. One possibility to improve the hydrogenation activity is to use instead of the quasicrystals their approximant phases with a higher TM content as it was demonstrated in Sect. 36.1 in the study of the AlPd compound. Another possible remedy of this problem is to reveal the subsurface TM sites. In the second TM-rich layer, there are many possibly active sites. This can be done, for example, by dissolving the surface Al atoms from the top plane by an etching process [9].

It is interesting to look also at the d-Al–Co–Ni quasicrystal related to the studied $Al_{13}Co_4$ compound. Unfortunately, neither the bulk nor the surface structure of this quasicrystal is so far well understood. As a representative for the surface of the d-Al–Co–Ni quasicrystal we have considered the surface of the W-approximant phase. In our earlier studies of the surface of the d-Al–Co–Ni quasicrystal [10], the chemical Al–TM ordering was based on the original proposal of Sugiyama et al. [11]. We note that a recent structure refinement [12] proposed a slightly modified chemical ordering. The bulk structure consists of two kinds of atomic planes—flat (F) and puckered (P). It can be expected that the more Al-rich P plane will be preferably exposed at the surface. A short inspection of the structure and the Al–TM ordering of the P plane shows that at this surface there are plenty of TM sites with a pentagonal environment; see Fig. 1 in [10]. The TM sites with the pentagonal environment can be found in the centers and in the vertices of a superposed pentagonal tiling. In a half of the TM sites, the TM atoms are slightly buried in the surface, but in the other half they are slightly protruding. This quasicrystal thus has a potential to be a good hydrogenation catalyst and deserves further theoretical investigations.

Acknowledgements This work has been supported by the Austrian Ministry for Education, Science and Art through the Center for Computational Materials Science. M.K. thanks also for support from the Grant Agency for Science of Slovakia (No. 2/0111/11), from CEX FUN-MAT, and from the Slovak Research and Development Agency (Grant No. APVV-0647-10, APVV-0076-11).

References

1. Kovnir K, Armbrüster M, Teschner D, Venkov TV, Jentoft FC, Knop-Gericke A, Grin Yu, Schlögl R (2007) Sci Technol Adv Mater 8:420
2. Armbrüster M, Kovnir K, Grin J, Schlögl R, Gille P, Heggen M, Feuerbacher M (2009) European patent application No 09157875.7
3. Krajčí M, Hafner J (2011) J Catal 278:200
4. Krajčí M, Hafner J (2012) J Phys Chem C 116:6307
5. Krajčí M, Windisch M, Hafner J, Kresse G, Mihalkovič M (1995) Phys Rev B 51:17355
6. Krajčí M, Hafner J (2005) Phys Rev B 71:054202
7. Krajčí M, Hafner J, Ledieu J, McGrath R (2006) Phys Rev B 73:024202
8. Unal B, Jenks CJ, Thiel PA (2008) Phys Rev B 77:195419
9. Tanabe T, Kameoka S, Tsai AP (2010) Appl Catal A, Gen 384:241–251
10. Krajčí M, Hafner J, Mihalkovič M (2006) Phys Rev B 73:134203
11. Sugiyama K, Nishimura S, Hiraga K (2002) J Alloys Compd 342:65
12. Strutz A, Yamamoto A, Steurer W (2010) Phys Rev B 82:064107

Chapter 37
Effect of Leaching on Surface Microstructure and Chemical Composition of Al-Based Quasicrystals

T.P. Yadav, M. Lowe, R. Tamura, R. McGrath, and H.R. Sharma

Abstract We have studied the effect of leaching treatments on the surface microstructure and chemical composition of Al-based quasicrystals. The high symmetry surfaces of single grain icosahedral (*i*-) Al–Cu–Fe and decagonal (*d*-) Al–Ni–Co quasicrystals and a polygrain *i*-Al–Pd–Re quasicrystal with random surface orientation were leached with NaOH solution at varying times and the resulting surfaces were characterized by scanning electron microscopy, energy dispersive X-ray analysis and X-ray photoelectron spectroscopy. The leaching treatments preferentially remove Al producing nanoparticles of the transition metals and their oxides. The leached fivefold surface of *i*-Al–Cu–Fe exhibits micron sized dodecahedral cavities on which the nanoparticles are precipitated. However, no specific microstructure has been observed on the tenfold surface of *d*-Al–Ni–Co and the polygrain *i*-Al–Pd–Re. The quasicrystalline surface can be regained after polishing the leached layer, indicating that leaching occurs only in a limited depth from the surface. This was revealed by low energy electron diffraction after the surface was prepared under ultra high vacuum conditions. These results provide important information for preparation of model catalysts of nanoparticles of catalytically active metals on quasicrystal surfaces.

37.1 Introduction

Quasicrystals (QCs) have orientationally ordered structures with no periodicity. They are intermetallic compounds with a specific chemical composition and often possess classically forbidden rotational symmetries such as fivefold and ten-

T.P. Yadav · M. Lowe · R. McGrath · H.R. Sharma (✉)
Surface Science Research Centre and Department of Physics, The University of Liverpool, Liverpool L69 3BX, UK
e-mail: H.R.Sharma@liverpool.ac.uk

H.R. Sharma
e-mail: h.r.sharma@liv.ac.uk

R. Tamura
Department of Materials Science and Technology, Tokyo University of Science, Noda 278-8510, Japan

S. Schmid et al. (eds.), *Aperiodic Crystals*, DOI 10.1007/978-94-007-6431-6_37,

fold. For the revolutionary discovery of QC, the 2011 Nobel Prize in chemistry was awarded to Danny Shechtman [1]. Quasicrystalline structure was discovered in 1982 on rapidly solidified Al–Mn alloys, which exhibited sharp diffraction peaks with icoshahedral symmetry [1]. QCs show physical properties different from those of their metallic and amorphous counterparts such as low coefficient of friction, high hardness, low surface energy, good wear-resistance, high thermal and electric resistivity [2], which can be exploited for industrial applications [3].

Many quasicrystals contain catalytically active elements and they are stable at high temperatures, and thus they have the potential to be used as catalysts where high thermal stability is necessary [4–9]. Tsai et al. studied the catalytic activity of leached quasicrystals in steam reforming of methanol [4, 5]. In their early experiments, the powder of polygrain icosahedral *i*-Al–Cu–Fe quasicrystal was leached with NaOH aqueous solutions. The thus-prepared powder shows higher reactivity and thermal stability than usual metal catalysts in steam reforming of methanol [8, 9]. Al is selectively dissolved during leaching yielding Cu and Fe nanoparticles on top of the quasicrystalline surface. These nanoparticles are believed to be responsible for the catalytic reaction. Iron species are homogeneously distributed in the leached layer, which suppresses the aggregation of Cu [9].

Recently, the same authors found that leaching of *i*-Al–Cu–Fe followed by calcination at 600 °C in air [10] drastically increases the catalytic activity. This is attributed to the formation of composite materials with fine Cu nanoparticles dispersed in a spinal matrix of $(Fe,Al)_3O_4$ [11]. Furthermore, the catalytic behaviour of QCs and related crystalline materials with composition similar to that of QCs, the so-called approximants, has been compared. It was found that QCs show higher catalytic activity and durability than approximants [9]. The dissolution rate of Al in approximants is higher than in QCs and produces a layer of Cu, instead of homogenous distribution of Cu and Fe [9], which deteriorates the activity.

Because of their complex nature, it has been difficult to gain atomic scale understanding of catalytic activity of leached quasicrystals. Additionally, the role of underlying quasicrystals in the catalytic activity is yet to be understood. In order to achieve this information, we have attempted to create a simple model catalyst of nanoparticles on quasicrystalline surfaces by leaching well defined surfaces of single grain quasicrystals. As the first step of these studies, we present here the effect of leaching treatments on surface morphology and chemical composition of different quasicrystals studied by scanning electron microscopy (SEM), energy dispersive X-ray (EDX) analysis and X-ray photoelectron spectroscopy (XPS). The studied systems include the fivefold surface of *i*-Al–Cu–Fe and tenfold surface of *d*-Al–Ni–Co. The results from the single grains QCs will be compared with a leached polygrain *i*-Al–Pd–Re QC with random surface orientations.

37.2 Experimental Procedure

Single grains of *i*-$Al_{63}Cu_{24}Fe_{13}$ and *d*-$Al_{72}Ni_{11}Co_{17}$ were grown by the Bergman method [12]. Single grain nature of the samples was confirmed by low energy electron diffraction (LEED), prior to leaching experiments. The *i*-Al–Pd–Re sample was synthesized by spark plasma sintering method using a mixture of raw material powders of Al, Pd and Re with a nominal composition of *i*-$Al_{70.5}Pd_{21}Re_{8.5}$ [13]. The as-cast ingots was annealed at 923 K for 48 hours for better homogeneity and the phase purity of the samples was examined by X-ray diffraction (XRD). All the peaks in the XRD pattern obtained from the surface of the disc can be indexed by icosahedral phase and the intensity distribution is similar to that obtained from *i*-$Al_{70.5}Pd_{21}Re_{8.5}$ QC powder. This reveals that the surface is composed of random grain orientations.

The surfaces were polished before each leaching treatment using diamond paste of successively smaller size of 6 to 0.25 μm. Leaching was performed in ambient conditions by placing droplets of NaOH aqueous solution with 10 mole concentration on the surfaces using a pipette. Leaching was performed at various times from 30 min to 8 hours. After leaching, the surfaces were thoroughly washed with methanol and distilled water in ultrasonic bath until no alkali was detected. It was found by XPS that 4–5 times of washing with water for 15 minutes each is needed to completely remove alkali from the surface. The leached surfaces were characterized using JEOL-JSM-6610 SEM, EDX and XPS.

37.3 Results and Discussions

The effect of leaching on surface was first examined by an optical camera. In the inset of Fig. 37.1(a), we show optical micrographs of fivefold surface of *i*-Al–Cu–Fe. The surface is mirror-like after polishing. Leaching turns the surface to reddish brown color with a reduced reflectivity. The colour and roughness indicates the presence of Cu and Fe metal particles on the surface after leaching.

Figure 37.1(a) shows an SEM of the polished fivefold *i*-Al–Cu–Fe surface, revealing the flatness of the surface. Some scratches were noticed, which were created during polishing the surface. The average elemental composition determined by EDX taken at different parts of the polished surface is Al: 63.4 at%, Cu: 23.2 at% and Fe: 13.1 at%, which is close to the nominal composition of the bulk. Small amount of oxygen (0.3 at%) was also detected as expected for an air-exposed surface.

Figure 37.1(b) shows an LEED pattern from the *i*-Al–Cu–Fe surface after several cycles of sputtering with 2 keV argon ions (8 μA ion current) for 30 min and annealing at 870 K for 4 hours. The LEED pattern exhibits fivefold symmetry with diffraction spots located at τ-scaling distances, τ is an irrational number characteristic of quasiperiodic symmetry, as expected from bulk symmetry of a single grain sample. After confirming single grain nature, the surface was taken out from ultra high vacuum (UHV) and leached by NaOH.

Figure 37.2 shows SEM images of the fivefold *i*-Al–Cu–Fe surface after leaching at different times. Leaching at 0.5 hours creates shallow cracks on the surface (Fig. 37.2(a)). After longer exposure to NaOH, the surface is leached deeper and

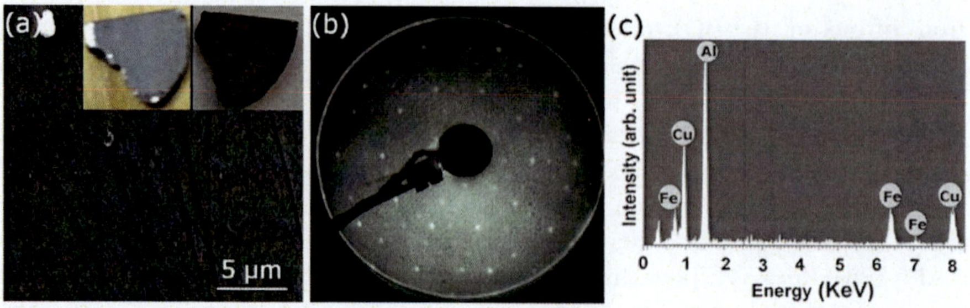

Fig. 37.1 (**a**) SEM micrograph of the fivefold *i*-Al–Cu–Fe surface after polishing. Inset: optical photographs of polished and leached surface (leaching time 4 hours). (**b**) LEED pattern from the same surface after ion bombardment and annealing under ultra high vacuum. (**c**) EDX spectrum from the polished surface

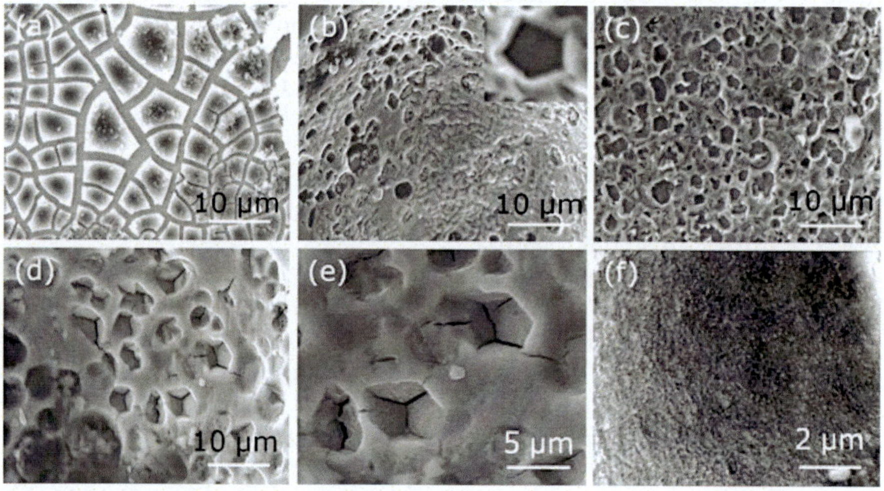

Fig. 37.2 (**a**) SEM micrograph of the fivefold *i*-Al–Cu–Fe surface after leaching at different times: (**a**) 0.5, (**b**) 2, (**c**) 4, (**d**) and (**e**) 8, (**f**) 1 hour

pentagonal facets are developed (Figs. 37.2(b)–(c)). The edge length of the pentagons varies from 3–5 μm for all leaching treatments. The pentagons have the same orientation. Their density increases with leaching time. After leaching at 8 hours, dodecahedral cavities are formed (Fig. 37.2(d)). A close up view of dodecahedrons is shown in Fig. 37.2(e). The boundary between the pentagonal facets are leached deeper. Possibly, dissolution of next layer begins from these deep boundaries.

High magnification of SEM images show nanoparticles which are precipitated atop of the pentagonal facets. The exact determination of particle size was not possible with the available microscope. We estimate the size to be 100 nm or less for the surface after leaching at 1 hour (Fig. 37.2(f)).

The local chemical information of the surface is achieved by EDX. It reveals that the nanoparticles observed on the pentagonal facets are composed of Cu and Fe,

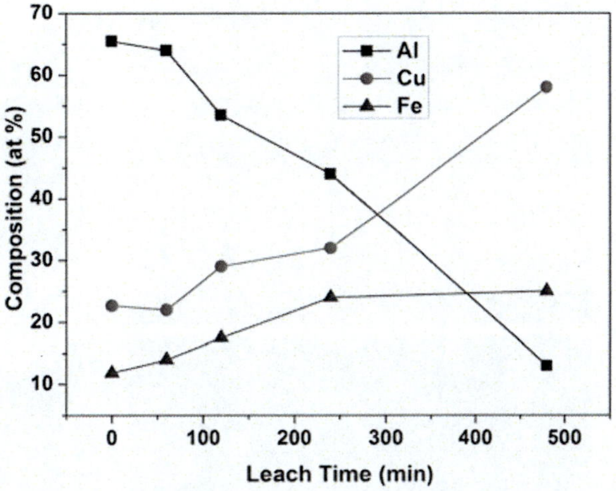

Fig. 37.3 Change in chemical composition of the fivefold *i*-Al–Cu–Fe surface as a function of leaching time determined by EDX

which are homogeneously distributed on the surface. Cu and Fe are also present on the dip boundaries of the pentagonal facets or elsewhere on the surface. EDX is unable to provide information about oxidation states of Cu and Fe and it is less surface sensitive. Therefore, we studied the surface by XPS, which is more surface sensitive and capable of providing oxidation states. XPS from the surface after leaching at 2 hours show the existence of copper and iron oxides and a trace of metallic iron. The observed chemical composition is very similar to that observed on polygrain *i*-Al–Cu–Fe QC powders after leaching [10, 11].

A trend of composition change with leaching time is shown in Fig. 37.3. The composition was determined by EDX taken at different locations of the surface. As seen in the graph, Al is selectively removed, while Cu and Fe content increases. The dissolution rate of Al is almost linear after 2 hours of leaching. The dissolution is slower for the first 2 hours. This is probably because Al oxide layer formed on the surface prior to leaching is hard to be dissolved compared to pure quasicrystal or it may take a while to establish a good contact between the surface and solution. After 8 hours of leaching, Cu content increases faster than Fe and thus the surface is more Cu rich than Fe.

The surface was polished after leaching experiments and checked by LEED in UHV. LEED patterns after UHV treatments were identical to those observed before the leaching experiments. This suggests that the leaching effect was limited to a certain depth from the surface.

Now we come back to the facets observed after leaching. At early stage of leaching, most of the facets are of fivefold symmetry and are parallel to the surface. Longer leaching produces dodecahedral cavities, which are composed of facets along other fivefold planes, which are inclined with respect to the surface. We note that an icosahedral quasicrystal has six fivefold axes inclined at 63.4° from each other. The formation of the dodecahedral cavities can be explained if the surface is preferentially leached along all fivefold planes, but not limited to the fivefold plane

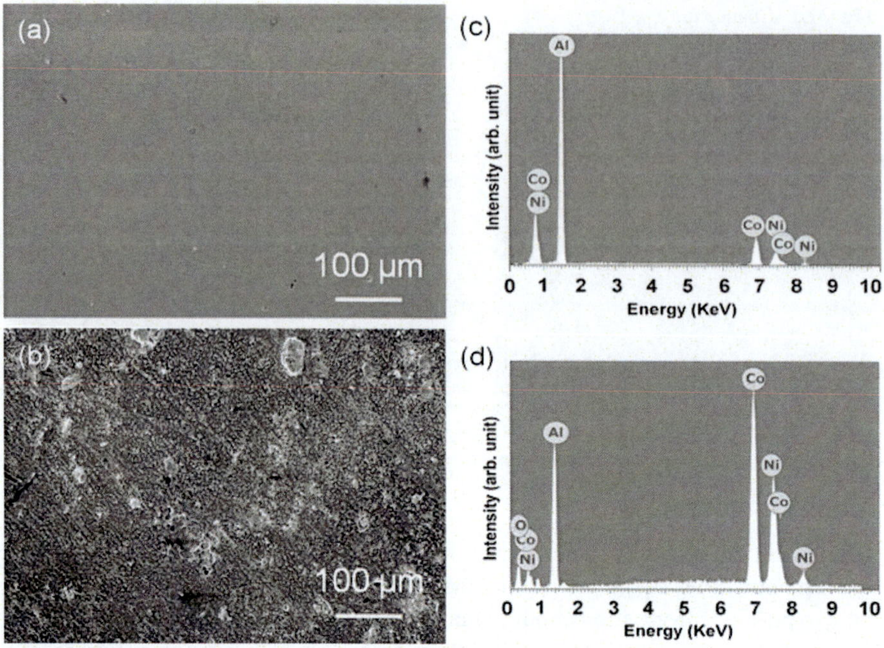

Fig. 37.4 SEM micrograph of the tenfold *d*-Al–Ni–Co surface after polishing (**a**) and after leaching (**b**). Corresponding EDX spectra of polished (**c**) and leached (**d**) surface. Leaching time was 2 hours

parallel to the surface. This further implies that the formation of pentagonal facets is not dictated by the surface orientation.

In order to check whether the facets were induced by surface matrix, we studied the influence of leaching on microstructure of tenfold surface of single grain *d*-Al–No–Co QC. The tenfold surface did not show any decagonal facets facets (refer to below), which may suggest that shape of facets is not induced by surface orientation but by crystallographic directions. However, results from leached *i*-Al–Pd–Re polygrain samples are somewhat different, which will be discussed later. Pentagonal facets are commonly observed in *i*-Al–Cu–Fe and *i*-Al–Pd–Mn systems. The *i*-Al–Cu–Fe QC grows in dodecahedron shape [14]. The same system after oxidation under UHV shows pentagonal micropores [15]. The fivefold surface of *i*-Al–Pd–Mn after cleavage [17] or after cleaning under UHV also exhibits similar features [16].

The tenfold surface of *d*-Al–Ni–Co was leached at selective time (2 hours) with NaOH solution in the same way as for *i*-Al–Cu–Fe. Figure 37.4 shows SEM image and EDX spectrum from the leached surface. The corresponding images from the polished surface are also given for comparison. The composition of the polished surface determined from EDX is Al: 72.5 at%, Ni: 10.0 at% and Co: 17.5 at%, which is close to the nominal composition of the bulk. As in *i*-Al–Cu–Fe, Al is preferentially removed by leaching and Co and Ni content is evolved (compare Figs. 37.4(c) and (d)). Small amount of oxygen was also detected. Unlike in *i*-Al–Cu–Fe, no

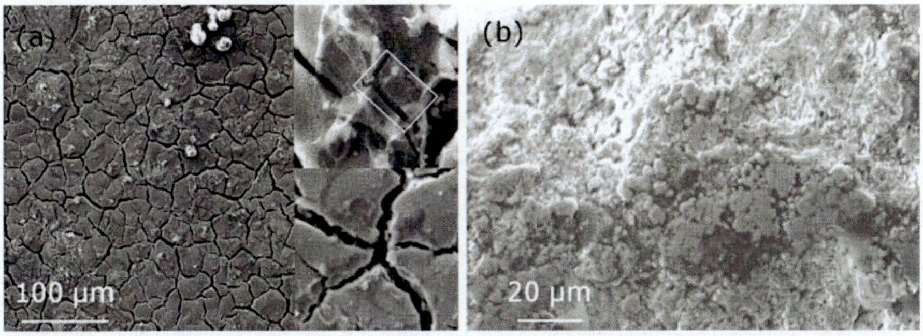

Fig. 37.5 (**a**) SEM micrograph of the polygrain *i*-Al–Pd–Re surface after leaching at 2 hours. *Inset*: magnified views of SEM images showing elongated features (marked) and grain boundaries. (**b**) High magnification SEM image of the same surface after leaching at 4 hours

decagonal features of the substrate symmetry were observed. Instead, liner features are seen.

Figure 37.5(a) shows an SEM image from the polygrain *i*-Al–Pd–Re surface. The leaching effect is enhanced at grain boundaries, which appear as dark thread like features in the image. A high magnification SEM image within a single grain is shown in Fig. 37.5(b), demonstrating nanoparticles. These particles are observed after 4 hours leaching. EDX confirms that the concentration of Al is quite low compared to Pd and Re. The surface becomes more Pd-rich than Re.

Unlike in *i*-Al–Cu–Fe, no pentagonal facets have been observed within a grain or inside grain boundaries, indicating that this surface is not leached along the fivefold axis. The leaching kinetic in *i*-Al–Pd–Re may not be same as in *i*-Al–Cu–Fe, possibly due to more covalent nature of the former, and thus may produce different surface microstructures. The same *i*-Al–Pd–Re sample studied by STM showed that twofold surface is most stable [13]. Occasionally, we observed elongated features appearing perpendicular to each others after leaching the *i*-Al–Pd–Re surface as shown in Fig. 37.5(a), inset. However, it is not clear at this stage whether higher stability of the twofold surface is related to these features. More experiments are underway in this direction.

37.4 Conclusions

The surface morphology and chemical composition of three different quasicrystals are studied after leaching with NaOH solution using different experimental techniques, namely, SEM, EDX and XPS. Investigated systems are the fivefold surface of *i*-Al–Cu–Fe and tenfold surface of *d*-Al–Ni–Co single grain quasicrystals and a polygrain *i*-Al-Pd-Re quasicrystal with random surface orientations. It was found that in all systems Al is preferentially dissolved producing nanoparticles of transition metals and their oxides. The fivefold *i*-Al–Cu–Fe surface is leached selectively along the fivefold planes producing micron sized dodecahedral cavities. However, no specific microstructure has been observed on the tenfold surface of *d*-Al–Ni–Co and the polygrain *i*-Al–Pd–Re under the given leaching conditions. Leaching occurs only in the top surface layers and the underlying quasicrystalline structure can be regained after polishing the surface. These preliminary results provide ideas about the preparation of model catalysts comprising nanoparticles of catalytically active metals on quasicrystal surfaces.

Acknowledgements H.R.S. is grateful to Engineering and Physical Sciences Research Council (EPSRC) for funding (Grants No. EP/D071828/1). T.P.Y. thanks the Department of Science and Technology (DST)-India for BOYACAST fellowship for this work. M.L. is grateful to EPSRC for support through Doctoral Training Grant. The authors would like to thank Dr. T. Joyce in the University of Liverpool for SEM facility. The quasicrystal samples were grown by T.A. Lograsso and A.R. Ross in Ames Laboratory, Iowa State University, USA. We thank A.P. Tsai in Tohoku University and V. Fournée and J. Ledieu at CNRS Nancy for fruitful discussions and J. Smerdon for critical reading of the manuscript.

References

1. Shechtman D, Bleeh I, Gratias D, Cahn JW (1984) Phys Rev Lett 53:1951–1953
2. Stadnik Z (1999) Physical properties of quasicrystals. Springer series in solid state sciences, vol 126. Springer, Berlin
3. Dubois J (2011) Isr J Chem 51:1168
4. Tsai AP, Yoshimura M (2001) Appl Catal A, Gen 214:237
5. Yoshimura M, Tsai AP (2002) J Alloys Compd 342:451
6. Ngoc B, Geantet C, Aouine M, Bergeret G, Raffy S, Marlin S (2008) Int J Hydrog Energy 33:1000
7. Jenk C, Thiel P (1998) J Mol Catal A, Chem 131:301
8. Kameoka S, Tanabe T, Tsai AP (2004) Catal Today 93–95:23
9. Tanabe T, Kameoka S, Tsai AP (2006) Catal Today 111:153
10. Tanabe T, Kameoka S, Tsai AP (2010) Appl Catal A, Gen 384:241
11. Tanabe T, Kameoka S, Tsai AP (2011) J Mater Sci 46:2242
12. Fisher IR, Kramer M, Islam Z, Ross R, Kracher A, Weiner T, Sailer M, Goldman A, Canfield P (1999) Philos Mag B 79:425
13. Tamura R, Yadav T, McGrath R, Sharma H (2012) Phys Rev B, submitted
14. Tsai AP, Inoue A, Masumoto T (1987) Jpn J Appl Phys 26:1505–1507
15. Rouxel D, Gil-Gavatz M, Pigeat P, Weber B (2005) J Non-Cryst Solids 351:802
16. Cappello G, Schmithusen F, Chevrier J, Comin F, Stierle A, Formoso V, de Boissieu M, Bourdard M, Lograsso T, Jenks C et al. (2000) Mater Sci Eng A 294–296:822
17. Suzuki S, Waseda Y, Urban K (1998) Mater Trans, JIM 39:314

Index

S. Schmid et al. (eds.), *Aperiodic Crystals*, DOI 10.1007/978-94-007-6431-6,